LONDON MATHEMATICAL SOCIETY LECTURE NOTE SERIES

Managing Editor: Professor Endre Süli, Mathematical Institute, University of Oxford,
Woodstock Road, Oxford OX2 6GG, United Kingdom

The titles below are available from booksellers, or from Cambridge University Press at
www.cambridge.org/mathematics

London Mathematical Society Lecture Note Series: 475

Effective Results and Methods for Diophantine Equations over Finitely Generated Domains

JAN-HENDRIK EVERTSE
Leiden University

KÁLMÁN GYŐRY
University of Debrecen

CAMBRIDGE
UNIVERSITY PRESS

CAMBRIDGE
UNIVERSITY PRESS

Shaftesbury Road, Cambridge CB2 8EA, United Kingdom

One Liberty Plaza, 20th Floor, New York, NY 10006, USA

477 Williamstown Road, Port Melbourne, VIC 3207, Australia

314–321, 3rd Floor, Plot 3, Splendor Forum, Jasola District Centre, New Delhi – 110025, India

103 Penang Road, #05–06/07, Visioncrest Commercial, Singapore 238467

Cambridge University Press is part of Cambridge University Press & Assessment, a department of the University of Cambridge.

We share the University's mission to contribute to society through the pursuit of education, learning and research at the highest international levels of excellence.

www.cambridge.org
Information on this title: www.cambridge.org/9781009005852

DOI: 10.1017/9781009042109

First published 2022

A catalogue record for this publication is available from the British Library

Library of Congress Cataloging-in-Publication data
Names: Evertse, J. H., author. | Győry, Kálmán, author.
Title: Effective results and methods for diophantine equations over finitely / Jan-Hendrik Evertse, Kálmán Győry.
Description: Cambridge, United Kingdom ; New York, NY : Cambridge University Press, 2022. | Includes bibliographical references and index.
Identifiers: LCCN 2021059210 | ISBN 9781009005852 (paperback)
| ISBN 9781009042109 (ebook)
Subjects: LCSH: Diophantine equations. | Diophantine analysis.
| Number theory. | BISAC: MATHEMATICS / Number Theory
Classification: LCC QA242 .E939 2022 | DDC 512.7/2–dc23/eng/20220128
LC record available at https://lccn.loc.gov/2021059210

ISBN   978-1-009-00585-2   Paperback

To our families

Contents

Preface

This book is devoted to Diophantine equations where the solutions are taken from an integral domain of characteristic 0 that is finitely generated over $\mathbb{Z}$, which is a domain of the shape $\mathbb{Z}[z_1, \ldots, z_r]$ with a quotient field of characteristic 0, where the generators $z_1, \ldots, z_r$ may be algebraic or transcendental over $\mathbb{Q}$. For instance, the ring of integers and the rings of S-integers of a number field are finitely generated domains where all generators are algebraic. Our aim is to prove effective finiteness results for certain classes of Diophantine equations, i.e., results that not only show that the equations from the said classes have only finitely many solutions, but whose proofs provide methods to determine the solutions in principle.

There is an extensive literature on Diophantine equations with solutions taken from the ring of rational integers $\mathbb{Z}$, or from more general domains, containing theorems on the finiteness of the set of solutions of such equations. Most of the finiteness theorems over $\mathbb{Z}$, and more generally over rings of integers and S-integers of number fields are ineffective. Their proofs are mainly based on techniques from Diophantine approximation (e.g., the Thue–Siegel–Roth–Schmidt theory) often combined with algebra and arithmetic geometry. These techniques yield the finiteness of the number of solutions but do not enable one to determine the solutions. Lang (1960) and others used certain specialization arguments to extend several ineffective finiteness results to the even more general case when the solutions are taken from an arbitrary integral domain of characteristic 0 that is finitely generated over $\mathbb{Z}$.

Since the 1960s, a great number of ineffective finiteness theorems over number fields were made effective and new theorems were obtained in effective form by means of A. Baker's effective theory of logarithmic forms. These results give effective upper bounds for the solutions, and thereby make it possible, at least in principle, to find all the solutions of the equations under consideration. Analogous theorems were established by Mason (1984) and others over function fields of characteristic 0 as well, which provide effective upper bounds for the heights of the solutions, but do not imply the finiteness of the number of solutions.

Győry (1983, 1984b) was the first to extend effective Diophantine results over number fields to the finitely generated case and proved effective finiteness theorems over certain restricted classes of finitely generated integral domains over $\mathbb{Z}$ of zero characteristic. He developed an effective specialization method, reducing the initial equations to the number field and function field cases, and using the corresponding effective results over number fields and function fields, he derived effective bounds for the solutions of the initial equations.

In the paper Evertse and Győry (2013), Győry's specialization method was extended to the case of arbitrary finitely generated domains of characteristic 0 over $\mathbb{Z}$. The crucial new tool in this extension was the work of Aschenbrenner (2004) on effective commutative algebra. Evertse's and Győry's general specialization method may be viewed as a "machine," which takes as input an effective Diophantine finiteness result concerning S-integral solutions over number fields together with an effective analogue over function fields, and produces as output a corresponding effective result over finitely generated domains. This general specialization method led to effective finiteness results for various classes of Diophantine equations over arbitrary domains of characteristic 0 that are finitely generated over $\mathbb{Z}$: Evertse and Győry (2013, 2014, 2015), Bérczes, Evertse, and Győry (2014), Bérczes (2015a, 2015b), and Koymans (2016, 2017) established general effective finiteness theorems over finitely generated domains of characteristic 0 for several classical equations, including *unit equations in two unknowns, Thue equations, hyper- and superelliptic equations*, and the *Catalan equation*. An important feature of these results is their quantitative nature, i.e., they give upper bounds for the sizes (suitable measures) of the solutions in terms of defining parameters for the domain from which the solutions are taken and for the Diophantine equation under consideration.

Our book provides the first comprehensive treatment of effective results and methods for Diophantine equations over finitely generated domains. Similarly to the above-mentioned literature, most of the results in our book are proved in quantitative form, giving effective bounds for the sizes of the solutions. Apart from the results mentioned above, our book contains new material, concerning *decomposable form equations* over finitely generated domains. Here, we have adapted the method of Győry (1973, 1980a) and Győry and Papp (1978) to reduce the decomposable form equations under consideration to systems of unit equations in two unknowns. Here again, we give effective upper bounds for the sizes of the solutions, and for this purpose, we had to work out new effective procedures. As a special case, we get back the results on *discriminant equations* from Evertse and Győry (2017a, 2017b).

We believe that the results in this book do not exhaust the possibilities of our techniques. Hopefully, they will inspire further investigations to obtain new effective results for other classes of Diophantine equations over finitely generated domains.

This book is aimed at anyone (graduate student and expert) with basic knowledge of algebra (groups, commutative rings, fields, Galois theory) and elementary algebraic number theory. No further specialized knowledge of commutative algebra or algebraic geometry is presupposed.

Acknowledgments

We are very grateful to Yann Bugeaud, Rob Tijdeman, and an anonymous referee for carefully reading and critically commenting on some chapters of our book, to Csanád Bertók for his typing of a considerable part of the book, and to the staff of Cambridge University Press, in particular David Tranah, Tom Harris, and Clare Dennison, for their suggestions and assistance with the final preparation of the manuscript.

The research of the second named author was supported in part by grants K115479 and K128088 from the Hungarian National Foundation for Scientific Research (OTKA) and from the Austrian–Hungarian joint project ANN130909 (FWF-NKFIH).

Glossary of Frequently Used Notation

General Notation

$|\mathcal{A}|$	cardinality of a finite set $\mathcal{A}$
$\log^* x$	$\max(1, \log x)$, $\log^* 0 := 1$
$\ll, \gg$	Vinogradov symbols; $A(x) \ll B(x)$ or $B(x) \gg A(x)$ means that there is a constant $c > 0$ such that $|A(x)| \leq cB(x)$ for all x in the specified domain. The constant c may depend on certain specified parameters independent of x
$\ll_{a,b,\ldots}$	the positive constants implied by $\ll_{a,b,\ldots}$ depend only on $a, b, \ldots$ and are effectively computable
$O(\cdot)$	$c \times$ the expression between the parentheses, where c is an effectively computable positive absolute constant. The c may be different at each occurrence of $O(\cdot)$
$\mathbb{Z}, \mathbb{Z}_{>0}, \mathbb{Z}_{\geq 0}$	integers, positive integers, non-negative integers
$\mathbb{Q}, \mathbb{R}, \mathbb{C}$	rational numbers, real numbers, complex numbers
gcd	greatest common divisor
$D(f)$	discriminant of a polynomial $f(X)$
$\overline{K}$	algebraic closure of a field K
A	integral domain (i.e., commutative ring with 1 and without divisors of 0)
A^*	unit group (multiplicative group of invertible elements) of A
A_G	integral closure of A in an extension G of the quotient field of A
$A[X_1, \ldots, X_n]$	ring of polynomials in n variables with coefficients in A

$A[\alpha_1,\ldots,\alpha_n]$ $\{f(\alpha_1,\ldots,\alpha_n) : f \in A[X_1,\ldots,X_r]\}$, A-algebra generated by $\alpha_1,\ldots,\alpha_n$

$\xi + \mathcal{M}$ $\{\xi + \eta : \eta \in \mathcal{M}\}$, $\mathcal{M}$-coset, where $\mathcal{M}$ is an A-module and ξ belongs to an A-module containing $\mathcal{M}$

$\mathcal{M}'/\mathcal{M}$ quotient A-module of two A-modules $\mathcal{M}',\mathcal{M}$, where $\mathcal{M}' \supseteq \mathcal{M}$; $\mathcal{M}'/\mathcal{M}$ consists of the $\mathcal{M}$-cosets $\xi + \mathcal{M}$ with $\xi \in \mathcal{M}'$, and is endowed with addition $(\xi_1 + \mathcal{M}) + (\xi_2 + \mathcal{M}) := (\xi_1 + \xi_2) + \mathcal{M}$ and scalar multiplication $a \cdot (\xi + \mathcal{M}) := a\xi + \mathcal{M}$, for $\xi_1,\xi_2,\xi \in \mathcal{M}'$ and $a \in A$

$H(Q), L(Q)$ maximum of the absolute values resp. the sum of the absolute values of the coefficients of $Q \in \mathbb{Z}[X_1,\ldots,X_n]$

$\deg Q, h(Q)$ the total degree of $Q \in \mathbb{Z}[X_1,\ldots,X_n]$, resp. the logarithmic height $\log H(Q)$ of Q

$s(Q)$ $\max(1,\deg Q, h(Q))$, the size of Q

Finite Étale Algebras over Fields

Ω/K finite étale algebra over a field K, i.e., a direct product $L_1 \times \cdots \times L_q$ of finite separable field extensions of K

$[\Omega : K]$ $\dim_K \Omega$

$x \mapsto x^{(i)}$ nontrivial K-algebra homomorphisms $\Omega \to \overline{K}$

$D_{\Omega/K}(\alpha)$ discriminant of $\alpha \in \Omega$ over K

A_Ω integral closure of an integral domain A with quotient field K in a finite étale K-algebra Ω

$\mathcal{O}$ A-order of Ω, i.e., a subring of A_Ω containing A and generating Ω as a K-vector space

Algebraic Number Fields

$\mathrm{ord}_p(a)$ exponent of a prime number p in the unique prime factorization of $a \in \mathbb{Q}$, and $\mathrm{ord}_p(0) = \infty$

$|a|_p$ $p^{-\mathrm{ord}_p(a)}$, p-adic absolute value of $a \in \mathbb{Q}$

$|a|_\infty$ $\max(a,-a)$, ordinary absolute value of $a \in \mathbb{Q}$

$\mathbb{Q}_p$ p-adic completion of $\mathbb{Q}$, $\mathbb{Q}_\infty = \mathbb{R}$

$\mathcal{M}_\mathbb{Q}$ $\{\infty\} \cup \{\text{primes}\}$, set of places of $\mathbb{Q}$

$\mathcal{O}_K, D_K, h_K, R_K$ ring of integers, discriminant, class number, regulator of a number field K

$\mathfrak{p}, \mathfrak{a}$ nonzero prime ideal, fractional ideal of $\mathcal{O}_K$

$[\alpha] = \alpha\mathcal{O}_K$	fractional ideal generated by α
$\mathrm{ord}_{\mathfrak{p}}(\mathfrak{a})$	exponent of $\mathfrak{p}$ in the unique prime ideal factorization of $\mathfrak{a}$
$\mathrm{ord}_{\mathfrak{p}}(\alpha)$	exponent of $\mathfrak{p}$ in the unique prime ideal factorization of (α) for $\alpha \in K$, with $\mathrm{ord}_{\mathfrak{p}}(0) := \infty$.
$N_K(\mathfrak{a})$	absolute norm of a fractional ideal $\mathfrak{a}$ of $\mathcal{O}_K$ (written as $N(\mathfrak{a})$ if it is clear which is the underlying number field)
$\mathcal{M}_K$	set of places of a number field K
$\lvert\cdot\rvert_v \ (v \in \mathcal{M}_K)$	normalized absolute values of K, satisfying the product formula, with $\lvert\alpha\rvert_v := N_K(\mathfrak{p})^{-\mathrm{ord}_{\mathfrak{p}}(\alpha)}$ if $\alpha \in K$ and $\mathfrak{p}$ is the prime ideal of $\mathcal{O}_K$ corresponding to the finite place v
K_v	completion of K at v
S_∞	set of infinite (archimedean) places
S	finite set of places of K, containing S_∞
$\mathcal{O}_S$	$\{\alpha \in K : \lvert\alpha\rvert_v \le 1 \text{ for } v \in \mathcal{M}_K\backslash S\}$, ring of S-integers, written as $\mathbb{Z}_S$ if $K = \mathbb{Q}$
$\mathcal{O}_S^*$	$\{\alpha \in K : \lvert\alpha\rvert_v = 1 \text{ for } v \in \mathcal{M}_K\backslash S\}$, group of S-units, written as $\mathbb{Z}_S^*$ if $K = \mathbb{Q}$
$N_S(\alpha)$	$\prod_{v\in S}\lvert\alpha\rvert_v$, S-norm of $\alpha \in K$
R_S	S-regulator
P_S, Q_S	$\max\{N_K(\mathfrak{p}_1),\ldots,N_K(\mathfrak{p}_t)\}$, $\prod_{i=1}^{t} N_K(\mathfrak{p}_i)$, where $\mathfrak{p}_1,\ldots,\mathfrak{p}_t$ are the prime ideals of $\mathcal{O}_K$ corresponding to the finite places of S
$\lvert\mathbf{x}\rvert_v \ (v \in \mathcal{M}_K)$	$\max_i \lvert x_i\rvert_v$, v-adic norm of $\mathbf{x} = (x_1,\ldots,x_n) \in K^n$
$H^{\mathrm{hom}}(\mathbf{x})$	$\left(\prod_{v\in\mathcal{M}_K}\lvert\mathbf{x}\rvert_v\right)^{1/[K:\mathbb{Q}]}$, absolute homogeneous height of $\mathbf{x} \in K^n$
$H(\mathbf{x})$	$\left(\prod_{v\in\mathcal{M}_K}\max(1,\lvert\mathbf{x}\rvert_v)\right)^{1/[K:\mathbb{Q}]}$, absolute height of $\mathbf{x} \in K^n$
$H(\alpha)$	$\left(\prod_{v\in\mathcal{M}_K}\max(1,\lvert\alpha\rvert_v)\right)^{1/[K:\mathbb{Q}]}$, absolute height of $\alpha \in K$
$h^{\mathrm{hom}}(\mathbf{x}), h(\mathbf{x}), h(\alpha)$	$\log H^{\mathrm{hom}}(\mathbf{x})$, $\log H(\mathbf{x})$, $\log H(\alpha)$, absolute logarithmic heights ($\mathbf{x} \in K^n$, $\alpha \in K$)
$h(P)$	$h(\mathbf{x}_P)$, $\mathbf{x}_P$ vector consisting of the nonzero coefficients of a polynomial $P \in K[X_1,\ldots,X_n]$

Function Fields

$\Bbbk$	field of constants (always algebraically closed)
$\Bbbk((z))$	field of Laurent series in z

$g_{K/\Bbbk}$	genus of function field K with constant field $\Bbbk$ ($K/\Bbbk$ is always assumed to be of transcendence degree 1)
$\mathcal{M}_K$	set of (normalized discrete) valuations of K, trivial on $\Bbbk$
$v(\mathbf{x})$ $(v \in \mathcal{M}_K)$	$\min_i v(x_i)$, v-adic norm of $\mathbf{x} = (x_1,\ldots,x_n) \in K^n$
$H_K^{\mathrm{hom}}(\mathbf{x})$	$-\sum_{v \in \mathcal{M}_K} v(\mathbf{x})$, homogeneous height of $\mathbf{x} \in K^n$
$H_K(x)$	$\sum_{v \in \mathcal{M}_K} \max(0, -v(x))$, height of $x \in K$
S	a finite subset of $\mathcal{M}_K$
$\mathcal{O}_S$	$\{\alpha \in K : v(\alpha) \geq 0 \text{ for } v \in \mathcal{M}_K \backslash S\}$, ring of S-integers
$\mathcal{O}_S^*$	$\{\alpha \in K : v(\alpha) = 0 \text{ for } v \in \mathcal{M}_K \backslash S\}$, group of S-units

Finitely Generated Domains

$A = \mathbb{Z}[z_1,\ldots,z_r]$	$\{f(z_1,\ldots,z_r) : f \in \mathbb{Z}[X_1,\ldots,X_r]\}$, finitely generated integral domain over $\mathbb{Z}$ with quotient field $K = \mathbb{Q}(z_1,\ldots,z_r)$
$A \simeq \mathbb{Z}[X_1,\ldots,X_r]/\mathcal{I}$	$\mathcal{I} := \{f \in \mathbb{Z}[X_1,\ldots,X_r] : f(z_1,\ldots,z_r) = 0\}$, finitely generated ideal in $\mathbb{Z}[X_1,\ldots,X_r]$
$\mathcal{I} = (f_1,\ldots,f_M)$	ideal representation for A
$\tilde{\alpha} \in \mathbb{Z}[X_1,\ldots,X_r]$	representative for $\alpha \in A$ if $\alpha = \tilde{\alpha}(z_1,\ldots,z_r)$
A effectively given	if an ideal representation $(f_1,\ldots,f_M)$ for A is given
$\alpha \in A$ effectively given (computable)	if a representative for α is given (can be computed)
$\{z_1 = X_1,\ldots,z_q = X_q\}$	transcendence basis for $K = \mathbb{Q}(z_1,\ldots,z_r)$ over $\mathbb{Q}$
$A_0 = \mathbb{Z}[X_1,\ldots,X_q]$	subring of A with unique factorization
$\deg \alpha$, $h(\alpha)$ for $\alpha \in A_0$	the total degree and logarithmic height of α
$K_0 = \mathbb{Q}(X_1,\ldots,X_q)$	quotient field of A_0
$K = K_0(w)$	where $w \in A$, integral over A_0 with degree D over K_0
$\overline{\deg} \, \alpha$ $(\alpha \in K)$	$\max(\deg P_{\alpha,0},\ldots, \deg P_{\alpha,D-1}, \deg Q_\alpha)$, where $P_{\alpha,0},\ldots,P_{\alpha,D-1},Q_\alpha \in A_0$ are relatively prime, and $\alpha = Q_\alpha^{-1} \sum_{j=0}^{D-1} P_{\alpha,j}\omega^j$
$\overline{h}(\alpha)$ $(\alpha \in K)$	$\max(h(P_{\alpha,0}),\ldots,h(P_{\alpha,D-1}),h(Q_\alpha))$

History and Summary

First, we give a brief historical overview of the equations treated in our book, and then outline the contents of the book.

We start with **ineffective** results. Thue (1909) developed an ingenious method for approximation of algebraic numbers by rationals. As an application, he proved that if $F \in \mathbb{Z}[X,Y]$ is a binary form (i.e., a homogeneous polynomial) of degree at least 3, which is irreducible over $\mathbb{Q}$ and δ is a nonzero integer, then the equation

$$F(x,y) = \delta \text{ in } x,y \in \mathbb{Z} \tag{1}$$

(nowadays called a *Thue equation*) has only finitely many solutions. Thue's approximation result was later considerably improved and generalized by many people including Siegel, Mahler, Dyson, Gel'fond, Roth, Schmidt, and Schlickewei.

Thue's finiteness theorem concerning equation (1) has many generalizations. Siegel (1921) generalized it for the number field case when the ground ring, i.e., the ring from which the solutions are taken, is the ring of integers $\mathcal{O}_K$ of a number field K. Mahler (1933) extended Thue's theorem to the case of ground rings of the form $\mathbb{Z}[(p_1, \ldots, p_s)^{-1}]$, where $p_1, \ldots, p_s$ are primes, while Parry (1950) gave a common generalization of the results of Siegel and Mahler to the case where the ground ring is the ring of S-integers of a number field.

Siegel's theorem has the following important consequence, which was not stated explicitly by Siegel but was implicitly proved by him. Denote by $\mathcal{O}_K^*$ the group of units of $\mathcal{O}_K$, and let α and β be nonzero elements of the number field K. Using the fact that $\mathcal{O}_K^*$ is finitely generated, it is easy to deduce from Siegel's theorem that the equation

$$\alpha x + \beta y = 1 \tag{2}$$

in $x, y \in \mathcal{O}_K^*$ has only finitely many solutions. Similarly, it follows from the results of Mahler and Parry that equation (2) has finitely many solutions even in S-units of K; these are elements of K composed of prime ideals from a finite, possibly empty set S of prime ideals of $\mathcal{O}_K$. Nowadays equation (2) is called a *unit equation* (when S is empty) resp. *S-unit equation* otherwise, or more precisely a unit equation and S-unit equation in two unknowns.

Further important equations are

$$f(x) = \delta y^m \text{ in } x, y \in \mathbb{Z}, \tag{3}$$

where $f \in \mathbb{Z}[X]$ is a polynomial of degree n and $\delta \in \mathbb{Z}\setminus\{0\}$. Equation (3) is called *elliptic* if $n = 3$ and $m = 2$, more generally *hyperelliptic* if $n \geq 3$ and $m = 2$, and *superelliptic* if $n \geq 2$ and $m \geq 3$. If m or n is at least 3 and f has no multiple zeros, equation (3) has only finitely many solutions. This was proved in the elliptic case by Mordell (1922a, 1922b, 1923), in the hyperelliptic case by Siegel (1929), and in the superelliptic case by Siegel (1929). LeVeque (1964) considered (3) in the more general case when f may have multiple zeros, and gave a finiteness criterion for (3) over the ring of integers of a number field.

A celebrated theorem of Siegel (1929) states that if $F(X, Y)$ is a polynomial with coefficients in a number field K, which is irreducible over $\overline{K}$, and the affine curve $F(x, y) = 0$ is of genus ≥ 1, then this curve has only finitely many points with integral coordinates in K. This theorem implies the above-mentioned finiteness results on Thue equations, unit equations, and hyperelliptic/superelliptic equations over number fields.

Lang (1960) generalized Siegel's theorem to what we call the **finitely generated** case, when the solutions are taken from an arbitrary integral domain of characteristic 0 that is finitely generated as a $\mathbb{Z}$-algebra, that is, a domain of the shape

$$\mathbb{Z}[z_1, \ldots, z_r] = \{f(z_1, \ldots, z_r) : f \in \mathbb{Z}[X_1, \ldots, X_r]\},$$

where $z_1, \ldots, z_r$ may be algebraic or transcendental over $\mathbb{Q}$. Recall that both the ring of integers of a number field K and the rings of S-integers of K are of this shape, with $z_1, \ldots, z_r$ all algebraic. In his proof, Lang used a specialization argument, reducing the theorem to the case of number fields and function fields of one variable, and then applied Siegel's theorem (1929) and its function field analogue from Lang (1960). As a consequence, Lang extended the earlier finiteness results concerning Thue equations, unit equations, and hyperelliptic/superelliptic equations to the finitely generated case.

Multivariate generalizations of Thue equations that have attracted much attention are the *decomposable form equations*

$$F(x_1, \ldots, x_m) = \delta \text{ in } x_1, \ldots, x_m \in \mathbb{Z}, \tag{4}$$

where $\delta \in \mathbb{Z}\backslash\{0\}$ and $F(X_1,\ldots,X_m)$ is a decomposable form of degree $n > m$ in $m \geq 2$ variables with coefficients in $\mathbb{Z}$, i.e., a homogeneous polynomial, which factorizes into linear forms with coefficients in the algebraic closure $\overline{\mathbb{Q}}$. Further important types of decomposable form equations are *norm form equations*, *discriminant form equations*, and *index form equations*, which are of basic importance in algebraic number theory. Schmidt (1971, 1972) developed a multidimensional generalization of Roth's theorem on the approximation of algebraic numbers, eventually leading to his famous Subspace Theorem, and from the latter he deduced a finiteness criterion for norm form equations. Evertse and Győry (1988b) proved a general finiteness criterion for decomposable form equations of the form (4). Their proof depends on the following finiteness result on *multivariate unit equations* of the form

$$\alpha_1 x_1 + \cdots + \alpha_m x_m = 1 \text{ in } x_1,\ldots,x_m \in \mathcal{O}_K^*, \qquad (5)$$

where K is a number field and $\alpha_1,\ldots,\alpha_m$ are nonzero elements of K. A solution of (5) is called *degenerate* if there is a vanishing subsum on the left hand side of (5). In this case (5) has infinitely many solutions if $\mathcal{O}_K^*$ is infinite. As a generalization of Siegel's theorem on equation (2), van der Poorten and Schlickewei (1982) and Evertse (1984) proved independently of each other that equation (5) has only finitely many non-degenerate solutions. This theorem was extended by Evertse and Győry (1988a) and van der Poorten and Schlickewei (1991) to the finitely generated case, when K is a finitely generated extension of $\mathbb{Q}$ and $\mathcal{O}_K^*$ is replaced by a finitely generated multiplicative subgroup of K^*. As a consequence, the above-mentioned general finiteness criterion for (4) was proved in Evertse and Győry (1988b) in a more general form, over finitely generated domains of characteristic 0.

In the 1960s, Baker developed an effective method in transcendence theory, providing nontrivial effective lower bounds for linear forms in logarithms of algebraic numbers. This furnished a very powerful tool to prove **effective** finiteness results for Diophantine equations over $\mathbb{Z}$ and more generally over number fields that enabled one to determine, at least in principle, all solutions of the equations under consideration. Using his method, Baker (1968b, 1968c, 1969) derived explicit upper bounds among others for the solutions of Thue equations and hyperelliptic/superelliptic equations. Győry (1974, 1979) used Baker's theory of logarithmic forms to obtain explicit upper bounds for the solutions of unit equations and S-unit equations in two unknowns (Evertse, Győry, Stewart and Tijdeman, 1988b). With the help of his bounds, Győry proved effective finiteness theorems for *discriminant equations* for polynomials

$$D(f) = \delta \text{ in monic polynomials } f \in \mathbb{Z}[X] \qquad (6)$$

and for elements

$$D(\alpha) = \delta \text{ in algebraic integers } \alpha. \tag{7}$$

Here, $D(\ )$ denotes the discriminant of a polynomial f resp. of an algebraic integer α, and δ is a nonzero integer. Two monic polynomials $f, f' \in \mathbb{Z}[X]$ are called *strongly $\mathbb{Z}$-equivalent* if $f'(X) = f(X + a)$ for some $a \in \mathbb{Z}$. Similarly, two algebraic integers α and α' are said to be *strongly $\mathbb{Z}$-equivalent* if $\alpha' - \alpha \in \mathbb{Z}$. Clearly, strongly $\mathbb{Z}$-equivalent monic polynomials resp. algebraic integers have the same discriminant.

Győry (1973) proved that there are only finitely many pairwise strongly $\mathbb{Z}$-inequivalent monic polynomials with the property (6). A similar finiteness theorem was proved for the solutions of (7) by Birch and Merriman (1972), and independently by Győry (1973). Győry's proofs for (6) and (7) are effective. These results, in less precise form, were generalized in Győry (1978a) for the number field case, and in Győry (1982) in an ineffective form, for the finitely generated case, subject to the condition that the ground ring is integrally closed. These results have many applications, among others, to power integral bases of ring extensions.

By using Győry's bounds on the solutions of unit equations in two unknowns, Győry (1976, 1980a) and Győry and Papp (1978) generalized Baker's effective theorem on Thue equations to equations in arbitrarily many unknowns. They derived explicit bounds for the solutions of a class of decomposable form equations over number fields, including discriminant form equations and certain norm form equations.

Tijdeman (1976) used Baker's theory of logarithmic forms to give an explicit upper bound for the solutions of the *Catalan equation*

$$x^m - y^n = 1 \text{ in positive integers } x, y, m, n \text{ with } m, n > 1 \text{ and } mn > 4. \tag{8}$$

Further, when in equation (3) m is also unknown and f has at least two distinct zeros, Schinzel and Tijdeman (1976) gave an effective upper bound for m. In this case, equation (3) is now called the *Schinzel–Tijdeman equation*. It is interesting to note that the effective theorems of Tijdeman (1976) and Schinzel and Tijdeman (1976) had no previously ineffective versions.

For Thue equations, unit equations, and hyper/superelliptic equations, analogous effective results were obtained by Mason (1981, 1983, 1984) and others over function fields of characteristic 0. The above-mentioned effective results over number fields and function fields were later improved and generalized by many people, and led to several further applications.

In Győry (1983, 1984b), the author extended the effective finiteness theorems concerning Thue equations, discriminant equations, and a class of decomposable form equations over number fields to similar such equations over restricted classes of finitely generated domains of characteristic 0, which may contain both algebraic and transcendental elements. To prove these extensions, Győry developed an effective specialization method to reduce the general equations under consideration to equations of the same type over number fields and function fields, and then used effective results concerning these reduced equations to derive effective bounds for the solutions of the initial equations.

Evertse and Győry (2013) refined the method of Győry and proved effective finiteness theorems for unit equations in two unknowns in full generality, over arbitrary finitely generated domains of characteristic 0 over $\mathbb{Z}$. In fact, they obtained their results by combining Győry's techniques with the work of Aschenbrenner (2004) concerning the effective resolution of systems of linear equations over polynomial rings $\mathbb{Z}[X_1, \ldots, X_n]$.

The general effective specialization method of Evertse and Győry led to effective finiteness results over finitely generated domains for several other classes of Diophantine equations, such as Thue equations, hyper/superelliptic equations, and the Schinzel–Tijdeman equation (Bérczes, Evertse and Győry 2014), a generalization of unit equations (Bérczes, 2015a, 2015b), and the Catalan equation (Koymans 2016, 2017). Further, generalizing another method of Győry (1973) and Győry and Papp (1978) applied over number fields, the present authors in Evertse and Győry (2017a, 2017b) and in Sections 2.6 and 2.8 of this book obtained effective finiteness theorems for decomposable form equations and discriminant equations over finitely generated domains. This other method is not based on specialization but instead uses a reduction of the equation under consideration to unit equations in two unknowns.

It is important to note that with the exception of discriminant equations and hyper- and superelliptic equations, both methods mentioned above provide quantitative results over finitely generated domains, giving effective bounds for the solutions. This is due to the effective and quantitative feature of the main tools from Chapters 4 to 8.

Major open problems are to make effective the general finiteness theorems of Siegel (1929) on integral points of curves and of van der Poorten and Schlickewei (1982) and Evertse (1984) on multivariate unit equations over number fields. Such effective versions could be extended to the finitely generated case, using existing analogues over function fields and applying our general effective specialization method.

We now outline the contents of our book. In Chapter 1, we present the most general ineffective finiteness results over finitely generated domains for Thue equations, unit equations in two unknowns, a generalization of unit equations, hyper- and superelliptic equations, curves of genus ≥ 1 with finitely many integral points, decomposable form equations, multivariate unit equations, and discriminant equations. Further, except for curves of genus ≥ 1 and multivariate unit equations, we cite the most general effective versions concerning the equations mentioned over number fields.

In Chapter 2, we state general effective finiteness theorems over finitely generated domains of characteristic 0 for unit equations in two unknowns, Thue equations, hyper- and superelliptic equations, the Schinzel–Tijdeman equation, the Catalan equation, decomposable form equations, and discriminant equations. As was mentioned above, apart from discriminant equations, the other results give also effective bounds for the solutions.

Chapter 3 is devoted to a short explanation of our general effective methods.

In Chapters 4 and 5, those effective results are collected on the above equations over number fields and function fields that are needed in Chapters 9 and 10, in the proofs of the general effective theorems stated in Chapter 2. We have skipped the complete proofs of the theorems in Chapters 4 and 5, which are rather technical. Instead, we sketch the proofs in simplified forms, which give sufficient insight into the main ideas.

Chapters 6–8 contain further important tools. In Chapter 6, we have collected results from effective commutative algebra; in Chapter 7, we give the detailed treatment of our effective specialization method; and in Chapter 8, we prove some useful results for "degree-height estimates," which may be viewed as an analogue of the naive height estimates of algebraic numbers for elements of the algebraic closure of a finitely generated field.

Lastly, in Chapters 9 and 10, the results and methods from Chapters 4 to 8 are combined to prove the general effective results presented in Chapter 2.

1

Ineffective Results for Diophantine Equations over Finitely Generated Domains

This book is about Diophantine equations where the solutions are taken from an integral domain of characteristic 0 that is finitely generated over $\mathbb{Z}$, that is, from a domain of the shape

$$\mathbb{Z}[z_1,\ldots,z_r] = \{f(z_1,\ldots,z_r) : f \in \mathbb{Z}[X_1,\ldots,X_r]\}$$

whose quotient field is of characteristic 0. The generators $z_1,\ldots,z_r$ may be either algebraic or transcendental over $\mathbb{Q}$.

For instance, let K be a number field and $\mathcal{O}_K$ its ring of integers. Let $\{\omega_1,\ldots,\omega_d\}$ be a $\mathbb{Z}$-module basis of $\mathcal{O}_K$. Then $\mathcal{O}_K = \mathbb{Z}[\omega_1,\ldots,\omega_d]$.

More generally, let K be a number field and with the notation introduced in Section 4.2, let S be a finite set of places of K, consisting of all infinite places of K and of the prime ideals $\mathfrak{p}_1,\ldots,\mathfrak{p}_t$ of $\mathcal{O}_K$. Then the ring of S-integers of K, denoted by $\mathcal{O}_S$, is given by the set of all elements α of K such that there are non-negative integers $k_1,\ldots,k_t$ with $\alpha\mathfrak{p}_1^{k_1}\cdots\mathfrak{p}_t^{k_t} \subseteq \mathcal{O}_K$. In the particular case that S consists only of the infinite places of K, the ring $\mathcal{O}_S$ is just equal to $\mathcal{O}_K$. We may express $\mathcal{O}_S$ otherwise as

$$\mathcal{O}_S = \mathbb{Z}[\omega_1,\ldots,\omega_d,\pi^{-1}],$$

where again, $\{\omega_1,\ldots,\omega_d\}$ is a $\mathbb{Z}$-module basis of $\mathcal{O}_K$ and where $\pi\mathcal{O}_K = (\mathfrak{p}_1,\ldots,\mathfrak{p}_t)^{h_K}$ with h_K being the class number of K. Thus, both the ring of integers and the rings of S-integers of a number field are domains finitely generated over $\mathbb{Z}$, with algebraic generators.

In general, we will consider Diophantine equations over integral domains $\mathbb{Z}[z_1,\ldots,z_r]$ where some of the generators, say $z_1,\ldots,z_q$, are algebraically independent of $\mathbb{Q}$, and the other generators are algebraic over $\mathbb{Q}(z_1,\ldots,z_q)$.

In this chapter, we present the most important **ineffective** finiteness theorems for integral solutions of various classes of Diophantine equations,

1

including *Thue equations, unit equations, hyper- and superelliptic equations, equations involving integral points on curves, decomposable form equations, and discriminant equations.* We consider these classes of equations in separate sections. For each class, we state the finiteness results in their most general form, over an arbitrary integral domain of characteristic 0 that is finitely generated over $\mathbb{Z}$, and give an account of the earlier special cases, leading to the general result. Over $\mathbb{Z}$ or more generally over the rings of integers or S-integers of number fields, these results were proved mostly by the powerful Thue–Siegel–Roth–Schmidt method, while in the finitely generated case, the equations are reduced either to the number field and function field cases by means of some specialization arguments or to such equations for which the finiteness of the number of solutions is already proved; see, e.g., Lang (1960), Győry (1982), Evertse and Győry (1988a, 1988b), and van der Poorten and Schlickewei (1991). At the end of each section, we make a mention to the corresponding **effective** results over $\mathbb{Z}$ or over number fields whose general versions over finitely generated domains will be presented in Chapter 2.

The above-mentioned equations have been studied very extensively, and they have many important generalizations, analogues, and applications. For details, we refer, e.g., to the books Lang (1962, 1978, 1983), Borevich and Shafarevich (1967), Mordell (1969), Baker (1975), Győry (1980b), Evertse (1983), Mason (1984), Shorey and Tijdeman (1986), Schmidt (1991), Sprindžuk (1993), Bombieri and Gubler (2006), Zannier (2009), Evertse and Győry (2015, 2017a), Bugeaud (2018), and the survey papers of Evertse, Győry, Stewart, and Tijdeman (1988b), and Győry (1984a, 1992, 2002).

1.1 Thue Equations

Let A denote an integral domain of characteristic 0 that is finitely generated over $\mathbb{Z}$. Let K denote the quotient field of A and fix an algebraic closure $\overline{K}$ of K. We first consider the equation

$$F(x,y) = \delta \quad \text{in} \quad x, y \in A \tag{1.1.1}$$

over A, where $F(X,Y)$ is a binary form of degree n with coefficients in A and $\delta \in A \setminus \{0\}$.

The following result is a consequence of the more general Theorem 1.4.1, which will be stated in Section 1.4.

Theorem 1.1.1 *Assume that F has at least three pairwise nonproportional linear factors over $\overline{K}$. Then equation (1.1.1) has only finitely many solutions.*

The condition in the theorem is obviously satisfied if F has degree at least 3 and its discriminant is nonzero. This theorem cannot be extended to binary forms F with fewer than three pairwise nonproportional linear factors; for instance, the Pell equation $x^2 - dy^2 = 1$ over $\mathbb{Z}$, where d is a positive integer not being a square, has infinitely many solutions.

In the classical case $A = \mathbb{Z}$, Theorem 1.1.1 was proved by Thue (1909). In fact, Thue proved it for irreducible F, but the general case can be easily reduced to the irreducible one. The proof of Thue's theorem is based on his result concerning approximations of algebraic numbers by rationals. After Thue, equations of the shape (1.1.1) are named *Thue equations*.

Thue's theorem has been generalized by many people. Siegel (1921) extended it to the case when A is the ring of integers of a number field, and Mahler (1933) extended it to rings of the shape $\mathbb{Z}[(p_1, \ldots, p_s)^{-1}]$, where $p_1, \ldots, p_s$ are distinct primes. Parry (1950) gave a common generalization of the results of Siegel and Mahler to rings of S-integers of a number field. In the above general form, Theorem 1.1.1 is due to Lang (1960).

We would like to mention another equivalent formulation of Theorem 1.1.1. First, we recall a result of Mahler (1933). Let $F \in \mathbb{Z}[X, Y]$ be a binary form with at least three pairwise nonproportional linear factors over $\overline{\mathbb{Q}}$, and let $p_1, \ldots, p_s$ be distinct prime numbers. Then the equation

$$F(x, y) = \pm p_1^{z_1} \cdots p_s^{z_s} \text{ in } x, y, z_1, \ldots, z_s \in \mathbb{Z} \text{ with } \gcd(x, y) = 1 \qquad (1.1.2)$$

has only finitely many solutions. If we drop the restriction $\gcd(x, y) = 1$, then we can construct infinite classes of solutions by multiplying (x, y) with products of powers of $p_1, \ldots, p_s$. Thus, it is easily seen that Mahler's result can be translated as follows. Let $S = \{p_1, \ldots, p_s\}$ be a finite set of primes, $\mathbb{Z}_S = \mathbb{Z}[(p_1, \ldots, p_s)^{-1}]$ the corresponding ring of S-integers, and $\mathbb{Z}_S^* = \{\pm p_1^{z_1} \cdots p_s^{z_s} : z_1, \ldots, z_s \in \mathbb{Z}\}$ the group of units of $\mathbb{Z}_S$. Then the solutions of

$$F(x, y) \in \mathbb{Z}_S^* \text{ in } (x, y) \in \mathbb{Z}_S^2 \qquad (1.1.3)$$

lie in finitely many $\mathbb{Z}_S^*$-cosets, where a $\mathbb{Z}_S^*$-coset is a set of solutions of the shape $\{u \cdot (x_0, y_0) : u \in \mathbb{Z}_S^*\}$, with $(x_0, y_0) \in \mathbb{Z}_S^2$ fixed.

We now generalize this last equation to arbitrary finitely generated domains of characteristic 0 that are finitely generated over $\mathbb{Z}$. Let A be such a domain and denote by A^* its unit group, i.e., the group of invertible elements. Further, let $F \in A[X, Y]$ be a binary form and δ a nonzero element of A, and consider the following generalization of (1.1.3):

$$F(x, y) \in \delta A^* \text{ in } (x, y) \in A^2. \qquad (1.1.4)$$

Because of its connection with (1.1.2), equation (1.1.4) is called a *Thue–Mahler equation*. Just like above, we can divide the solutions $(x, y) \in A^2$ of (1.1.4) into A^*-*cosets* $A^*(x_0, y_0) = \{u \cdot (x_0, y_0) : u \in A^*\}$.

The following assertion is equivalent to Theorem 1.1.1.

Theorem 1.1.2 *Assume again that F has at least three pairwise nonproportional linear factors over $\overline{K}$. Then equation (1.1.4) has only finitely many A^*-cosets of solutions.*

Theorem 1.1.1$\Rightarrow$Theorem 1.1.2 Assume Theorem 1.1.1. According to a theorem of Roquette (1957), the unit group A^* is finitely generated. Let $\{v_1, \ldots, v_s\}$ be a set of generators for A^*, and define $\mathcal{U} := \{v_1^{m_1} \cdots v_s^{m_s} : m_1, \ldots, m_s \in \{0, \ldots, n-1\}\}$. Then every element of A^* can be expressed as $u_1 u_2^n$, where $u_1 \in \mathcal{U}$ and $u_2 \in A^*$. Clearly, if $(x, y) \in A^2$ satisfies (1.1.4), then $F(x, y) = \delta u_1 u_2^n$ for some $u_1 \in \mathcal{U}$, $u_2 \in A^*$, and so $F(x', y') = \delta u_1$, where $(x', y') = u_2^{-1}(x, y)$. Hence, every A^*-coset of solutions of (1.1.4) contains (x', y') with $F(x', y') = \delta u_1$ with some $u_1 \in \mathcal{U}$, and Theorem 1.1.1 implies that for each $u_1 \in \mathcal{U}$, there are only finitely many possibilities for (x', y'). This implies Theorem 1.1.2.

Theorem 1.1.2$\Rightarrow$Theorem 1.1.1 Assume Theorem 1.1.2. Let $A^*(x_0, y_0)$ be one of the finitely many A^*-cosets of solutions of (1.1.4) and pick those solutions from it that satisfy (1.1.1). These solutions are all of the shape $u(x_0, y_0)$ with $u^n = F(x_0, y_0)/\delta$, and there are only finitely many of those. Hence, (1.1.1) has only finitely many solutions. $\qquad\qquad\qquad\qquad\qquad\qquad\qquad\qquad\qquad\qquad\qquad$ $\square$

Equation (1.1.1) has many further generalizations, see, e.g., equation (1.4.1) in Section 1.4, equations (1.5.1), (1.5.2), and (1.5.4) in Section 1.5, and Evertse and Győry (2015, Chapter 9).

In the case $A = \mathbb{Z}$, the first general **effective** result for equation (1.1.1) was established by Baker (1968b). He gave an explicit upper bound for the solutions by means of his effective method based on lower bounds for linear forms in logarithms. Coates (1969) extended Baker's result to the case of ground rings of the type $A = \mathbb{Z}[(p_1 \cdots p_s)^{-1}]$, and later, Kotov and Sprindžuk (1973) extended that to the case when A is the ring of S-integers of a number field. Győry (1983), using his effective specialization method, generalized the above results for a wide but special class of finitely generated domains that may contain both algebraic and transcendental elements. In Chapter 2, Theorem 2.3.1 gives an effective version of Theorem 1.1.1 in quantitative form over an arbitrary integral domain of characteristic 0 that is finitely generated over $\mathbb{Z}$. Its proof uses a precise effective version of Theorem 1.1.1 over rings of S-integers of number fields; see Theorem 4.4.1 in Chapter 4, as well as an effective version over function fields, see Theorem 5.4.1 in Chapter 5, which is a slight variation of a result of Mason (1981, 1984).

1.2 Unit Equations in Two Unknowns

Let again A be an integral domain of characteristic 0 that is finitely generated over $\mathbb{Z}$ and K its quotient field. Further, let a and b be the nonzero elements of K. Consider the *unit equation*

$$ax + by = 1 \quad \text{in} \quad x, y \in A^*, \tag{1.2.1}$$

where A^* denotes the unit group of A, i.e., the multiplicative group of invertible elements of A.

By a theorem of Roquette (1957), the group A^* is finitely generated. Lang (1960) proved the following general result.

Theorem 1.2.1 *Equation* (1.2.1) *has only finitely many solutions.*

The first finiteness result for equation (1.2.1) was implicitly proved by Siegel (1921) in the case where K is a number field and A is the ring of integers of K. For the case when A is of the type $\mathbb{Z}[(p_1 \cdots p_s)^{-1}]$ with distinct primes $p_1, \ldots, p_s$, the finiteness of the number of solutions was obtained by Mahler (1933), while a common generalization of the results of Siegel and Mahler follows from Parry (1950).

In fact, in Lang (1960), the following more general version of Theorem 1.2.1 is established.

Theorem 1.2.2 *Let K be a field of characteristic* 0, *a and b the nonzero elements of K, and Γ a finitely generated multiplicative subgroup of K^*. Then the equation*

$$ax + by = 1 \quad \text{in} \quad x, y \in \Gamma \tag{1.2.2}$$

has only finitely many solutions.

Proof Using an argument due to Siegel (1921), the theorem can be easily reduced to Theorem 1.1.1. Indeed, suppose that equation (1.2.2) has infinitely many solutions. Let n be an integer ≥ 3. Since Γ is finitely generated, the quotient group Γ/Γ^n is finite. Hence, there is a solution (x_0, y_0) of (1.2.2) such that there are infinitely many solutions x, y such that $x \in x_0\Gamma^n$ and $y \in y_0\Gamma^n$. Each of these solutions x, y can be written in the forms $x = x_0 u^n$ and $y = y_0 v^n$ with some $u, v \in \Gamma$. Denoting by A the ring generated by Γ over $\mathbb{Z}$, it follows that the Thue equation

$$(ax_0)u^n + (by_0)v^n = 1$$

has infinitely many solutions $u, v \in A$. This contradicts Theorem 1.1.1. $\quad\square$

We note that, conversely, Thue equations can be reduced to finitely many appropriate unit equations; see, e.g., Evertse and Győry (2015). In other words, Thue equations and unit equations in two unknowns are, in fact, equivalent. This was (implicitly) pointed out by Siegel (1926).

Theorem 1.2.2 has several generalizations, see, e.g., Theorem 1.5.4 in Section 1.5, Lang (1960, 1983), and Evertse and Győry (2015). Here we present one of them.

Lang (1960) extended his result concerning equation (1.2.2) to equations of the shape

$$F(x,y) = 0 \quad \text{in} \quad x,y \in \Gamma, \tag{1.2.3}$$

where Γ is again a finitely generated multiplicative subgroup of K, and where $F \in A[X,Y]$ is a nonconstant polynomial. He proved the following.

Theorem 1.2.3 *Let $F \in A[X,Y]$ be a nonconstant polynomial that is not divisible by any polynomial of the shape*

$$X^m Y^n - \alpha \quad \text{or} \quad X^m - \alpha Y^n, \tag{1.2.4}$$

with $\alpha \in \Gamma$ and with non-negative integers m and n, not both zero. Then equation (1.2.3) has only finitely many solutions.

It is easy to see that the exceptions described in Theorem 1.2.3 must be excluded.

Lang (1965a, 1965b) conjectured that Theorem 1.2.3 remains valid if one replaces Γ by its division group $\overline{\Gamma}$, which consists of those $\gamma \in \overline{K}^*$ such that $\gamma^k \in \Gamma$ for some positive integer k. Hence, in this case, the solutions x, y do not necessarily belong to K. Lang's conjecture has been proved by Liardet (1974, 1975) who obtained the following.

Theorem 1.2.4 *Let $F \in A[X,Y]$ be a nonconstant polynomial that is not divisible by any polynomial of the shape (1.2.4) with $\alpha \in \overline{\Gamma}$ and with non-negative integers m and n, not both zero. Then equation (1.2.3) has only finitely many solutions even in $x, y \in \overline{\Gamma}$.*

The first general **effective** results for equation (1.2.1) over the ring of integers of algebraic number fields were proved in Győry (1972, 1973, 1974, 1976), over rings of S-integers of an algebraic number field in Győry (1979), and independently, in a less precise form, in Kotov and Trelina (1979). Using Baker's method concerning linear forms in logarithms, effective upper bounds were given for the solutions. These bounds were improved later by several authors; see, e.g., Bugeaud and Győry (1996a), Győry and Yu (2006), and Győry (2019).

Over algebraic number fields, Bombieri and Gubler (2006) gave an effective version of Lang's theorem on equation (1.2.3), which was made explicit by Bérczes, Evertse, Győry, and Pontreau (2009). These results are proved under a slightly stronger condition than (1.2.4), with $\alpha \in \overline{K}$ used in place of $\alpha \in \Gamma$.

In the number field case, an effective version of Liardet's theorem for linear polynomials F is due to Bérczes, Evertse, and Győry (2009), and for the general case to Bérczes, Evertse, Győry, and Pontreau (2009).

In Section 2.2, we present effective versions of Theorems 1.2.1 and 1.2.2 in quantitative form over an arbitrary integral domain of characteristic 0 that is finitely generated over $\mathbb{Z}$; see Theorems 2.2.1 and 2.2.3. In its proof, we use the result of Győry and Yu (2006) concerning equation (1.2.1) for the group of S-units of a number field, as well as the Mason–Stothers abc-theorem for function fields (as in Mason (1984)), see Theorem 5.2.2 in Chapter 5. Further, we formulate some effective generalizations for equation (1.2.3), due to Bérczes (2015a, 2015b), see Theorems 2.2.4 and 2.2.5.

1.3 Hyper- and Superelliptic Equations

Now consider the equation

$$f(x) = \delta y^m \quad \text{in} \quad x, y \in A, \tag{1.3.1}$$

where A is again an integral domain of characteristic 0 that is finitely generated over $\mathbb{Z}$, $f \in A[X]$ is a polynomial of degree $n \geq 2$, $\delta \in A \setminus \{0\}$, and $m \geq 2$ is an integer. Equation (1.3.1) is called *elliptic* if $n = 3$ and $m = 2$, *hyperelliptic* if $n \geq 3$ and $m = 2$, and *superelliptic* if $n \geq 2$ and $m \geq 3$.

The following theorem follows from the general ineffective Theorem 1.4.1 of Lang.

Theorem 1.3.1 *Suppose that in* (1.3.1) *m or n is at least* 3 *and that f has no multiple zeros. Then* (1.3.1) *has only finitely many solutions.*

Under the assumptions of Theorem 1.3.1, the affine curve $f(x) - \delta y^m = 0$ has genus ≥ 1. Thus, Theorem 1.3.1 is a consequence of the general Theorem 1.4.1 stated below on the finiteness of the number of intregral points on algebraic curves. The example of Pell equations shows that (1.3.1) may have infinitely many solutions if $m = 2$ and $n = 2$.

In the special case $A = \mathbb{Z}$, Mordell (1922a, 1922b, 1923) proved the finiteness of the numbers of solutions of elliptic equations for which the polynomial f has no multiple zeros. In particular, this implies that for every nonzero integer k, the *Mordell equation* $x^3 + k = y^2$ has only finitely many solutions. Mordell's finiteness results were extended by Siegel (1926) to hyperelliptic equations, by reducing such equations to unit equations. LeVeque (1964) considered (1.3.1), where f may have multiple zeros, and gave a finiteness criterion for the equation (1.3.1) when A is the ring of integers of a number field. The proofs of Mordell, Siegel, and LeVeque are ineffective.

Over $\mathbb{Z}$, Baker (1968b, 1968c, 1969) was the first to give **effective** upper bounds for the solutions of (1.3.1) in the case when f has at least three simple zeros if $m = 2$ and at least two simple zeros if $m \geq 3$. Brindza (1984) made LeVeque's theorem effective and extended it to S-integral solutions from a number field.

Schinzel and Tijdeman (1976) considered equation (1.3.1) in the more general situation when m is also unknown. In the case that $A = \mathbb{Z}$ and that f has at least two distinct zeros, they derived an effective upper bound for m. Equation (1.3.1) with m also unknown is nowadays called the *Schinzel–Tijdeman equation*. All the effective results mentioned above depend on Baker's method.

In Chapter 2, we present effective versions of Theorem 1.3.1 and the Schinzel–Tijdeman theorem in quantitative form, over an arbitrary integral domain of characteristic 0 that is finitely generated over $\mathbb{Z}$; see Theorems 2.4.1 and 2.4.2. These results follow from similar effective results over number fields (see Theorems 4.5.1–4.5.3) and function fields (see Theorems 5.5.1 and 5.5.2).

1.4 Curves with Finitely Many Integral Points

Let K be a finitely generated extension of $\mathbb{Q}$ and A a subring of K that is finitely generated over $\mathbb{Z}$. The following finiteness theorem is of fundamental importance in Diophantine number theory.

Theorem 1.4.1 *Let $F \in K[X,Y]$ be a polynomial irreducible over $\overline{K}$ such that the affine curve $F(x,y) = 0$ is of genus ≥ 1. Then this curve has only finitely many points with coordinates in A.*

In other words, under the above assumptions, the equation

$$F(x,y) = 0 \quad \text{in} \quad x,y \in A \qquad (1.4.1)$$

has only finitely many solutions.

In the case when K is a number field and A its ring of integers, this celebrated theorem was proved by Siegel (1929). Further, Siegel described the cases when the curve has genus 0 and has infinitely many points with coordinates in A. Mahler (1934) conjectured that a similar statement holds for rational points with coordinates having only finitely many fixed primes in their denominators, and proved this for curves of genus 1. In the above general form, Theorem 1.4.1 is due to Lang (1960); see also Lang (1962, 1983). In this proof, Lang used a specialization argument, reducing Theorem 1.4.1 to the case of number fields resp. function fields of one variable, and then applied Siegel's theorem and its analogue over function fields from Lang (1960).

Confirming Mordell's (1922a) famous conjecture on rational points on curves, Faltings (1983) proved, first for number fields K and later for finitely generated extensions K of $\mathbb{Q}$; (see Faltings and Wüstholz [1984, p. 205, Theorem 3]), that if the above curve has genus ≥ 2, then it has only finitely many points even with coordinates in K as well. Except for the genus 1 case, Faltings' theorem contains Theorem 1.4.1.

All known proofs of Theorem 1.4.1 and those of Faltings are ineffective. As was mentioned in Sections 1.1–1.3, Theorem 1.4.1 has been made **effective** in a couple of important special cases. Further, in the case when K is a number field, an effective version of Theorem 1.4.1 for genus 1 curves was obtained by Baker and Coates (1970).

It is a **major open problem** to give an effective version of Theorem 1.4.1 in full generality.

1.5 Decomposable Form Equations and Multivariate Unit Equations

Let K be a finitely generated extension field of $\mathbb{Q}$ and $F \in K[X_1, \ldots, X_m]$ a decomposable form in $m \geq 2$ variables, i.e., F factorizes into linear forms over an extension of K, which we may choose to be a given algebraic closure $\overline{K}$ of K. Let $\delta \in K^*$ and A be a subring of K that is finitely generated over $\mathbb{Z}$. As a generalization of the Thue equation, we consider the *decomposable form equation*

$$F(\mathbf{x}) = \delta \ \text{ in } \mathbf{x} = (x_1, \ldots, x_m) \in A^m. \tag{1.5.1}$$

Let $\mathcal{L}_0$ be a maximal set of pairwise linearly independent linear factors of F. That is, we can express F as $c\ell_1^{e_1} \cdots \ell_n^{e_n}$, where $\mathcal{L}_0 = \{\ell_1, \ldots, \ell_n\}$, $c \in K^*$, and $e_1, \ldots, e_n$ are positive integers. For applications, it is convenient to consider the following generalization of equation (1.5.1). Let $\mathcal{L} \supseteq \mathcal{L}_0$ be a finite set of pairwise linearly independent linear forms of $X_1, \ldots, X_m$, with coefficients in $\overline{K}$, and consider now the equation

$$F(\mathbf{x}) = \delta \ \text{ in } \mathbf{x} = (x_1, \ldots, x_m) \in A^m \text{ with } \ell(\mathbf{x}) \neq 0 \text{ for all } \ell \in \mathcal{L}. \tag{1.5.1a}$$

For $\mathcal{L} = \mathcal{L}_0$, equation (1.5.1a) gives (1.5.1).

To state the main results, we need some definitions. Given a nonzero linear subspace V of the K-vector space K^m and linear forms $\ell_1, \ldots, \ell_r$ in $\overline{K}[X_1, \ldots, X_m]$, we say that $\ell_1, \ldots, \ell_r$ are linearly dependent on V if there are $c_1, \ldots, c_r \in \overline{K}$, not all 0, such that $c_1 \ell_1 + \cdots + c_r \ell_r$ vanishes identically on V. Otherwise, we say that $\ell_1, \ldots, \ell_r$ are linearly independent on V.

We say that a nonzero linear subspace V of K^m is $\mathcal{L}$-*nondegenerate* if $\mathcal{L}$ contains $r \geq 3$ linear forms $\ell_1, \ldots, \ell_r$ that are linearly dependent on V, while each pair ℓ_i, ℓ_j ($i \neq j$) is linearly independent of V. Otherwise, the space V is called $\mathcal{L}$-*degenerate*. That is, V is $\mathcal{L}$-degenerate precisely if there are $\ell_1, \ldots, \ell_r \in \mathcal{L}$ such that $\ell_1, \ldots, \ell_r$ are linearly independent of V, while each other's linear form $\ell \in \mathcal{L}$ is linearly dependent on V to one of $\ell_1, \ldots, \ell_r$. In particular, V is $\mathcal{L}$-degenerate if V has dimension 1.

Lastly, we call V $\mathcal{L}$-*admissible* if no linear form in $\mathcal{L}$ vanishes identically on V.

The following general finiteness criterion was proved by Evertse and Győry (1988b).

Theorem 1.5.1 *The following two statements are equivalent:*

(i) *Every $\mathcal{L}$-admissible linear subspace of K^m of dimension ≥ 2 is $\mathcal{L}_0$-nondegenerate;*

(ii) *For every subring A of K that is finitely generated over $\mathbb{Z}$ and for every $\delta \in K^*$, equation* (1.5.1a) *has only finitely many solutions.*

For $\mathcal{L} = \mathcal{L}_0$, this theorem gives a finiteness criterion for equation (1.5.1). It relates a statement (cf. (ii)) about the finiteness of the number of solutions to a condition (cf. (i)) that can be formulated in terms of linear algebra. It can be shown that (i) is effectively decidable once K, $\mathcal{L}_0$, and $\mathcal{L}$ are given in some explicit form; see Evertse and Győry (2015, Theorem 9.1.1) for an equivalent formulation of (i) for which the effective decidability is clear.

In the case $m = 2$ and $\mathcal{L} = \mathcal{L}_0$, Theorem 1.5.1 gives immediately Theorem 1.1.1 on Thue equations. For a more general version of Theorem 1.5.1, see Evertse and Győry (2015, Chapter 9).

Decomposable form equations are of basic importance in Diophantine number theory. Besides Thue equations (when $m = 2$), important classes of decomposable form equations are norm form equations, discriminant form equations, and index form equations.

Let us start with norm form equations. Let $\alpha_1 = 1, \alpha_2, \ldots, \alpha_m \in \overline{K}$ and suppose they are linearly independent over K. Put $K' := K(\alpha_1, \ldots, \alpha_m)$. Assume that K' is of degree $n \geq 3$ over K. Put $\ell(\mathbf{X}) := \alpha_1 X_1 + \cdots + \alpha_m X_m$ and denote by $\ell^{(i)}(\mathbf{X}) := \alpha_1^{(i)} X_1 + \cdots + \alpha_m^{(i)} X_m$ ($i = 1, \ldots, n$) the conjugates of $\ell(\mathbf{X})$ with respect to K'/K. Then

$$N_{K'/K}(\alpha_1 X_1 + \cdots + \alpha_m X_m) := \prod_{i=1}^{n} \ell^{(i)}(\mathbf{X})$$

is a decomposable form of degree n with coefficients in K. Such a form is called a *norm form* over K (or with respect to K'/K) and, for $\delta \in K^*$,

$$N_{K'/K}(\alpha_1 x_1 + \cdots + \alpha_m x_m) = \delta \quad \text{in} \quad x_1, \ldots, x_m \in A \qquad (1.5.2)$$

a *norm form equation*.

Let $\mathcal{V}$ be the K-vector space generated by $\alpha_1, \ldots, \alpha_m$ in K'. We say that $\mathcal{V}$ is *degenerate* if there exist a $\mu \in K'^*$ and an intermediate number field K'' with $K \subsetneq K'' \subsetneq K'$, such that $\mu K'' \subseteq \mathcal{V}$. The following finiteness criterion is a consequence of Theorem 1.5.1; see Evertse and Győry (1988b).

Corollary 1.5.2 *The following two statements are equivalent:*

(i) $\mathcal{V}$ is nondegenerate.
(ii) For all $\delta \in K^*$ and all subrings A of K that are finitely generated over $\mathbb{Z}$, equation (1.5.2) has only finitely many solutions.

For $K = \mathbb{Q}$, $A = \mathbb{Z}$, Schmidt (1971) gave a criterion for equation (1.5.2) to have only finitely many solutions for every $\delta \in \mathbb{Q}^*$. Then Schmidt (1972) proved that all solutions of (1.5.2) over $\mathbb{Z}$ belong to finitely many so-called families of solutions. These results were later extended by Schlickewei (1977) to the case of arbitrary finitely generated subrings A of $\mathbb{Q}$ and by Laurent (1984) to the above general case. As a generalization of Schmidt's (1972) result, Győry (1993) showed that all solutions of equation (1.5.1) belong to finitely many so-called wide families of solutions.

Next, consider discriminant form equations. Let again $K' = K(\alpha_1, \ldots, \alpha_m)$ be an extension of degree $n \geq 3$ of K, where now $1, \alpha_1, \ldots, \alpha_m$ are K-linearly independent elements of K'. Let $\ell^{(i)}(\mathbf{X}) = X_0 + \alpha_1^{(i)} X_1 + \cdots + \alpha_m^{(i)} X_m$, $i = 1, \ldots, n$, be the conjugates of $\ell(\mathbf{X}) = X_0 + \alpha_1 X_1 + \cdots + \alpha_m X_m$ with respect to K'/K. Then the decomposable form

$$D_{K'/K}(\alpha_1 X_1 + \cdots + \alpha_m X_m) = \prod_{1 \leq i < j \leq n} (\ell^{(i)}(\mathbf{X}) - \ell^{(j)}(\mathbf{X}))^2$$

has its coefficients in K and is independent of X_0. It is called a *discriminant form*, while, for $\delta \in K^*$,

$$D_{K'/K}(\alpha_1 x_1 + \cdots + \alpha_m x_m) = \delta \quad \text{in} \quad x_1, \ldots, x_m \in A \qquad (1.5.3)$$

is called a *discriminant form equation*.

The following finiteness result is due to Győry (1982). It can be deduced from Theorem 1.5.1 as well.

Theorem 1.5.3 *Under the above assumptions, equation* (1.5.3) *has only finitely many solutions.*

This theorem and its various versions have several important applications, among others, to *index form equations* and *power integral bases*; for references, see, e.g., Győry (1980b) and Evertse and Győry (2017a).

The above results concerning equations (1.5.1), (1.5.1a), (1.5.2), and (1.5.3) have been extended to equations of the form

$$F(\mathbf{x}) \in \delta A^* \quad \text{in} \quad \mathbf{x} = (x_1, \dots, x_m) \in A^m. \tag{1.5.4}$$

The set of solutions of equation (1.5.4) can be divided into A^*-cosets $\mathbf{x}_0 A^*$, where $\mathbf{x}_0 = (x_1, \dots, x_m)$ is a solution of (1.5.4). As was already mentioned, by a theorem of Roquette (1957) A^* is finitely generated. Hence, (1.5.4) can be reduced to finitely many equations of the form (1.5.1).

The proof of Theorem 1.5.1 and its variant concerning (1.5.4) depend on the following finiteness result on *multivariate unit equations* of the form

$$a_1 x_1 + \cdots + a_m x_m = 1 \quad \text{in} \quad x_1, \dots, x_m \in A^* \quad \text{resp. in} \quad \Gamma, \tag{1.5.5}$$

where K is a field of characteristic 0, $a_1, \dots, a_m$ are nonzero elements of K, A is a subring of K that is finitely generated over $\mathbb{Z}$, and Γ is a finitely generated subgroup of K^*. A solution $x_1, \dots, x_m$ of (1.5.5) is called *degenerate* if there is a vanishing subsum on the left-hand side of (1.5.5). In this case (1.5.5) has obviously infinitely many solutions if A^* resp. Γ is infinite. The following theorem was proved by van der Poorten and Schlickewei (1982) and Evertse (1984) in the number field case, and by Evertse and Győry (1988a) and van der Poorten and Schlickewei (1991) in the finitely generated case.

Theorem 1.5.4 *Equation* (1.5.5) *has only finitely many nondegenerate solutions.*

As is pointed out in Evertse and Győry (1988b), Theorem 1.5.4 and the implication (i)$\Rightarrow$(ii) of Theorem 1.5.1 are equivalent statements; see also Evertse and Győry (2015, Chapter 9).

The afore mentioned results are all ineffective. In certain important cases, they have **effective** versions. Concerning the discriminant form equation (1.5.3) over $\mathbb{Z}$, the first effective finiteness result was obtained by Győry (1976). This was extended to the number field case by Győry and Papp (1977) and Győry (1981a). Győry and Papp (1978) over $\mathbb{Z}$ and Győry (1981a) over arbitrary number fields established effective finiteness theorems for equations (1.5.1), (1.5.1a), and (1.5.2), for some classes of decomposable forms and norm forms, including binary forms and discriminant forms. As was mentioned in

Section 1.2, the first effective finiteness results for bivariate unit equations, i.e., equations (1.5.5) in $m = 2$ unknowns over number fields were given by Győry (1972, 1973, 1974).

Győry (1983) extended his effective results on equations (1.5.1), (1.5.2) and (1.5.3) over number fields to a class of finitely generated ground domains over $\mathbb{Z}$, which may contain both algebraic and transcendental elements over $\mathbb{Q}$. In Chapter 2, we present a further extension, in a slightly more general form, to the case of arbitrary ground domains of characteristic 0 that are finitely generated over $\mathbb{Z}$; see Theorem 2.6.1. However, apart from the case of general discriminant form equations (1.5.3), it remains a **major open problem** to make Theorem 1.5.1, Corollary 1.5.2, and Theorem 1.5.4 effective in full generality.

1.6 Discriminant Equations for Polynomials and Integral Elements

Let again A be an integral domain of characteristic 0 that is finitely generated over $\mathbb{Z}$ and K its quotient field. Take a finite extension G of K. Let $n \geq 2$ be an integer, δ a nonzero element of A, and consider the *discriminant equation for polynomials*

$$D(f) = \delta \text{ in monic } f \in A[X] \text{ of degree } n \text{ having all its zeros in } G, \quad (1.6.1)$$

where $D(f)$ denotes the discriminant of f. Two monic polynomials $f, f' \in A[X]$ are called *strongly A-equivalent* if $f'(X) = f(X + a)$ for some $a \in A$.[1] In this case, f and f' have the same discriminant. Hence, the solutions of (1.6.1) can be divided into strong A-equivalence classes.

Denote by A_K the integral closure of A in K.

Theorem 1.6.1 *Let $n \geq 2$ be an integer and A an integral domain of characteristic 0, finitely generated over $\mathbb{Z}$, with a quotient field K such that the quotient A-module*

$$(\tfrac{1}{n}A \cap A_K)/A \quad \text{is finite.} \tag{1.6.2}$$

Further, let G be a finite extension of K and δ a nonzero element of A. Then the set of monic polynomials $f \in A[X]$ satisfying (1.6.1) is a union of finitely many strong A-equivalence classes.

---

[1] With our definitions of strong A-equivalence and A-equivalence (sec), which is considered below, we follow Győry (1982) and Evertse and Győry (2017b).

This was proved in Evertse and Győry (2017b) in an effective form; see also Theorem 2.8.4 in Section 2.8.

The class of domains A with (1.6.2) contains, among others, all finitely generated subrings of $\overline{\mathbb{Q}}$ and, more generally, all finitely generated domains over $\mathbb{Z}$ of characteristic 0 that are integrally closed. In the latter case, Győry (1982) proved the following more precise result, without fixing the degrees of the polynomials under consideration.

Theorem 1.6.2 *Let A be an integrally closed integral domain of characteristic 0 that is finitely generated over $\mathbb{Z}$ and G a finite extension of the quotient field of A. Then the set of solutions of (1.6.1) is a union of finitely many strong A-equivalence classes.*

We do not know if condition (1.6.2) is the weakest possible condition. As is pointed out in Evertse and Győry (2017b), Theorem 1.6.1 is not true for arbitrary finitely generated domains of characteristic 0.

We also consider discriminant equations where the unknowns are elements of orders of finite étale K-algebras. Let Ω be a *finite étale K-algebra*, i.e., $\Omega = K[X]/(P) = K[\theta]$, where $P \in K[X]$ is some separable polynomial and $\theta := X \pmod P$. If, in particular, P is irreducible over K, then Ω is a finite extension field of K. Writing $[\Omega : K] := \dim_K \Omega$, we have $[\Omega : K] = \deg P$. Let $\overline{K}$ be an algebraic closure of K. By a K-*homomorphism* from Ω to $\overline{K}$, we mean a nontrivial K-algebra homomorphism. There are precisely $n := [\Omega : K]$ K-homomorphisms from Ω to $\overline{K}$ that map θ to the n distinct zeros of P in $\overline{K}$. We denote these by $x \mapsto x^{(i)}$ $(i = 1, \ldots, n)$. The *discriminant* of $\alpha \in \Omega$ over K is given by

$$D_{\Omega/K}(\alpha) = \prod_{1 \le i < j \le n} (\alpha^{(i)} - \alpha^{(j)})^2,$$

where $\alpha^{(i)}$ denotes the image of α under $x \mapsto x^{(i)}$. This is an element of K. It is easy to see that $D_{\Omega/K}(\alpha + a) = D_{\Omega/K}(\alpha)$ for $\alpha \in \Omega$, $a \in K$. Further, $D_{\Omega/K}(\alpha)$ is different from zero if and only if $\Omega = K[\alpha]$.

Consider now *discriminant equations for integral elements* of the shape

$$D_{\Omega/K}(\xi) = \delta \quad \text{in } \xi \in \mathcal{O}, \tag{1.6.3}$$

where δ is a nonzero element of A and $\mathcal{O}$ is an A-*order* of Ω, i.e., an A-subalgebra of Ω that spans Ω as a K-vector space and that is finitely generated as an A-module. Then $\mathcal{O}$ is, in fact, an A-subalgebra of the integral closure of A in Ω. As was mentioned above, A_Ω is finitely generated as an A-module.

If $\xi \in \mathcal{O}$ is a solution of (1.6.3), then so is $\xi + a$ for every $a \in A$. Thus, the solution of (1.6.3) splits into A-*cosets* $\xi + A = \{\xi + a : a \in A\}$.

The following theorem was established by Evertse and Győry (2017b) in an effective form; see also Corollary 2.8.3.

Theorem 1.6.3 *Let A be an integral domain of characteristic 0 that is finitely generated over* $\mathbb{Z}$. *Further, let K be the quotient field of A,* Ω *a finite étale K-algebra,* $\mathcal{O}$ *an A-order of* Ω, *and* δ *a nonzero element of A. Then the following two assertions are equivalent:*

(i) *The quotient A-module* $(\mathcal{O} \cap K)/A$ *is finite.*

(ii) *For every nonzero* $\delta \in A$, *the set of* $\xi \in \mathcal{O}$ *with* (1.6.3) *is a union of finitely many A-cosets.*

The implication (ii)$\Rightarrow$(i) is obvious. Suppose (i) does not hold. Pick $\xi_0 \in \mathcal{O}$ with $\Omega = K[\xi_0]$ and let $\delta := D_{\Omega/K}(\xi_0)$. Then $\delta \neq 0$. The $\mathcal{O} \cap K$-coset $\xi_0 + \mathcal{O} \cap K$ is contained in the set of solutions of (1.6.3), and this $\mathcal{O} \cap K$-coset is clearly the union of infinitely many A-cosets.

We note that $\mathcal{O} \cap K = A$ if A is integrally closed. Hence, Theorem 1.6.3 immediately gives the following.

Corollary 1.6.4 *Let A be an integrally closed integral domain of characteristic 0 that is finitely generated over* $\mathbb{Z}$, *K its quotient field,* Ω *a finite étale K-algebra,* $\mathcal{O}$ *an A-order of* Ω, *and* δ *a nonzero element of A. Then the set of* $\xi \in \mathcal{O}$ *with* (1.6.3) *is a union of finitely many A-cosets.*

As was mentioned above, A_Ω is finitely generated as an A-module. Taking for Ω a finite extension L of K and for $\mathcal{O}$ the integral closure A_L of A in L, we get the following important special case which is due to Győry (1982).

Corollary 1.6.5 *Let A be an integrally closed integral domain of characteristic 0 that is finitely generated over* $\mathbb{Z}$, *K its quotient field, L a finite extension of K, and* $\delta \in A \setminus \{0\}$. *Then the set of solutions of the equation*

$$D_{L/K}(\xi) = \delta \quad in \quad \xi \in A_L \tag{1.6.4}$$

is a union of finitely many A-cosets.

The following more general versions of equations (1.6.1) and (1.6.3) are also important for applications:

$$D(f) \in \delta A^* \text{ in monic } f \in A[X] \text{ of degree } n \geq 2$$
$$\text{having all its zeros in } G \tag{1.6.1a}$$

and

$$D_{\Omega/K}(\xi) \in \delta A^* \quad in \quad \xi \in \mathcal{O}. \tag{1.6.3a}$$

The solutions of (1.6.1a) can be partitioned into the so-called A-equivalence classes, where two monic polynomials $f, f' \in A[X]$ of degree n are called *A-equivalent* if $f'(X) = u^n f(u^{-1}X + a)$ for some $u \in A^*, a \in A$. Combining Theorem 1.6.1 with Roquette's (1957) theorem that A^* is finitely generated, it follows that under the assumption (1.6.2) the polynomials f with (1.6.1a) lie only in finitely many A-equivalence classes. For integrally closed A, this finiteness result was proved by Győry (1982) in a more general form, without fixing the degree of the polynomials f under consideration.

Similarly, the solutions of (1.6.3a) can be divided into A-equivalence classes, where two elements α and α' of $\mathcal{O}$ are called *A-equivalent* if $\alpha' = u\alpha + a$ with some $u \in A^*$ and $a \in A$. Together with Roquette's theorem, Theorem 1.6.3 implies that under the condition (i) of Theorem 1.6.3, equation (1.6.3a) has only finitely many A-equivalence classes of solutions. In the case of integrally closed A, this was obtained in Evertse and Győry (2017a, Chapter 5).

A further important application is as follows. If

$$\mathcal{O} = A[\xi] \tag{1.6.5}$$

for some $\xi \in \mathcal{O}$ and ξ' is A-equivalent to ξ, then also $\mathcal{O} = A[\xi']$. The above result concerning (1.6.3a) implies that under the condition (i) of Theorem 1.6.3, the set of ξ with (1.6.5) is a union of finitely many A-equivalence classes. For integrally closed A, see Evertse and Győry (2017a, Chapter 5), and if, in addition, Ω is a finite extension of K, see Győry (1982).

Over $\mathbb{Z}$ and more generally over number fields, the first finiteness results concerning equations (1.6.1), (1.6.1a), (1.6.3), (1.6.3a), and (1.6.5) were proved by Győry in an **effective form**. He proved in Győry (1973) for $A = \mathbb{Z}$ that given a nonzero $\delta \in \mathbb{Z}$, there are only finitely many strong $\mathbb{Z}$-equivalence classes of monic $f \in \mathbb{Z}[X]$ with discriminant δ, and a full set of representatives of these equivalence classes can be effectively determined. Here neither the degree n nor the splitting field G of the polynomials f is fixed. This result implied the first effective finiteness theorem for equation (1.6.4) with $A = \mathbb{Z}$. Further, in Győry (1976), it is proved in an effective form that if L is a number field with ring of integers O_L, then there are only finitely many $\mathbb{Z}$-equivalence classes of $\alpha \in \mathcal{O}_L$ with $\mathcal{O}_L = \mathbb{Z}[\alpha]$.

It follows from finiteness results of Győry (1978a, 1978b, 1984b) that if A is the ring of integers or S-integers of a number field, then the finiteness results presented above on equations (1.6.1), (1.6.1a), (1.6.3), (1.6.3a), and (1.6.5) are valid in an effective form. Moreover, these versions of Theorem 1.6.2 remain true without fixing the number field G or the degree n of the polynomials f. Perhaps such a finiteness result without fixing G can be extended to certain finitely generated integral domains of low transcendence degree. But extending this to arbitrary finitely generated domains over $\mathbb{Z}$ seems to be very hard.

For a class of finitely generated ground domains over $\mathbb{Z}$ that may contain both algebraic and transcendental elements over $\mathbb{Q}$, effective versions of Theorem 1.6.2 and Corollary 1.6.5 were obtained by Győry (1984b). These were extended, in a slightly more general form, by Evertse and Győry (2017b) to the case of arbitrary finitely generated ground domains over $\mathbb{Z}$. These will be presented in Chapter 2, Section 2.8.

2

Effective Results for Diophantine Equations over Finitely Generated Domains: The Statements

In this chapter, general **effective** finiteness theorems are presented for Diophantine equations over finitely generated integral domains of characteristic 0, including *unit equations, Thue equations, hyper- and superelliptic equations*, the *Schinzel–Tijdeman equation*, the *Catalan equation, decomposable form equations*, and *discriminant equations*. Apart from discriminant equations, the other theorems are established in *quantitative form*, providing effective bounds for the solutions. The results presented make it possible to solve, at least in principle, the equations under consideration. Their proofs are given in Chapters 9 and 10.

2.1 Notation and Preliminaries

To make sense of statements such as "a particular Diophantine equation can be solved effectively over a given finitely generated domain," we need an *explicit representation* for this *domain*, as well as for its *elements*. We start with the necessary definitions. A more detailed treatment can be found in Chapter 6, and in particular in Section 6.3.

Let $A = \mathbb{Z}[z_1, \ldots, z_r]$ be a finitely generated integral domain of characteristic 0. Assume that $r > 0$ and let

$$\mathcal{I} := \{f \in \mathbb{Z}[X_1, \ldots, X_r] : f(z_1, \ldots, z_r) = 0\}.$$

Then $\mathcal{I}$ is an ideal of $\mathbb{Z}[X_1, \ldots, X_r]$, which, by Hilbert's Basis Theorem is finitely generated; that is, we have

$$A \cong \mathbb{Z}[X_1, \ldots, X_r]/\mathcal{I}, \quad \text{with} \quad \mathcal{I} = (f_1, \ldots, f_M) \qquad (2.1.1)$$

for a finite set of polynomials $f_1, \ldots, f_M \in \mathbb{Z}[X_1, \ldots, X_r]$. We call the tuple $(f_1, \ldots, f_M)$ an *ideal representation* for A. Recall that a necessary and sufficient condition for A to be a domain of characteristic 0 is that $\mathcal{I}$ be a prime ideal of $\mathbb{Z}[X_1,\ldots,X_r]$ with $\mathcal{I} \cap \mathbb{Z} = (0)$. Given a set of generators $(f_1,\ldots,f_M)$ for $\mathcal{I}$, this can be checked effectively, for example, by using Aschenbrenner (2004, Lemma 4.8, Corollary 4.9) and the comments in Chapter 6 of the present work.

To perform computations in A, it will be necessary to be able to decide whether for any given $f \in \mathbb{Z}[X_1,\ldots,X_r]$ and any given ideal $\mathcal{I} = (f_1,\ldots,f_M)$ of $\mathbb{Z}[X_1,\ldots,X_r]$ we have $f \in \mathcal{I}$, that is, whether there exist $g_1,\ldots,g_M \in \mathbb{Z}[X_1,\ldots,X_r]$ such that $f = g_1 f_1 + \cdots + g_M f_M$. An algorithm performing this task is called an *ideal membership algorithm for* $\mathbb{Z}[X_1,\ldots,X_r]$. Several such algorithms have been developed since the 1960s; we mention only the algorithm of Simmons (1970), and the more precise algorithm of Aschenbrenner (2004), which plays an important role in our work; see Corollary 6.1.6 in Chapter 6.

Denote by K the quotient field of A. For $\alpha \in A$ we call f a *representative* for α or say that f *represents* α if $f \in \mathbb{Z}[X_1,\ldots,X_r]$ and $\alpha = f(z_1,\ldots,z_r)$. With the notation (2.1.1), this means that α corresponds to the residue class f mod $\mathcal{I}$. Further, for $\alpha \in K$, we call (f,g) a *pair of representatives* for α or say that (f,g) *represents* α if $f,g \in \mathbb{Z}[X_1,\ldots,X_r]$, $g \notin \mathcal{I}$, and $\alpha = f(z_1,\ldots,z_r)/g(z_1,\ldots,z_r)$. Note that $g \notin \mathcal{I}$ can be verified by means of an ideal membership algorithm for $\mathbb{Z}[X_1,\ldots,X_r]$. A representative for a tuple $(x_1,\ldots,x_m) \in A^m$ is a tuple $(\widetilde{x}_1,\ldots,\widetilde{x}_m)$ with elements from $\mathbb{Z}[X_1,\ldots,X_r]$ such that $\widetilde{x}_i$ represents x_i, for $i = 1,\ldots,m$. Finally, a representative for a polynomial F with coefficients in A is a polynomial $\widetilde{F}$ with coefficients in $\mathbb{Z}[X_1,\ldots,X_r]$ that represent the corresponding coefficients of F.

We say that the domain A is *effectively given* if an ideal representation as in (2.1.1) for it is given. Further, we say that an element α of A, resp. of K is *effectively given/computable* if a representative, resp. a pair of representatives for α is given/can be computed.

Given (pairs of) representatives for two elements of A or K, it is clear how to compute a (pair of) representative(s) for their sum, difference, product, or quotient. Using an ideal membership algorithm for $\mathbb{Z}[X_1,\ldots,X_r]$, we can decide whether two given f and $f' \in \mathbb{Z}[X_1,\ldots,X_r]$ represent the same element of A (i.e., whether $f - f' \in \mathcal{I}$) and whether two given pairs (f,g) and (f',g') in $\mathbb{Z}[X_1,\ldots,X_r]$ represent the same element of K (this amounts to $g \cdot g' \notin \mathcal{I}$ and $fg' - f'g \in \mathcal{I}$).

Suppose we know somehow that a particular system of polynomial Diophantine equations

$$F_1(\mathbf{x}) = 0,\ldots,F_s(\mathbf{x}) = 0 \text{ in } \mathbf{x} = (x_1,\ldots,x_m) \in A^m, \qquad (2.1.2)$$

where $F_1,\ldots,F_s \in A[Y_1,\ldots,Y_m]$ has only finitely many solutions. Then determining the solutions of (2.1.2) effectively means finding a finite list of tuples

$\widetilde{\mathbf{x}} = (\widetilde{x}_1, \ldots, \widetilde{x}_m) \in \mathbb{Z}[X_1, \ldots, X_r]^m$ that represents all the solutions in A^m of (2.1.2) and such that no two tuples in the list represent the same solution.

Rather than merely showing that the set of solutions of (2.1.2) can be determined effectively, it is sometimes possible to obtain more precise quantitative statements by estimating the *sizes* of the coordinates of the tuples representing the solutions. In fact, we define the size of a nonzero polynomial $f \in \mathbb{Z}[X_1, \ldots, X_r]$ by

$$s(f) := \max(1, \deg f, h(f)),$$

where $\deg f$ denotes the *degree*, that is, total degree, of f and $h(f)$ denotes the *logarithmic height* of f, that is, the logarithm of the maximum of the absolute values of the coefficients of f. Further, we define $s(0) := 1$. Clearly, there are only finitely many polynomials in $\mathbb{Z}[X_1, \ldots, X_r]$ with size below a given bound, and these can be determined effectively.

Proposition 2.1.1 *Let the domain A have an ideal representation $f_1, \ldots, f_M$ as in (2.1.1), and let $F_1, \ldots, F_s \in A[Y_1, \ldots, Y_m]$ be given by representatives $\widetilde{F}_1, \ldots, \widetilde{F}_s$, which are polynomials in the variables $Y_1, \ldots, Y_m$ with coefficients in $\mathbb{Z}[X_1, \ldots, X_r]$. Suppose we can compute C in terms of $f_1, \ldots, f_M, \widetilde{F}_1, \ldots, \widetilde{F}_s$ such that every solution of the system (2.1.2) has a representative $\widetilde{\mathbf{x}} = (\widetilde{x}_1, \ldots, \widetilde{x}_m)$ with*

$$\widetilde{x}_i \in \mathbb{Z}[X_1, \ldots, X_r], \quad s(\widetilde{x}_i) \leq C, \ for \ i = 1, \ldots, m.$$

Then we can effectively determine the solutions of (2.1.2).

Proof We enumerate all m-tuples consisting of elements in $\mathbb{Z}[X_1, \ldots, X_r]$ of size at most C. By means of an ideal membership algorithm for $\mathbb{Z}[X_1, \ldots, X_r]$, we check for each of these tuples $\widetilde{\mathbf{x}}$ whether $\widetilde{F}_i(\widetilde{\mathbf{x}}) \in \mathcal{I}$ for $i = 1, \ldots, s$ and make a list of the tuples passing this test. This list contains at least one representative for each solution of (2.1.2). Subsequently, we check, for any two tuples $\widetilde{\mathbf{x}}_1$ and $\widetilde{\mathbf{x}}_2$ from this list, whether there is an index i such that the difference of their ith coordinates is not in $\mathcal{I}$. If there is not such an index i, then $\widetilde{\mathbf{x}}_1$ and $\widetilde{\mathbf{x}}_2$ represent the same solution of (2.1.2) so we may remove one of them from our list. What remains is a list with precisely one representative for each solution. $\square$

In Sections 2.2–2.5, we present effective finiteness results in quantitative form, i.e., with bounds for the sizes for representatives of their solutions, for unit equations, a generalization of unit equations, Thue equations, hyper- and superelliptic equations, the Schinzel–Tijdeman equation, and the Catalan equation over finitely generated domains. These results have been proved by means of the effective method of Evertse and Győry (2013), reducing the equations to the number field and function field cases, applying effective specializations.

As will be pointed out in Chapters 3 and 7, this is an improved version of the effective specialization method of Győry (1983, 1984b).

Sections 2.6–2.8 are devoted to effective finiteness results concerning decomposable form equations, norm form equations, and discriminant equations over finitely generated domains. Here, following Győry's method, the equations are reduced to unit equations, and then the general effective results concerning unit equations are used. The proofs use several effective results from commutative algebra and some new, effective, so-called "degree–height estimates" from Chapter 8 for elements of $\overline{K}$. In Section 2.9, we mention some Diophantine problems that can be solved effectively over number fields but for which as yet no effective analogue over finitely generated domains could be established. We recall that, except for Section 2.5 dealing with the Catalan equation, the earlier less general effective theorems were mentioned in Chapter 1, in the corresponding sections. They will be also referred to in Chapter 4 where we treat effective results over number fields. Hence, apart from Section 2.5, no mention will be made in this chapter on earlier effective results over number fields or special finitely generated domains.

Throughout this work, we shall use $O(\cdot)$ to denote a quantity that is an effectively computable positive absolute constant times the expression between the parentheses. The constant may be different at each occurrence of the O-symbol.

2.2 Unit Equations in Two Unknowns

In what follows, A will denote an integral domain of characteristic 0 that is finitely generated over $\mathbb{Z}$. We assume that $A \cong \mathbb{Z}[X_1,\ldots,X_r]/(f_1,\ldots,f_M)$, where $f_1,\ldots,f_M$ are given elements of $\mathbb{Z}[X_1,\ldots,X_r]$.

We start with *unit equations in two unknowns*; these are equations of the form

$$ax + by = c \quad \text{in } x, y \in A^*, \tag{2.2.1}$$

where a, b, c are nonzero elements of A. Let $\widetilde{a}, \widetilde{b}, \widetilde{c}$ be representatives in $\mathbb{Z}[X_1,\ldots,X_r]$ for a, b, c, respectively.

The following effective result was established by Evertse and Győry (2013).

Theorem 2.2.1 *Assume that $f_1,\ldots,f_M$ and $\widetilde{a}, \widetilde{b}, \widetilde{c}$ all have degree at most d and logarithmic height at most h, where $d \geq 1, h \geq 1$. Then for each solution x, y of (2.2.1), there are representatives $\widetilde{x}, \widetilde{x}', \widetilde{y}, \widetilde{y}'$ of x, x^{-1}, y, y^{-1}, respectively, such that*

$$s(\widetilde{x}), s(\widetilde{x}'), s(\widetilde{y}), s(\widetilde{y}') \leq \exp\left((2d)^{\exp O(r)} h\right).$$

The exponential dependence of the upper bound on d and h is a consequence of the use of Baker's method in the proof of Theorem 4.3.1 on unit equations over number fields. The doubly exponential dependence on r comes from the effective commutative algebra for polynomial rings over fields and over $\mathbb{Z}$, which is used in the specialization method of Evertse and Győry (2013); see also Chapters 6–9.

We deduce the following effective version of Lang's Theorem 1.2.1.

Corollary 2.2.2 *Equation* (2.2.1) *has only finitely many solutions. Further, if A and a, b, c are effectively given, then all solutions of* (2.2.1) *can be determined effectively.*

Proof of Corollary 2.2.2 Notice that equation (2.2.1) is equivalent to the system

$$ax + by = c, \quad x \cdot x' = 1, \quad y \cdot y' = 1 \quad \text{in } x, x', y, y' \in A.$$

Apply Proposition 2.1.1 with C being the upper bound occurring in Theorem 2.2.1. □

We present a variation on Theorem 2.2.1. Let $\gamma_1, \ldots, \gamma_s$ be multiplicatively independent elements of K^*. We mention that Proposition 7.5.2 in Chapter 7 provides a method to check whether the elements $\gamma_1, \ldots, \gamma_s$ given by pairs of representatives are multiplicatively independent; see also Lemma 7.2 from Evertse and Győry (2013). Again, let a, b, c be nonzero elements of A and consider the equation

$$a\gamma_1^{u_1} \ldots \gamma_s^{u_s} + b\gamma_1^{v_1} \ldots \gamma_s^{v_s} = c \quad \text{in} \quad u_1, \ldots, u_s, v_1, \ldots, v_s \in \mathbb{Z}. \quad (2.2.2)$$

Theorem 2.2.3 *Let $\widetilde{a}, \widetilde{b}, \widetilde{c}$ be representatives for a, b, c, and for $i = 1, \ldots, s$, let $(g_{i,1}, g_{i,2})$ be a pair of representatives for γ_i. Suppose that $f_1, \ldots, f_M, \widetilde{a}, \widetilde{b}, \widetilde{c}$, and $g_{i,1}, g_{i,2}$ $(i = 1, \ldots, s)$ all have degree at most d and logarithmic height at most h, where $d \geq 1, h \geq 1$. Then for each solution $(u_1, \ldots, v_s)$ of* (2.2.2)*, we have*

$$\max(|u_1|, \ldots, |u_s|, |v_1|, \ldots, |v_s|) \leq \exp\left((2d)^{\exp O(r+s)} h\right).$$

An immediate consequence of Theorem 2.2.3 is that for given $f_1, \ldots, f_M$, a, b, c, and $\gamma_1, \ldots, \gamma_s$, all solutions of (2.2.2) can be determined effectively.

Since the unit group A^* is finitely generated, equation (2.2.1) may be viewed as a special case of (2.2.2). But no general effective algorithm is known to find a finite system of generators for A^*; hence, we cannot deduce an effective result for (2.2.1) from Theorem 2.2.3. In fact, in Chapter 9, we shall argue the reverse and prove Theorem 2.2.3 by combining Theorem 2.2.1 with an effective result for equations of the type $\gamma_1^{u_1} \ldots \gamma_s^{u_s} = \gamma_0$ in integers $u_1, \ldots, u_s$, where $\gamma_0, \gamma_1, \ldots, \gamma_s \in K^*$.

Finally, we mention that using the effective method of Evertse and Győry (2013), Bérczes (2015a, 2015b) made effective, in full generality, the results of Lang and Liardet, resp., presented in Section 1.2 for the equations

$$F(x, y) = 0 \quad \text{in} \quad x, y \in A^*, \tag{2.2.3}$$

$$F(x, y) = 0 \quad \text{in} \quad x, y \in \overline{\Gamma}, \tag{2.2.4}$$

where Γ is a finitely generated multiplicative subgroup of K^*, $\overline{\Gamma}$ is its division group, and $F \in A[X, Y]$. In Bérczes' results, F has to satisfy a slightly stronger condition than those in Theorems 1.2.3 and 1.2.4, that is, $F \in A[X, Y]$ is a nonconstant polynomial that is not divisible by any polynomial of the form

$$X^m Y^n - \alpha \quad \text{or} \quad X^m - \alpha Y^n,$$

with $\alpha \in \overline{K}$ and with non-negative integers m, n, not both zero. As was pointed out by Bérczes, this condition can be checked effectively once pairs of representatives for the coefficients of F are given.

Assume that $f_1, \dots, f_M$ and a set of representatives for the coefficients of F have degree at most d and logarithmic height at most h, with $d > 1, h > 1$. Further, denote by N the total degree of F. Bérczes (2015a, Theorem 2.1) proved in a more precise form the following.

Theorem 2.2.4 *Under the above assumptions, there is an effectively computable number C_1 depending only on r, d, h, and N such that for every solution $x, y \in A^*$ of (2.2.3), there are representatives $\widetilde{x}, \widetilde{x}', \widetilde{y}, \widetilde{y}'$ of x, x^{-1}, y, y^{-1}, respectively, such that*

$$s(\widetilde{x}), s(\widetilde{x}'), s(\widetilde{y}), s(\widetilde{y}') \leq C_1.$$

Let the generators $\gamma_1, \dots, \gamma_s$ of Γ be given by pairs of representatives $(g_1, h_1), \dots, (g_s, h_s)$. Assume now that $f_1, \dots, f_M, (g_1, h_1), \dots, (g_s, h_s)$, and a set of representatives for the coefficients of F have degree at most d and logarithmic height at most h, with $d, h > 1$. Then Bérczes (2015b, Theorem 2.1) obtained the following.

Theorem 2.2.5 *There is an effectively computable number C_2 depending only on r, s, d, h, and N such that for every solution x, y of (2.2.4) in $\overline{\Gamma}$, we have*

$$x^k = \gamma_1^{k_{1,x}} \cdots \gamma_s^{k_{s,x}}, \quad y^k = \gamma_1^{k_{1,y}} \cdots \gamma_s^{k_{s,y}},$$

where $k, k_{1,x}, \dots, k_{s,x}$, and $k_{1,y}, \dots, k_{s,y}$ are the integers with $k \geq 1$ and

$$k, |k_{1,x}|, \dots, |k_{s,x}|, |k_{1,y}|, \dots, |k_{s,y}| \leq C_2.$$

Theorems 2.2.4 and 2.2.5 imply, in an effective form, the finiteness of the number of solutions of equations (2.2.3) and (2.2.4). We have not included the rather technical proofs of Theorems 2.2.4 and 2.2.5 in our book.

2.3 Thue Equations

As before, A denotes an integral domain of characteristic 0 that is finitely generated over $\mathbb{Z}$, and we assume that $A \cong \mathbb{Z}[X_1, \ldots, X_r]/\mathcal{I}$, where $\mathcal{I} = (f_1, \ldots, f_M)$, with $f_1, \ldots, f_M$, being the given elements of $\mathbb{Z}[X_1, \ldots, X_r]$. We consider the *Thue equation*

$$F(x, y) = \delta \text{ in } (x, y) \in A^2 \tag{2.3.1}$$

over A, where

$$F(X, Y) = a_0 X^n + a_1 X^{n-1} Y + \cdots + a_n Y^n \in A[X, Y]$$

is a binary form of degree $n \geq 3$ with the nonzero discriminant D_F, i.e., with n being the pairwise nonproportional linear factors, and $\delta \in A \backslash \{0\}$. We choose representatives

$$\widetilde{a_0}, \widetilde{a_1}, \ldots, \widetilde{a_n}, \widetilde{\delta} \in \mathbb{Z}[X_1, \ldots, X_r]$$

of $a_1, \ldots, a_n, \delta$, respectively, where $\widetilde{\delta} \notin \mathcal{I}$ and the discriminant $D_{\widetilde{F}}$ of $\widetilde{F} :=$ $\sum_{j=0}^n \widetilde{a}_j X^{n-j} Y^j$ is not in $\mathcal{I}$. These conditions on $\widetilde{\delta}$ and $D_{\widetilde{F}}$ can be checked by means of an ideal membership algorithm for $\mathbb{Z}[X_1, \ldots, X_r]$.

The next theorem is due to Bérczes, Evertse, and Győry (2014).

Theorem 2.3.1 *Assume that* $f_1, \ldots, f_M$ *and* $\widetilde{a}_0, \ldots, \widetilde{a}_n, \widetilde{\delta}$ *all have degree at most* d *and logarithmic height at most* h, *where* $d \geq 1, h \geq 1$. *Then every solution* (x, y) *of equation* (2.3.1) *has a representative* $(\widetilde{x}, \widetilde{y})$ *such that*

$$s(\widetilde{x}), s(\widetilde{y}) \leq \exp\{n!(nd)^{\exp O(r)} h\}. \tag{2.3.2}$$

Combining Theorem 2.3.1 with Proposition 2.1.1, one immediately obtains the following.

Corollary 2.3.2 *Equation* (2.3.1) *has only finitely many solutions. Further, if* $A, a_1, \ldots, a_n$, *and* δ *are effectively given, then all solutions of* (2.3.1) *can be determined effectively.*

2.4 Hyper- and Superelliptic Equations, the Schinzel–Tijdeman Equation

We keep our notation that A is an integral domain of characteristic 0 that is finitely generated over $\mathbb{Z}$, satisfying $A \cong \mathbb{Z}[X_1, \ldots, X_r]/\mathcal{I}$, where $\mathcal{I} = (f_1, \ldots, f_M)$, with $f_1, \ldots, f_M$ being the given elements of $\mathbb{Z}[X_1, \ldots, X_r]$. We now consider the equation

$$F(x) = \delta y^m \text{ in } x, y \in A, \tag{2.4.1}$$

where

$$F(X) = a_0 X^n + a_1 X^{n-1} + \cdots + a_n \in A[X],$$

$\delta \in A \backslash \{0\}$, and F has n distinct roots in an algebraic closure of the quotient field of A, i.e., a_0 and the discriminant of F are different from zero. We choose representatives

$$\widetilde{a_0}, \widetilde{a_1}, \ldots, \widetilde{a_n}, \widetilde{\delta} \in \mathbb{Z}[X_1, \ldots, X_r]$$

for $a_0, a_1, \ldots, a_n, \delta$, respectively, where $\widetilde{\delta}, \widetilde{a_0}$ and the discriminant of $\widetilde{F} :=$ $\sum_{j=0}^{n} \widetilde{a_j} X^{n-j}$ are not in $\mathcal{I}$. We assume that

$$\text{either} \quad m = 2 \text{ and } n \geq 3 \quad \text{or} \quad m \geq 3 \text{ and } n \geq 2.$$

The following theorems are due to Bérczes, Evertse, and Győry (2014).

Theorem 2.4.1 *Assume that* $f_1, \ldots, f_M$ *and* $\widetilde{a_0}, \ldots, \widetilde{a_n}, \widetilde{\delta}$ *have degree at most* d *and logarithmic height at most* h, *where* $d \geq 1$, $h \geq 1$. *Then every solution of equation (2.4.1) has representatives* $\widetilde{x}, \widetilde{y}$ *such that*

$$s(\widetilde{x}), s(\widetilde{y}) \leq \exp\{m^3 (nd)^{\exp O(r)} h\}. \tag{2.4.2}$$

When combined with Proposition 2.1.1, this provides a method to determine effectively the solutions of (2.4.1), i.e., it provides an effective version of Theorem 1.3.1 concerning hyperelliptic/superelliptic equations.

Our next result concerns the *Schinzel–Tijdeman equation*

$$F(x) = \delta y^m \quad \text{in} \quad x, y \in A, \quad m \in \mathbb{Z}_{\geq 2}. \tag{2.4.3}$$

Keeping the above notation, we have

Theorem 2.4.2 *Assume that in (2.4.3)* F *has degree* $n \geq 2$ *and a nonzero discriminant. Let* $x, y \in A$ *and* $m \geq 2$ *integer be a solution of (2.4.3). Then with the same notation as in Theorem 2.4.1, we have*

$$m \leq \exp\{(nd)^{\exp O(r)} h\}$$

$$\text{if } y \in \overline{\mathbb{Q}}, \ y \neq 0, \text{ and } y \text{ is not a root of unity,} \tag{2.4.4}$$

$$m \leq (nd)^{\exp O(r)} \quad \text{if } y \notin \overline{\mathbb{Q}}. \tag{2.4.5}$$

2.5 The Catalan Equation

As before, A is an integral domain of characteristic 0 that is finitely generated over $\mathbb{Z}$, satisfying $A \cong \mathbb{Z}[X_1, \ldots, X_r]/\mathcal{I}$, where $\mathcal{I} = (f_1, \ldots, f_M)$, with $f_1, \ldots, f_M$ being the given elements of $\mathbb{Z}[X_1, \ldots, X_r]$. Consider the *Catalan equation*

$$x^m - y^n = 1 \quad \text{in } x, y \in A \backslash \{0\}$$

$$\text{and } m, n \in \mathbb{Z} \text{ with } m, n > 1 \text{ and } mn > 4. \tag{2.5.1}$$

In contrast with the other equations from this chapter, we cite here the most important earlier results concerning equation (2.5.1). In the classical case $A = \mathbb{Z}$, Catalan (1844) conjectured that $3^2 - 2^3 = 1$ is the only solution of the equation in the positive integers x, y, m, n, with $m, n > 1$. In this case, Tijdeman (1976), using Baker's method, gave an effectively computable but very large upper bound for the solutions of equation (2.5.1). Brindza, Győry, and Tijdeman (1986) and Brindza (1987) generalized Tijdeman's result for the case when x and y are integers resp. S-integers of a given number field. Subsequently, Brindza (1993) further generalized this for the case of the restricted class of finitely generated ground domains A considered in Győry (1983, 1984b).

Mihailescu (2004) used methods from pure algebraic number theory to prove Catalan's conjecture over $\mathbb{Z}$.

Strengthening earlier results of Brindza (1987, 1993) on equation (2.5.1), Koymans (2016, 2017) proved the following theorem using the method of Evertse and Győry (2013).

Theorem 2.5.1 *Assume that $f_1, \ldots, f_M$ have degree at most d and logarithmic height at most h, where $d \geq 1, h \geq 1$. Let x, y, m, n be a solution of (2.5.1) such that x and y are not roots of unity. Then*

$$\max(m, n) \leq \exp \exp \exp \{ (2d)^{\exp O(r)} h \} \quad \textit{if } x, y \in \overline{\mathbb{Q}}, \tag{2.5.2}$$

$$\max(m, n) \leq (2d)^{\exp O(r)} \quad \textit{if } x, y \notin \overline{\mathbb{Q}}. \tag{2.5.3}$$

It is easy to see that in Theorem 2.5.1, the conditions $m, n > 1, mn > 4$ are necessary.

By combining Theorem 2.5.1 with Theorem 2.4.1, it follows immediately that equation (2.5.1) has only finitely many solutions with x, y not being roots of unity, and combining this with Proposition 2.1.1, we obtain that from a given ideal representation $(f_1, \ldots, f_M)$, all solutions can be determined effectively.

In his master's thesis, Koymans (2016) proved an analogue of Theorem 2.5.1 over finitely generated domains of positive characteristic.

2.6 Decomposable Form Equations

Again, let A be an integral domain of characteristic 0 that is finitely generated over $\mathbb{Z}$ such that $A \cong \mathbb{Z}[X_1, \ldots, X_r]/\mathcal{I}$, with $\mathcal{I} = (f_1, \ldots, f_M)$ for some given polynomials $f_1, \ldots, f_M \in \mathbb{Z}[X_1, \ldots, X_r]$. We denote by K the quotient field of A and by $\overline{K}$ an algebraic closure of K.

Pick linear forms

$$\ell_i = \alpha_{i,1}X_1 + \cdots + \alpha_{i,m}X_m \in \overline{K}[X_1,\ldots,X_m] \quad (i = 1,\ldots,n) \qquad (2.6.1)$$

in $m \geq 2$ variables. We allow that some of these linear forms are equal. Let $F = \ell_1 \cdots \ell_n$ be their product, $\delta \in \overline{K}^*$, and consider the *decomposable form equation*

$$F(\mathbf{x}) = \ell_1(\mathbf{x}) \cdots \ell_n(\mathbf{x}) = \delta \quad \text{in } \mathbf{x} = (x_1,\ldots,x_m) \in A^m. \qquad (2.6.2)$$

We do not require that F have its coefficients in K. Subject to certain conditions on $\ell_1,\ldots,\ell_n$ and $\mathbf{x}$, we will formulate an effective finiteness result on equation (2.6.2) in a quantitative form. To this end, we introduce some notation and explain the conditions imposed on the ℓ_i.

Let us first introduce the A-module

$$\mathcal{Z}_{A,F} = \{\mathbf{x} \in A^m : \ell_1(\mathbf{x}) = \cdots = \ell_m(\mathbf{x}) = 0\}. \qquad (2.6.3)$$

Clearly, if $\mathbf{x}$ is a solution of (2.6.2), then so is $\mathbf{x} + \mathbf{y}$ for every $\mathbf{y} \in \mathcal{Z}_{A,F}$. Hence, the set of solutions of (2.6.2) falls apart into $\mathcal{Z}_{A,F}$-*cosets* $\mathbf{x} + \mathcal{Z}_{A,F}$,[1] and we want to determine representatives for these cosets.

If $\mathrm{rank}\{\ell_1,\ldots,\ell_n\} = m$, then $\mathcal{Z}_{A,F} = \{\mathbf{0}\}$, and the $\mathcal{Z}_{A,F}$-cosets are just single solutions. Had we been interested in noneffective finiteness results only, the generalization to the case $\mathrm{rank}\{\ell_1,\ldots,\ell_n\} < m$ and $\mathcal{Z}_{A,F} \neq \{\mathbf{0}\}$ would not have been necessary, but for certain *effective* applications this turned out to be of importance.

Given $\mathbf{x} = (x_1,\ldots,x_m) \in A^m$, a representative for $\mathbf{x}$ is a tuple $\widetilde{\mathbf{x}} = (\widetilde{x_1},\ldots,\widetilde{x_m})$ with

$$\widetilde{x_i} \in \mathbb{Z}[X_1,\ldots,X_r], \; x_i = \widetilde{x_i}(z_1,\ldots,z_r) \text{ for } i = 1,\ldots,m.$$

We define the size of this tuple by

$$s(\widetilde{\mathbf{x}}) := \max\left(s(\widetilde{x_1}),\ldots,s(\widetilde{x_m})\right)$$
$$= \max\left(1, \deg \widetilde{x_1}, h(\widetilde{x_1}),\ldots,\deg \widetilde{x_m}, h(\widetilde{x_m})\right).$$

Slightly diverging from its usual meaning, by a *representative* for a $\mathcal{Z}_{A,F}$-coset, we shall mean a tuple $\widetilde{\mathbf{x}} \in \mathbb{Z}[X_1,\ldots,X_r]^m$ that represents any element from this coset. Thus, in order to effectively determine a full system of representatives for the $\mathcal{Z}_{A,F}$-cosets of solutions, it suffices to compute a number C such that each of these cosets has a representative $\widetilde{\mathbf{x}} \in \mathbb{Z}[X_1,\ldots,X_r]^m$ with $s(\widetilde{\mathbf{x}}) \leq C$.

Next, we need some measures for the elements of $\overline{K}$. Let $\alpha \in \overline{K}$. We denote by $\deg_K \alpha$ the degree of α over K. A *tuple of representatives* for α is a tuple

---

[1] If $\mathcal{M}_1$ and $\mathcal{M}_2$ are the modules over a ring R, with $\mathcal{M}_1 \subset \mathcal{M}_2$, then by an $\mathcal{M}_1$-*coset* in $\mathcal{M}_2$, we mean a set of the shape $a + \mathcal{M}_1 = \{a + x : x \in \mathcal{M}_1\}$ with $a \in \mathcal{M}_2$.

$(g_0, \ldots, g_n)$, where $n = \deg_K \alpha$ and where $g_0, \ldots, g_n \in \mathbb{Z}[X_1, \ldots, X_r]$, $g_0 \notin \mathcal{I}$, such that

$$X^n + \frac{g_1(z_1, \ldots, z_r)}{g_0(z_1, \ldots, z_r)} X^{n-1} + \cdots + \frac{g_n(z_1, \ldots, z_r)}{g_0(z_1, \ldots, z_r)}$$

is the monic minimal polynomial of α over K. If $\alpha \in K$, then a tuple of representatives for α is up to sign a pair of representatives for α, as introduced in Section 2.1. We say that $(g_0, \ldots, g_n)$ has degree at most d and logarithmic height at most h, if each g_i $(i = 0, \ldots, n)$ has total degree at most d and logarithmic height at most h.

In order to formulate our effective results, we adopt some terminology from Győry and Papp (1978) and Győry (1981a, 1982, 1983). Let $\mathcal{L} = (\ell_1, \ldots, \ell_n)$ be the system of linear forms from (2.6.1). As said before, these linear forms need not be pairwise distinct. We define the *triangular graph* $\mathcal{G}(\mathcal{L})$ of $\mathcal{L}$ as follows:

$\mathcal{G}(\mathcal{L})$ has vertex system $\mathcal{L}$;

ℓ_i and ℓ_j with $i \neq j$ are connected by an edge if either ℓ_i, ℓ_j are linearly dependent on $\overline{K}$ or they are linearly independent of $\overline{K}$, and there is $q \neq i, j$ such that ℓ_i, ℓ_j, ℓ_q are linearly dependent on $\overline{K}$.
$\left.\right\}$ (2.6.4)

Let $\mathcal{L}_1, \ldots, \mathcal{L}_k$ denote the vertex systems of the connected components of $\mathcal{G}(\mathcal{L})$. When $k = 1$, we say that $\mathcal{L}$ or F is *triangularly connected*; see Győry and Papp (1978). For $j = 1, \ldots, k$, denote by $[\mathcal{L}_j]$ the $\overline{K}$-vector space generated by the linear forms from $\mathcal{L}_j$ and assume that

$$[\mathcal{L}_1] \cap \cdots \cap [\mathcal{L}_k] \neq (0). \tag{2.6.5}$$

This is in general a serious restriction that is not satisfied by most systems $\mathcal{L}$. In fact, it is much stronger than the criterion from Theorem 1.5.1.

In what follows, we want to consider solutions $\mathbf{x} \in A^m$ of (2.6.2) such that

$$\text{there is } \ell \in [\mathcal{L}_1] \cap \cdots \cap [\mathcal{L}_k] \text{ with } \ell(\mathbf{x}) \neq 0. \tag{2.6.6}$$

This is the set of solutions of (2.6.2) to which our effective method can be applied. Here, the linear form ℓ may vary with $\mathbf{x}$. We should note that if $\mathbf{x} \in A^m$ satisfies (2.6.6), then so does every element of the $\mathcal{Z}_{A,F}$-coset $\mathbf{x} + \mathcal{Z}_{A,F}$.

In the case that $\mathcal{L}$ is triangularly connected, i.e., $k = 1$, we have $\mathcal{L}_1 = \mathcal{L} = (\ell_1, \ldots, \ell_n)$; hence, (2.6.5) is satisfied. Further, if $\delta \neq 0$, then every solution of (2.6.2) automatically satisfies (2.6.6).

We are now ready to state our results. We first formulate a quantitative result, and then a corollary giving an effective finiteness statement.

Theorem 2.6.1 *Suppose the following:*

- *the given generators $f_1, \ldots, f_M$ of $\mathcal{I}$ have degree at most d and logarithmic height at most h;*
- *δ and the coefficients of $\ell_1, \ldots, \ell_n$ all have tuples of representatives of degree at most d and logarithmic height at most h;*
- *the coefficients of $\ell_1, \ldots, \ell_n$ all have degree at most v over K;*
- *$[\mathcal{L}_1] \cap \cdots \cap [\mathcal{L}_k] \neq (0)$.*

Then every $\mathcal{Z}_{A,F}$-coset of $\mathbf{x} \in A^m$ such that

$$F(\mathbf{x}) = \delta; \quad \text{there is } \ell \in [\mathcal{L}_1] \cap \cdots \cap [\mathcal{L}_k], \text{ with } \ell(\mathbf{x}) \neq 0 \qquad (2.6.7)$$

is represented by $\widetilde{\mathbf{x}} \in \mathbb{Z}[X_1, \ldots, X_r]^m$ with

$$s(\widetilde{\mathbf{x}}) \leq \exp\left((2mn \cdot v^{vmn}d)^{\exp O(r)}h\right). \qquad (2.6.8)$$

We deduce from this an effective finiteness result. Assume that A and its quotient field K are given effectively. A finite extension G of K is said to be *given effectively*, if it is given in the form of $K[X]/(P)$, where P is an effectively given irreducible monic polynomial in $K[X]$. We note that for a given polynomial $P \in K[X]$, it can be decided effectively whether it is irreducible; see for instance Theorem 6.2.3 in Chapter 6. We may write $G = K(\theta)$, where $\theta := X \pmod{P}$. Thus, elements of G can be expressed uniquely as $\sum_{i=0}^{g-1} a_i \theta^i$, with $a_0, \ldots, a_{g-1} \in K$, where g denotes the degree of G over K. We say that an *element* of G is *given/can be determined effectively* if the corresponding $a_0, \ldots, a_{g-1}$ are given/can be determined effectively.

Corollary 2.6.2 *There are only finitely many $\mathcal{Z}_{A,F}$-cosets of $\mathbf{x} \in A^m$ with (2.6.7). Moreover, if δ and the coefficients of $\ell_1, \ldots, \ell_n$ all belong to a finite extension G of K and if A, K, G, δ, and the coefficients of $\ell_1, \ldots, \ell_n$ are given effectively, then one can determine effectively a set consisting of precisely one representative $\widetilde{\mathbf{x}} \in \mathbb{Z}[X_1, \ldots, X_r]^m$ for each of these cosets.*

The essence of the proofs of Theorem 2.6.1 and Corollary 2.6.2 is that, thanks to the condition (2.6.6), equation (2.6.7) can be reduced to a finite system of unit equations in two unknowns, but with units from a subring $A' \supset A$ of G that is finitely generated over $\mathbb{Z}$. Then, Theorem 2.6.1 and Corollary 2.6.2 are deduced by applying Theorem 2.2.1, with A' instead of A. In the course of the proof of Theorem 2.6.1, we use so-called "degree–height estimates" for elements of $\overline{K}$; see Chapter 8. The proofs of Theorems 2.6.1 and Corollary 2.6.2 are given in Section 10.1 of Chapter 10.

In this generality, Theorem 2.6.1 and Corollary 2.6.2 are new. The finiteness statement of Corollary 2.6.2 was first proved in Győry (1982), but with slightly stronger conditions instead of (2.6.5) and (2.6.6): instead of (2.6.5) Győry assumed that $X_m \in [\mathcal{L}_1] \cap \cdots \cap [\mathcal{L}_k]$, and instead of (2.6.6) he assumed that $x_m \neq 0$; see also Evertse and Győry (2015, Chapter 9).

Győry (1983) established a quantitative result comparable to Theorem 2.6.1 but only for a restricted class of finitely generated integral domains A; see also Győry (1984a). Over number fields, more precise quantitative versions were obtained in Győry (1998) and Győry and Yu (2006).

We now discuss some applications. Let $F \in A[X,Y]$ be a nonzero binary form and δ a nonzero element of A, and consider again the Thue equation

$$F(x,y) = \delta \text{ in } (x,y) \in A^2. \tag{2.3.1}$$

We can factorize F as

$$F = \ell_1 \cdots \ell_n, \text{ with linear forms } \ell_i \in \overline{K}[X,Y].$$

Assume that at least three among the linear forms $\ell_1,\ldots,\ell_n$ are pairwise non-proportional over $\overline{K}$. Then it is easily verified that $\mathcal{L} = (\ell_1,\ldots,\ell_n)$ is triangularly connected and that $\mathcal{Z}_{A,F} = \{0\}$. Further, let δ be a nonzero element of A. Theorem 2.6.1 and Corollary 2.6.2 imply the following variation on Theorem 2.3.1, with worse bounds.

Corollary 2.6.3 *Assume that the given generators $f_1,\ldots,f_M$ of $\mathcal{I}$ have degree at most d and logarithmic height at most h, and that δ and the coefficients of F have representatives in $\mathbb{Z}[X_1,\ldots,X_r]$ of degree at most d and logarithmic height at most h, where $d \geq 1$ and $h \geq 1$.*

Then each solution $(x,y) \in A^2$ of (2.3.1) is represented by a pair $(\widetilde{x},\widetilde{y})$ with

$$\widetilde{x},\widetilde{y} \in \mathbb{Z}[X_1,\ldots,X_r], \ s(\widetilde{x}), s(\widetilde{y}) \leq \exp\left((n^{n^2}d)^{\exp O(r)}h\right). \tag{2.6.9}$$

Consequently, the solutions $(x,y) \in A^2$ of (2.3.1) can be determined effectively.

The next application is to a system of double Pell equations

$$\gamma_1 x_1^2 - \gamma_2 x_2^2 = \beta_{1,2}, \ \gamma_1 x_1^2 - \gamma_3 x_3^2 = \beta_{1,3} \text{ in } (x_1,x_2,x_3) \in A^3 \tag{2.6.10}$$

where $\gamma_1, \gamma_2, \gamma_3, \beta_{1,2}, \beta_{1,3} \in A$ with

$$\gamma_1 \gamma_2 \gamma_3 \beta_{1,2} \beta_{1,3}(\beta_{1,2} - \beta_{1,3}) \neq 0. \tag{2.6.11}$$

From the two equations in (2.6.10), it follows that

$$\gamma_2 x_2^2 - \gamma_3 x_3^2 = \beta_{1,3} - \beta_{1,2}$$

and thus,

$$F(x_1,x_2,x_3) = \delta, \tag{2.6.12}$$

where

$$\delta = \beta_{1,2}\beta_{1,3}(\beta_{1,3} - \beta_{1,2}) \qquad (2.6.13)$$

and

$$
\begin{aligned}
F &= \left(\gamma_1 X_1^2 - \gamma_2 X_2^2\right)\left(\gamma_1 X_1^2 - \gamma_3 X_3^2\right)\left(\gamma_2 X_2^2 - \gamma_3 X_3^2\right) \\
&= (\sqrt{\gamma_1}X_1 + \sqrt{\gamma_2}X_2)(\sqrt{\gamma_1}X_1 - \sqrt{\gamma_2}X_2) \\
&\quad (\sqrt{\gamma_1}X_1 + \sqrt{\gamma_3}X_3)(\sqrt{\gamma_1}X_1 - \sqrt{\gamma_3}X_2) \\
&\quad (\sqrt{\gamma_2}X_2 + \sqrt{\gamma_3}X_3)(\sqrt{\gamma_2}X_2 - \sqrt{\gamma_3}X_3)
\end{aligned}
\qquad (2.6.14)
$$

with appropriate choices for the square roots. It is easy to verify that the linear factors of F form a triangularly connected system and that $\mathcal{Z}_{A,F} = \{0\}$. So Theorem 2.6.1 can be applied. This leads to the following:

Corollary 2.6.4 *Assume that the given generators $f_1,\ldots,f_M$ of $\mathcal{I}$ have degree at most d and logarithmic height at most h, and that γ_i ($i = 0, 1, 2$), $\beta_{1,2}$, $\beta_{1,3}$ have representatives in $\mathbb{Z}[X_1,\ldots,X_r]$ of degree at most d and logarithmic height at most h, where $d \geq 1$, $h \geq 1$. Assume (2.6.11).*

Then each solution $(x_1, x_2, x_3) \in A^3$ of (2.6.10) is represented by a triple $(\tilde{x}_1, \tilde{x}_2, \tilde{x}_3)$ with

$$\tilde{x}_i \in \mathbb{Z}[X_1,\ldots,X_r], \quad s(\tilde{x}_i) \leq \exp\left((2d)^{\exp O(r)} h\right) \text{ for } i = 1, 2, 3. \quad (2.6.15)$$

Consequently, the solutions $(x_1, x_2, x_3) \in A^3$ of (2.6.10) can be determined effectively.

2.7 Norm Form Equations

We keep the notation that A is an integral domain of characteristic 0 which is finitely generated over $\mathbb{Z}$ such that $A \cong \mathbb{Z}[X_1,\ldots,X_r]/\mathcal{I}$, with $\mathcal{I} = (f_1,\ldots,f_M)$ for some given polynomials $f_1,\ldots,f_M \in \mathbb{Z}[X_1,\ldots,X_r]$. As before, we denote by K the quotient field of A and by $\overline{K}$ an algebraic closure of K.

Let $\alpha_1 = 1, \alpha_2,\ldots,\alpha_m \in \overline{K}$ ($m \geq 2$) be linearly independent of K. Consider the *norm form equation*

$$N_{K'/K}(\alpha_1 x_1 + \cdots + \alpha_m x_m) = \delta \quad \text{in } (x_1,\ldots,x_m) \in A^m, \qquad (2.7.1)$$

where $K' = K(\alpha_1,\ldots,\alpha_m)$ and δ is a nonzero element of K. In Section 10.2, we deduce the following from Theorem 2.6.1 and Corollary 2.6.2:

Theorem 2.7.1 *Let $f_1,\ldots,f_M$ have degree at most d and logarithmic height at most h, and let δ, $\alpha_1,\ldots,\alpha_m$ be represented by tuples of degree at most d and logarithmic height at most h. Suppose that $[K': K] = n \geq 3$, that*

$\alpha_1, \ldots, \alpha_m$ *are linearly independent of K and that α_m is of degree ≥ 3 over* $K(\alpha_1, \ldots, \alpha_{m-1})$. *Then each solution* $\mathbf{x} = (x_1, \ldots, x_m) \in A^m$ *of* (2.7.1) *with* $x_m \neq 0$ *is represented by* $\widetilde{\mathbf{x}} \in \mathbb{Z}[X_1, \ldots, X_r]^m$ *with*

$$s(\widetilde{\mathbf{x}}) \leq \exp\left((n^{mn^2} d)^{\exp O(r)} h \right).$$

Corollary 2.7.2 *Suppose again that $\alpha_1, \ldots, \alpha_m$ are linearly independent of K and that α_m is of degree ≥ 3 over $K(\alpha_1, \ldots, \alpha_{m-1})$. Then equation (2.7.1) has only finitely many solutions with $x_m \neq 0$. Moreover, if A, K, K', $\alpha_1, \ldots, \alpha_m$, and δ are given effectively, then all solutions of (2.7.1) with $x_m \neq 0$ can be determined effectively.*

The norm form in (2.7.1) can be expressed as

$$N_{K'/K}(\alpha_1 X_1 + \cdots + \alpha_m X_m) = \ell_1 \cdots \ell_n$$
$$\text{with } \ell_i = \alpha_1^{(i)} X_1 + \cdots + \alpha_m^{(i)} X_m \ (i = 1, \ldots, m). \tag{2.7.2}$$

Let $\mathcal{L} = (\ell_1, \ldots, \ell_n)$, $\mathcal{G}(\mathcal{L})$ be the triangular graph of $\mathcal{L}$, and $\mathcal{L}_1, \ldots, \mathcal{L}_k$ be the vertex systems of the connected components of $\mathcal{L}$. The essential observation in the deduction of Theorem 2.7.1 and Corollary 2.7.2 is that by the assumptions on $\alpha_1, \ldots, \alpha_m$, we have $X_m \in [\mathcal{L}_1] \cap \cdots \cap [\mathcal{L}_m]$; see Section 10.2. So the solutions of (2.7.1) with $x_m \neq 0$ satisfy (2.6.6). Further, letting F denote the decomposable form from (2.7.2), we have $\mathcal{Z}_{A,F} = \{\mathbf{0}\}$, since $\alpha_1, \ldots, \alpha_m$ are linearly independent of K.

The finiteness statement of Corollary 2.7.2 was proved by Győry (1982) in full generality. In Győry (1983), he also established the effectivity statement of this corollary, but only for a restricted class of integral domains A. In Győry (1983), it is pointed out that Corollary 2.7.2 does not remain valid in general if we lower the bound 3 concerning the degree of α_m over $K(\alpha_1, \ldots, \alpha_{m-1})$. Further, under the assumptions of Corollary 2.7.2, equation (2.7.1) may have infinitely many solutions $(x_1, \ldots, x_m)$ with $x_m = 0$.

The following result is an easy consequence of Corollary 2.7.2.

Corollary 2.7.3 *Suppose that in (2.7.1) α_{i+1} is of degree ≥ 3 over $K(\alpha_1, \ldots, \alpha_i)$ for $i = 1, \ldots, m - 1$. Then (2.7.1) has only finitely many solutions. Moreover, if $A, K, K', \alpha_1, \ldots, \alpha_m$, and δ are given effectively, then all solutions of (2.7.1) can be determined effectively.*

2.8 Discriminant Form Equations and Discriminant Equations

Let Ω be a finite étale K-algebra. We represent Ω in the form $K[X]/(P)$, where $P \in K[X]$ is monic and separable. We view K as a subfield of Ω. The degree

$[\Omega : K] := \dim_K \Omega$ is equal to $\deg P$. We say that Ω is *given effectively* if P is given effectively. The separability of P can be checked for instance by determining the factorization of P into irreducible factors; see for instance Theorem 6.2.3. By the choice of our representation, we have $\Omega = K[\theta]$, where $\theta := X \pmod{P}$. Elements of Ω can be expressed uniquely as $\sum_{i=0}^{n-1} a_i \theta^i$, with $a_0, \ldots, a_{n-1} \in K$, where $n = [\Omega : K]$. We say that an element of Ω is *given/can be determined effectively* if $a_0, \ldots, a_{n-1}$ are given/can be determined effectively. Denote by G the splitting field of P over K. Then there are precisely n K-algebra homomorphisms from Ω to G, denoted by $\alpha \mapsto \alpha^{(i)}$ for $i = 1, \ldots, n$, mapping θ to the n distinct zeros of P in G. One can verify that if $\alpha \in \Omega$, then

$$\alpha \in K \iff \alpha^{(1)} = \cdots = \alpha^{(n)}. \qquad (2.8.1)$$

Assume that $[\Omega : K] = n \geq 2$. Let $\mathcal{M} \subset \Omega$ be a finitely generated A-module, i.e., there are $\omega_1, \ldots, \omega_m \in \mathcal{M}$ such that

$$\mathcal{M} = \left\{ \sum_{i=1}^{m} b_i \omega_i : b_1, \ldots, b_m \in A \right\}.$$

We do not require that $\mathcal{M}$ is free over A. We say that $\mathcal{M}$ is given effectively if such $\omega_1, \ldots, \omega_m$ are given effectively. Further, we say that an *element* α of $\mathcal{M}$ is *given/can be determined effectively*, if $b_1, \ldots, b_m \in A$ are given/can be determined effectively such that $\alpha = \sum_{i=1}^{m} b_i \omega_i$.

We consider the *discriminant equation for elements of $\mathcal{M}$*

$$D_{\Omega/K}(\xi) = \prod_{1 \leq i < j \leq n} \left(\xi^{(i)} - \xi^{(j)} \right)^2 = \delta \quad \text{in } \xi \in \mathcal{M}, \qquad (2.8.2)$$

where $\delta \in K^*$.

Assertion (2.8.1) implies that if ξ is a solution of (2.8.2), then so is $\xi + \eta$ for every $\eta \in \mathcal{M} \cap K$. Hence, the set of solutions of (2.8.2) is a union of $\mathcal{M} \cap K$-*cosets* $\xi + \mathcal{M} \cap K := \{\xi + \eta : \eta \in \mathcal{M} \cap K\}$. Our aim is to determine a full system of representatives for these cosets.

We can give an equivalent formulation of (2.8.2) in terms of *discriminant form equations*. Choosing a set of A-module generators $\{\omega_1, \ldots, \omega_m\}$ for $\mathcal{M}$ (this need not be an A-basis), we can express a solution $\xi \in \mathcal{M}$ of (2.8.2) as $\sum_{i=1}^{m} x_i \omega_i$, with $x_1, \ldots, x_m \in A$. Thus, (2.8.2) translates into the *discriminant form equation*

$$D_{\Omega/K}(x_1 \omega_1 + \cdots + x_m \omega_m) = \prod_{1 \leq i < j \leq n} \left(\sum_{k=1}^{m} x_k \left(\omega_k^{(i)} - \omega_k^{(j)} \right) \right)^2 = \delta$$

$$\text{in } \mathbf{x} = (x_1, \ldots, x_m) \in A^m, \qquad (2.8.3)$$

which is a decomposable form of equation. Let

$$\mathcal{Z}_{A,D} := \left\{ \mathbf{x} = (x_1, \ldots, x_m) \in A^m : \sum_{i=1}^{m} x_i \omega_i \in K \right\}. \tag{2.8.4}$$

Then the set of solutions in A^m of (2.8.3) is a union of $\mathcal{Z}_{A,D}$-cosets $\mathbf{x} + \mathcal{Z}_{A,D}$. By a representative for a $\mathcal{Z}_{A,D}$-coset, we mean a tuple $\widetilde{\mathbf{x}} \in \mathbb{Z}[X_1, \ldots, X_r]^m$ that is a representative for an element of this coset.

In Section 10.3, we deduce the following result from Theorem 2.6.1. The essential observation is that $D_{\Omega/K}(X_1 \omega_1 + \cdots + X_m \omega_m)$ is a decomposable form whose linear factors form a triangularly connected system.

Theorem 2.8.1 *Assume that $f_1, \ldots, f_M$ have degree at most d and logarithmic height at most h and that δ and $\omega_i^{(j)}$ $(i = 1, \ldots, m, j = 1, \ldots, n)$ have tuples of representatives of degree at most d and logarithmic height at most h. Then every $\mathcal{Z}_{A,D}$-coset of solutions of (2.8.3) has a representative $\widetilde{\mathbf{x}} \in \mathbb{Z}[X_1, \ldots, X_r]^m$ with*

$$s(\widetilde{\mathbf{x}}) \le \exp\left((n^{mn^4} d)^{\exp O(r)} h \right).$$

Recall that a finitely generated A-module $\mathcal{M} \subset \Omega$ is given effectively once a finite set of A-module generators $\{\omega_1, \ldots, \omega_m\}$ for $\mathcal{M}$ is given effectively. According to the definitions, determining a full system of representatives for the $\mathcal{M} \cap K$-cosets of solutions of (2.8.2) means the same as determining a full system of representatives for the $\mathcal{Z}_{A,D}$-cosets of (2.8.3). This leads to the following consequence.

Corollary 2.8.2 *Let Ω be a finite étale K-algebra with $[\Omega : K] \ge 2$, $\mathcal{M} \subset \Omega$ a finitely generated A-module and $\delta \in K^*$. Then equation (2.8.2) has only finitely many $\mathcal{M} \cap K$-cosets of solutions. Moreover, if A, Ω, δ and $\mathcal{M}$ are given effectively, then one can effectively determine a set consisting of precisely one element from each of these cosets.*

We mention here that once $\mathcal{M}$ is given effectively and $\alpha \in \Omega$ is given effectively, then it can be decided whether $\alpha \in \mathcal{M}$. Further, given $\alpha, \alpha' \in \mathcal{M}$, one can decide whether $\alpha - \alpha' \in K$; see Corollary 6.3.8.

We consider the special case that $\mathcal{M} = \mathcal{O}$ is an A-order in Ω, i.e., $\mathcal{O}$ is a subring of Ω such that $A \subseteq \mathcal{O} \subseteq \Omega$, $K\mathcal{O} = \Omega$ and Ω is finitely generated as an A-module.

By an A-coset, we mean a set $\xi + A := \{\xi + a : a \in A\}$.

Corollary 2.8.3 *Let $\delta \in A \setminus \{0\}$, and let $\mathcal{O}$ be an A-order in Ω such that the quotient A-module*

$$(\mathcal{O} \cap K)/A \quad \text{is finite.} \tag{2.8.5}$$

Then the set of $\xi \in \mathcal{O}$ *with*

$$D_{\Omega/K}(\xi) = \delta$$

is a union of finitely many A-cosets. Moreover, if A, Ω, δ and $\mathcal{O}$ are given effectively, then one can effectively determine a set consisting of precisely one element from each of these cosets.

Corollary 2.8.3 is an easy consequence of Corollary 2.8.2. It will be deduced in Section 10.3. In the deduction of Corollary 2.8.3, we use Corollary 6.3.9 from Chapter 6. Since for this latter result we do not have a quantitative version at our disposal, we were not able to deduce a quantitative version of Corollary 2.8.3 similar to Theorem 2.8.1.

We mention here that if $\mathcal{O}$ is given effectively, then it can be decided effectively whether it is an A-order or satisfies (2.8.5); see Corollary 6.3.9. Further, for any two given $\xi_1, \xi_2 \in \mathcal{O}$, it can be decided whether $\xi_1 - \xi_2 \in A$; see Corollary 6.3.8 and Theorem 6.3.2.

Corollary 2.8.2 has further consequences, among others for index form equations; we refer to Győry (1982) for ineffective finiteness results and Győry (1983, 1984b) for effective results over a class of finitely generated domains over $\mathbb{Z}$.

We now consider another type of discriminant equation. Let A be as above a finitely generated integral domain over $\mathbb{Z}$ of characteristic 0 and with the quotient field K, and let $n \geq 2$ be an integer, δ a nonzero element of A, and G a finite extension of K. We consider the *discriminant equation for polynomials*

$$D(f) = \delta \quad \text{in monic polynomials } f \in A[X]$$

$$\text{of degree } n \text{ having all their zeros in } G. \tag{2.8.6}$$

As in Section 1.6, two monic polynomials $f, f' \in A[X]$ are called *strongly A-equivalent* if there is $a \in A$ such that $f'(X) = f(X + a)$. We recall that strongly A-equivalent polynomials have the same discriminant, and so the solutions of equation (2.8.6) split into strong A-equivalence classes.

Assuming that A is effectively given in the above sense, we say that a *polynomial* with coefficients in A or K is *given/can be determined* effectively if its coefficients are given/can be determined effectively.

Denote by A_K the integral closure of A in K. The following theorem is an effective version of Theorem 1.6.1. In Evertse and Győry (2017b), this result was deduced directly from a general effective result for unit equations. In Section 10.3, we give another proof, taking Corollary 2.6.2 as a starting point.

Theorem 2.8.4 *Let $n \geq 2$ be an integer and A an integral domain of characteristic 0, finitely generated over $\mathbb{Z}$ with the quotient field K such that the quotient A-module*

$$\left(\tfrac{1}{n}A \cap A_K\right)/A \quad \text{is finite.} \tag{2.8.7}$$

Further, let G be a finite extension of K and δ a nonzero element of A. Then the set of monic polynomials $f \in A[X]$ with (2.8.6) is a union of finitely many strong A-equivalence classes.

Moreover, for any effectively given n, A, G, δ as above, a set consisting of precisely one element from each of these classes can be determined effectively.

For any effectively given integral domain A of characteristic 0 that is finitely generated over $\mathbb{Z}$, it can be decided effectively whether it satisfies (2.8.7); see Corollary 6.3.7 or Evertse and Győry (2017b).

The proof of Theorem 2.8.4 uses both Corollary 2.6.2 and Theorem 6.3.6 and Corollary 6.3.9 from Chapter 6. Since for the latter two results we do not have quantitative versions at our disposal, we were not able to deduce a quantitative version of Theorem 2.8.4 with estimates for the sizes of the coefficients of a polynomial solution f of (2.8.6) from each strong A-equivalent class.

For integrally closed domains A, Theorem 2.8.4 gives the following result of Evertse and Győry (2017a). We note that for an effectively given domain A of characteristic 0 that is finitely generated over $\mathbb{Z}$, it can be decided whether it is integrally closed; see e.g. Theorem 10.7.17 in Evertse and Győry (2017a) and the references given there.

Corollary 2.8.5 *Let $n \geq 2$ be an integer, A an integrally closed integral domain of characteristic 0 that is finitely generated over $\mathbb{Z}$ and G a finite extension of the quotient field of A. Then the solutions of (2.8.6) lie in finitely many strong A-equivalence classes. If moreover A, G, δ are given effectively, then a set consisting of precisely one element from each of these classes can be determined effectively.*

The finiteness part of this corollary was proved in a more general, but ineffective form in Győry (1982), without fixing the degree of the polynomials under consideration; see also Theorem 1.6.2. For a restricted class of integral domains A containing transcendental elements, the effective part was proved in Győry (1984b).

2.9 Open Problems

Let again A be a domain of characteristic 0 that is finitely generated over $\mathbb{Z}$, K its quotient field, and $\overline{K}$ an algebraic closure of K.

Let $F \in A[X,Y]$ be a binary form of degree n having at least three pairwise nonproportional linear factors over $\overline{K}$, and let $\delta \in A\backslash\{0\}$. In Section 1.1, it was explained that the *Thue–Mahler equation*

$$F(x,y) \in \delta A^* \text{ in } (x,y) \in A^2 \tag{2.9.1}$$

has at most finitely many A^*-cosets of solutions. As was explained there, if $\{v_1, \ldots, v_s\}$ is a set of generators for A^*, and $\mathcal{U} = \{v_1^{m_1} \cdots v_s^{m_s} : m_1, \ldots, m_s \in \{0, \ldots, n-1\}\}$, every A^*-coset of solutions of (2.9.1) contains a pair $(x, y) \in A^2$ with

$$F(x,y) = \delta u_1 \text{ for some } u_1 \in \mathcal{U}.$$

By applying the known results on Thue equations to the latter, it follows that (2.9.1) has only finitely many A^*-cosets of solutions.

Unfortunately, as yet no method is known that on input of an arbitrary domain A of characteristic 0 that is finitely generated over $\mathbb{Z}$ computes a set of generators for A^*. Consequently, it is therefore an **open problem** how to determine effectively the A^*-cosets of solutions of (2.9.1) for arbitrary A.

It should be mentioned that for a restricted class of domains A, it is possible to compute a set of generators for A^* and, thus, the A^*-cosets of solutions of (2.9.1), for instance, for localizations of polynomial rings $A = \mathcal{O}_S[X_1, \ldots, X_q, 1/g]$, where $\mathcal{O}_S$ is the ring of S-integers in a number field and $g \in \mathcal{O}_S[X_1, \ldots, X_q] \setminus \{0\}$; see Evertse and Győry (2017a, Lemma 10.6.2).

The same can be said about decomposable form equations in $m \geq 3$ unknowns

$$F(\mathbf{x}) \in \delta A^* \text{ in } \mathbf{x} \in A^m,$$

where F is a decomposable form of degree n. The solutions of the latter equation can be divided into A^*-cosets $A^* \mathbf{x}_0 = \{u \cdot \mathbf{x}_0 : u \in A^*\}$. Completely similarly as for Thue–Mahler equations, one can reduce the above equation to finitely many equations of the form

$$F(\mathbf{x}) = \delta u_1 \text{ with } u_1 \in \mathcal{U}.$$

Again, for arbitrary A, this reduction cannot be made effective since we cannot compute a set of generators for A^*. The same applies to discriminant equations for polynomials and integral elements,

$$D_{\Omega/K}(\xi) \in \delta A^* \text{ in } \xi \in \mathcal{M}, \tag{2.9.2}$$

$$D(f) \in \delta A^*, \tag{2.9.3}$$

respectively, where the solutions of the latter equations are monic polynomials having their zeros in a given finite extension G of K.

Lastly, we would like to discuss some open problems related to monogeneity of orders. Let $A, K, \overline{K}$ be as above, Ω a finite étale K-algebra with

$[\Omega : K] = n \geq 2$, and $\mathcal{O}$ an A-order in Ω such that the quotient A-module $(\mathcal{O} \cap K)/A$ is finite. Consider the equation

$$A[\xi] = \mathcal{O} \text{ in } \xi \in \mathcal{O}. \tag{2.9.4}$$

We call two elements ξ and ξ' of $\mathcal{O}$ *A-equivalent* if $\xi' = u\xi + a$ for some $u \in A^*$ and $a \in A$. Clearly, the set of solutions of (2.9.4) is a union of A-equivalence classes. It is as yet an **open problem** to effectively determine these for arbitrary finitely generated domains A. In what follows, we will discuss some of the obstacles. For more details, we refer to Evertse and Győry (2017a, Chapters 5 and 10).

First, we observe that ξ is a solution of (2.9.4) if and only if $\{1, \xi, \ldots, \xi^{n-1}\}$ is an A-basis of $\mathcal{O}$. So for (2.9.4) to be solvable, it is necessary that $\mathcal{O}$ be free. Suppose this is the case, and let $\{\omega_1, \ldots, \omega_n\}$ be an A-basis for $\mathcal{O}$. Define the discriminant of this basis:

$$\delta := D_{\Omega/K}(\omega_1, \ldots, \omega_n) = \left(\det(\omega_i^{(j)})_{i,j=1,\ldots,n} \right)^2.$$

Recall that the discriminant of an A-basis of $\mathcal{O}$ is uniquely determined up to multiplication with a factor from A^*; see Evertse and Győry (2017a, Subsection 5.4.4). Thus, $\xi \in \mathcal{O}$ satisfies (2.8.2) if and only if

$$D_{\Omega/K}(\xi) = D_{\Omega/K}(1, \xi, \ldots, \xi^{n-1}) \in \delta A^*. \tag{2.9.5}$$

Similarly, as mentioned above, from a set of generators for A^*, we can compute a finite set $\mathcal{U} \subset A^*$, such that every element of A^* can be expressed as $u_1 \cdot u_2^{n(n-1)}$, with $u_1 \in \mathcal{U}$, $u_2 \in A^*$. Then for every solution $\xi \in \mathcal{O}$ of (2.9.4) and thus of (2.9.5), there are $u_1 \in \mathcal{U}$, $u_2 \in A^*$ such that $\xi' = u_2^{-1}\xi$ satisfies

$$D_{\Omega/K}(\xi') = \delta u_1. \tag{2.9.6}$$

By Corollary 2.8.3, the solutions $\xi' \in \mathcal{O}$ of (2.9.6) lie in finitely many A-cosets. Hence, the solutions of (2.9.4) lie in finitely many A-equivalence classes.

As yet we do not know how to solve the following problems for arbitrary domains A finitely generated over $\mathbb{Z}$ with quotient field K of characteristic 0. The *first problem* is to decide whether a given A-order $\mathcal{O}$ in a given finite étale K-algebra Ω is a free A-module and, if so, to determine an A-basis for it. The *second problem*, as mentioned above, is to compute a set of generators for A^*, needed to get the set $\mathcal{U}$.

These two problems can be solved, and thus, the A-equivalence classes of solutions of (2.9.4) can be computed, for the class of domains A mentioned above, i.e., those of the shape $\mathcal{O}_S[X_1, \ldots, X_q, g^{-1}]$, where $\mathcal{O}_S$ is the ring of S-integers in some number field and g is a nonzero element of $\mathcal{O}_S[X_1, \ldots, X_q]$; see Evertse and Győry (2017a, Chapter 10).

3

A Brief Explanation of Our Effective Methods over Finitely Generated Domains

There are two **effective** methods for solving Diophantine equations over finitely generated integral domains over $\mathbb{Z}$ of characteristic 0, which may contain both algebraic and transcendental elements over $\mathbb{Q}$. The *first one*, reducing equations to the number field and function field cases by means of effective specializations, was introduced by Győry (1983, 1984b) for a restricted class of finitely generated integral domains over $\mathbb{Z}$ of characteristic 0. This can be regarded as an effective version of Lang's method (1960) mentioned in Section 1.4. Győry's method was later refined and extended by Evertse and Győry (2013) to arbitrary finitely generated integral domains of characteristic 0; see also Bérczes, Evertse, and Győry (2014), Evertse and Győry (2015), and Chapters 7 and 9 of this book.

Over number fields, the *second effective method*, reducing equations in two unknowns to unit equations, was extended by Győry to equations in an arbitrary number of unknowns, including discriminant equations and important classes of decomposable form equations; see, e.g., Győry (1973, 1976, 1980b) and Győry and Papp (1978). This was generalized in Győry (1982) in an ineffective way and in Evertse and Győry (2017a, 2017b) and Chapter 10 of this book in an effective form to the case of arbitrary finitely generated integral domains over $\mathbb{Z}$ of characteristic 0.

We note that both effective methods have quantitative versions as well that provide effective bounds for the solutions of the equations under consideration.

In this chapter, we briefly explain the two effective methods and illustrate their applications to Diophantine equations. Detailed presentations, quantitative versions, and applications are given in Chapters 2 and 7–10.

3.1 Sketch of the Effective Specialization Method

First, we briefly outline the *first method of* Győry (1983, 1984b), which enabled him to obtain effective finiteness results for some important classes of

Diophantine equations over a class of finitely generated domains of characteristic 0. The core of the method is to reduce the equations under consideration to equations of the same type over function fields and over number fields by means of an *effective specialization* procedure, and then to apply the existing effective results over number fields and over function fields. Győry applied his method to discriminant equations and decomposable form equations, including Thue equations, index form equations, and some norm form equations. Later, this method was applied by Brindza (1989) and Végső (1994) to hyper- and superelliptic equations and by Brindza (1993) to the Catalan equation over the class of domains considered by Győry.

Evertse and Győry (2013) refined Győry's method and extended it to arbitrary finitely generated domains. We present their *general method* and compare it with that of Győry. For convenience, we use here the notation of Chapters 7–10.

Let

$$A = \mathbb{Z}[z_1, \ldots, z_r]$$

be a finitely generated integral domain of characteristic 0 with quotient field K of characteristic 0. Denote by q the transcendence degree of K. We consider only the case that $q > 0$, since otherwise, K is algebraic over $\mathbb{Q}$, and no specialization argument is needed. We assume without loss of generality that $\{z_1, \ldots, z_q\}$ is a transcendence basis of K over $\mathbb{Q}$. Since $z_1, \ldots, z_q$ may be viewed as polynomial variables, we henceforth write X_i for z_i, for $i = 1, \ldots, q$. Let

$$A_0 := \mathbb{Z}[X_1, \ldots, X_q], \quad K_0 := \mathbb{Q}(X_1, \ldots, X_q).$$

Then $K = K_0(z_{q+1}, \ldots, z_r)$ is a finite extension of K_0. Given $\alpha \in A_0$, we denote by $\deg \alpha$, $h(\alpha)$ the total degree and logarithmic height of α. Recall that A_0 is a unique factorization domain with unit group $A_0^* = \{\pm 1\}$; hence, any finite set $a_1, \ldots, a_r$ of nonzero elements of A_0 has a greatest common divisor $\gcd(a_1, \ldots, a_r)$, which is up to sign uniquely determined, such that every element of A_0 that divides $a_1, \ldots, a_r$, in fact divides their gcd.

We first describe the approach of Győry (1983, 1984b). Take $w \in A$ such that $K = K_0(w)$ and w has minimal polynomial $\mathcal{F}(X) = X^D + \mathcal{F}_1 X^{D-1} + \cdots + \mathcal{F}_D$ over K_0 with coefficients in A_0. For every $\alpha \in K$, there are up to sign unique $P_{\alpha,0}, \ldots, P_{\alpha,D-1}, Q_\alpha \in A_0$ such that $\gcd(P_{\alpha,0}, \ldots, P_{\alpha,D-1}, Q_\alpha) = 1$ and

$$\alpha = Q_\alpha^{-1} \sum_{j=0}^{D-1} P_{\alpha,j} w^j. \tag{3.1.1}$$

Now define the measures

$$\overline{\deg}\, \alpha := \max(\deg P_{\alpha,0}, \ldots, \deg P_{\alpha,D-1}, \deg Q_\alpha)$$

and

$$\overline{h}(\alpha) := \max(h(P_{\alpha,0}),\dots,h(P_{\alpha,D-1}),h(Q_\alpha)).$$

Given any bounds C_1 and C_2, there are only finitely many elements in K with $\overline{\deg}$-value at most C_1 and $\overline{h}$-value at most C_2, and these can be effectively determined.

Next, let g be the product $\prod_{i=q+1}^{r} Q_{z_i}$ of the denominators of $z_{q+1},\dots,z_r$ in their representations of the form (3.1.1). Then $g \in A_0\setminus\{0\}$. Suppose that

$$\max(\deg \mathcal{F}_1,\dots,\deg \mathcal{F}_D,\deg g) \le d_0,$$
$$\max(h(\mathcal{F}_1),\dots,h(\mathcal{F}_D),h(g)) \le h_0.$$

Győry (1983, 1984b) considered the solutions of the equations under consideration from a larger domain than A, i.e., the overring B of A defined by

$$A \subseteq B := A_0[w,g^{-1}]$$

and gives explicit upper bounds for the $\overline{\deg}$-values and $\overline{h}$-values of the solutions from B, depending on d_0 and h_0 and on appropriate parameters of the equations. This implies in an effective form the finiteness of the number of solutions of the equations with coordinates in B. What remains is to select from these the solutions with coordinates in A. Győry was able to do this only if A is given in a special manageable form. This is the case, e.g., if $A = A_0$, if $A = B$ with the domain B just introduced, or if A is an A_0-module with a given basis, say $\{z_{q+1},\dots,z_r\}$. For arbitrary A, he could prove only the finiteness of the number of solutions.

We note that Győry (1983, 1984b) established effective results in the so-called *relative case* as well, when A is a finitely generated domain over a field of characteristic 0 instead of $\mathbb{Z}$, and he gave explicit upper bounds for the $\overline{\deg}$-values of the solutions.

We now outline the general effective method of Evertse and Győry (2013), which can be applied to the case of arbitrary finitely generated domains A over $\mathbb{Z}$. Further, we point out the refinements compared with Győry's method.

Evertse and Győry (2013) use the representation for $A = \mathbb{Z}[z_1,\dots,z_r]$ that we introduced in Section 2.1. That is, let $\mathcal{I}$ be the ideal

$$\mathcal{I} = \{f \in \mathbb{Z}[X_1,\dots,X_r] : f(z_1,\dots,z_r) = 0\}.$$

Then $\mathcal{I}$ is finitely generated, and so

$$A \cong \mathbb{Z}[X_1,\dots,X_r]/\mathcal{I}, \quad \mathcal{I} = (f_1,\dots,f_M)$$

for certain polynomials $f_1,\dots,f_M$. We call $(f_1,\dots,f_M)$ an *ideal representation* for A and say that A is *effectively given* if such a representation is given.

Further, $\alpha \in A$ is said to be *effectively given/computable* if a *representative* of α in $\mathbb{Z}[X_1, \ldots, X_r]$, say $\widetilde{\alpha}$, is given/computable such that $\alpha = \widetilde{\alpha}(z_1, \ldots, z_r)$. For $\alpha \in K$, we call (a, b) a *pair of representatives* for α if $a, b \in \mathbb{Z}[X_1, \ldots, X_r]$, $b \notin \mathcal{I}$, and $\alpha = a(z_1, \ldots, z_r)/b(z_1, \ldots, z_r)$.

We collect here, in a simplified form, those lemmas/propositions from Chapter 7 that together constitute our general specialization method and give a brief explanation how these can be used in Chapter 9 to prove the effective results formulated in Chapter 2. We assume again that $z_i = X_i$ $(i = 1, \ldots, q)$ form a transcendence basis of K and keep the notation $A_0 = \mathbb{Z}[X_1, \ldots, X_q]$ and $K_0 = \mathbb{Q}(X_1, \ldots, X_q)$. We recall that the notation $O(r)$, introduced in Chapter 2, denotes any expression of the type "effectively computable absolute constant times r," where at each occurrence of $O(r)$, the constant may be different.

The following lemma can be regarded as a modified and more explicit version of Győry's result on the overring B of A. To allow for more flexibility in applications, we have extended Győry's result a little further and prescribe that certain elements of K^* are units of B. Thus, let $\mathcal{A}$ be a possibly empty finite subset of K^*, and for $\alpha \in \mathcal{A}$, let (a_α, b_α) be a pair of representatives for α.

Let $d_1 \geq d \geq 1, h_1 \geq h \geq 1$ and assume that

$$\left. \begin{array}{l} \deg f_i \leq d, \quad h(f_i) \leq h, \text{ for } i = 1, \ldots, M, \\ \deg a_\alpha, \deg b_\alpha \leq d_1, \quad h(a_\alpha), h(b_\alpha) \leq h_1, \text{ for } \alpha \in \mathcal{A}. \end{array} \right\} \quad (3.1.2)$$

Lemma 3.1.1 *There are w and g with $w \in A$ and $g \in A_0 \backslash \{0\}$ such that*

$$A \subseteq B := A_0[w, g^{-1}] \text{ and } \mathcal{A} \subset B^*,$$

such that w has minimal polynomial $\mathcal{F}(X) = X^D + \mathcal{F}_1 X^{D-1} + \cdots + \mathcal{F}_D$ over K_0 of degree $D \leq d^{r-q}$ with

$$\mathcal{F}_i \in A_0, \quad \deg \mathcal{F}_i \leq (2d)^{\exp O(r)}, \quad h(\mathcal{F}_i) \leq (2d)^{\exp O(r)} h \quad \text{for } i = 1, \ldots, D,$$

and such that

$$\deg g \leq (k+1)(2d_1)^{\exp O(r)}, \quad h(g) \leq (k+1)(2d_1)^{\exp O(r)} h_1, \text{ where } k := |\mathcal{A}|.$$

Proof This is a combination of Corollary 3.4 and Lemma 3.6 of Evertse and Győry (2013); see also Propositions 7.2.5 and 7.2.7 in Chapter 7. One has to take $g := \prod_{i=q+1}^{r} Q_{z_i} \cdot \prod_{\alpha \in \mathcal{A}} Q_\alpha Q_{\alpha^{-1}}$. □

The next lemma is new in the method of Evertse and Győry (2013). It plays an important role in the extension of Győry's method to the case of arbitrary finitely generated domains.

Lemma 3.1.2 *Let $\alpha \in A \backslash \{0\}$.*

(i) Let $\widetilde{\alpha} \in \mathbb{Z}[X_1, \ldots, X_r]$ be a representative for α. Put

$$d_2 := \max(d, \deg \widetilde{\alpha}), \quad h_2 := \max(h, h(\widetilde{\alpha})).$$

Then

$$\overline{\deg} \, \alpha \le (2d_2)^{\exp O(r)}, \quad \overline{h}(\alpha) \le (2d_2)^{\exp O(r)} h_2.$$

(ii) Put

$$d_2' := \max(d, \overline{\deg} \, \alpha), \quad h_2' := \max(h, \overline{h}(\alpha)).$$

Then α has a representative $\widetilde{\alpha} \in \mathbb{Z}[X_1, \ldots, X_r]$ such that

$$\deg \widetilde{\alpha} \le (2d_2')^{\exp O(r \log^* r)} h_2', \quad h(\widetilde{\alpha}) \le (2d_2')^{\exp O(r \log^* r)} h_2'^{r+1}.$$

Proof This is a combination of Lemmas 3.5 and 3.7 of Evertse and Győry (2013); see also Lemmas 7.2.6 and 7.3.1 in Chapter 7. The proofs of these lemmas depend heavily on work of Aschenbrenner (2004). □

Both Győry (1983, 1984b) and Evertse and Győry (2013) embedded the Diophantine equations under consideration into appropriate function fields in a fixed algebraic closure $\overline{K}_0$ of K_0. The next lemma relates $\overline{\deg} \, \alpha$ to the function field height of α in such a function field. Let $\alpha \mapsto \alpha^{(j)}$ ($j = 1, \ldots, D$) denote the K_0-isomorphic embeddings of K in $\overline{K}_0$, $j = 1, \ldots, D$. For $i = 1, \ldots, q$, let $\Bbbk_i$ be the algebraic closure of $\mathbb{Q}(X_1, \ldots, X_{i-1}, X_{i+1}, \ldots, X_q)$ in $\overline{K}_0$ and $L_i = \Bbbk_i(X_i, w^{(1)}, \ldots, w^{(D)})$. Thus, K may be viewed as a subfield of $L_1, \ldots, L_q$. We recall that the height of $\alpha \in K$ relative to $L_i/\Bbbk_i$ is defined as

$$H_{L_i}(\alpha) := \sum_{v \in \mathcal{M}_{L_i}} \max(0, -v(\alpha)),$$

where $\mathcal{M}_{L_i}$ denotes the set of normalized discrete valuations on L_i that are trivial on $\Bbbk_i$. Further, we put $\Delta_i := [L_i : \Bbbk_i(X_i)]$.

A slightly different version of the following lemma was implicitly proved in Győry (1983, 1984b) with dependence on d_0 and h_0 instead of d and h. We recall that d and h are given by (3.1.2).

Lemma 3.1.3 *Let $\alpha \in K^*$. Then*

$$\overline{\deg} \, \alpha \le (2d)^{\exp O(r)} + r \cdot d^r \max_{i,j} \Delta_i^{-1} H_{L_i}(\alpha^{(j)})$$

and

$$\max_{i,j} \Delta_i^{-1} H_{L_i}(\alpha^{(j)}) \le 2d^r \, \overline{\deg} \, \alpha + (2d)^{\exp O(r)},$$

where the maxima are taken over $i = 1, \ldots, q$ and $j = 1, \ldots, D$.

Proof The first assertion is a consequence of Lemma 4.4 of Evertse and Győry (2013) and Lemma 3.1.1. The second assertion is Lemma 4.4 of Bérczes, Evertse, and Győry (2014); see also Lemmas 7.3.3 and 7.3.4 in Chapter 7.   □

The main idea of the specialization method is to construct ring homomorphisms from an overring of A to $\overline{\mathbb{Q}}$, with which one can reduce the equations under consideration over A to the number field case. We use a refined version, due to Evertse and Győry (2013), of the ring homomorphisms from Győry (1983, 1984b). Take the overring B from Lemma 3.1.1. Define

$$\mathcal{T} := \Delta_{\mathcal{F}} \mathcal{F}_D \cdot g,$$

where $\Delta_{\mathcal{F}}$ denotes the discriminant of $\mathcal{F}$. Clearly, $\mathcal{T} \in A_0 \setminus \{0\}$ and, by Lemma 3.1.1, the additivity of the total degree, and the "almost additivity" of the logarithmic height for products of polynomials, we have

$$\deg \mathcal{T} \le (k+1)(2d_1)^{\exp O(r)}, \quad h(\mathcal{T}) \le (k+1)(2d_1)^{\exp O(r)} h_1,$$

where d_1 and h_1 are given by (3.1.2).

Any $\mathbf{u} = (u_1, \ldots, u_q) \in \mathbb{Z}^q$ gives rise to a ring homomorphism $\varphi_{\mathbf{u}} \colon A_0 = \mathbb{Z}[X_1, \ldots, X_q] \to \mathbb{Z}$ by substituting u_i for X_i for $i = 1, \ldots, q$. We write $\alpha(\mathbf{u}) := \varphi_{\mathbf{u}}(\alpha)$ for $\alpha \in A_0$. The map $\varphi_{\mathbf{u}}$ can be extended to B in the following way. Choose $\mathbf{u} \in \mathbb{Z}^q$ such that

$$\mathcal{T}(\mathbf{u}) \ne 0.$$

Let $\mathcal{F}_{\mathbf{u}} := X^D + \mathcal{F}_1(\mathbf{u})X^{D-1} + \cdots + \mathcal{F}_D(\mathbf{u})$. By our choice of $\mathbf{u}$, the polynomial $\mathcal{F}_{\mathbf{u}}$ has a nonzero discriminant; hence, it has D distinct zeros, say $w_1(\mathbf{u}), \ldots, w_D(\mathbf{u}) \in \overline{\mathbb{Q}}$, which are all nonzero since $\mathcal{F}_D(\mathbf{u}) \ne 0$. Further, $g(\mathbf{u}) \ne 0$. Hence, each substitution

$$X_1 \mapsto u_1, \ldots, X_q \mapsto u_q, \, w \mapsto w_j(\mathbf{u}), \, (j = 1, \ldots, D),$$

defines a ring homomorphism $\varphi_{\mathbf{u},j} \colon B \to \overline{\mathbb{Q}}$. We write $\alpha_j(\mathbf{u}) := \varphi_{\mathbf{u},j}(\alpha)$ for $\alpha \in B, j = 1, \ldots, D$. It follows from $\alpha_i \in B^*$ that

$$\alpha_{i,j}(\mathbf{u}) \ne 0 \text{ for } i = 1, \ldots, k \text{ and } j = 1, \ldots, D. \tag{3.1.3}$$

The image $\varphi_{\mathbf{u},j}(B)$ of B is contained in the algebraic number field $K_{\mathbf{u},j} := \mathbb{Q}(w_j(\mathbf{u}))$, with $[K_{\mathbf{u},j} : \mathbb{Q}] \le D \le d^{r-q}$.

As usual, we denote by $h(\xi)$ the absolute logarithmic height of $\xi \in \overline{\mathbb{Q}}$. For $\mathbf{u} = (u_1, \ldots, u_q) \in \mathbb{Z}^q$, we write $|\mathbf{u}| = \max(|u_1|, \ldots, |u_q|)$. An earlier version of Lemma 3.1.4 stated was proved in Győry (1983, 1984b) with different bounds, depending on d_0 and h_0 instead of d and h.

Lemma 3.1.4 *Let d, h, d_1, h_1 be given by (3.1.2). Then for $\alpha \in B \setminus \{0\}$, we have the following:*

(i) Let $\mathbf{u} \in \mathbb{Z}^q$ with $\mathcal{T}(\mathbf{u}) \neq 0$ and let $j \in \{1, \ldots, D\}$. Then

$$h(\alpha_j(\mathbf{u})) \leq \overline{h}(\alpha) + (2d)^{\exp O(r)} (h + (\overline{\deg}\,\alpha + 1) \log \max(1, |\mathbf{u}|)).$$

(ii) There exists $\mathbf{u} \in \mathbb{Z}^q$, $j \in \{1, \ldots, D\}$ such that

$$|\mathbf{u}| \leq \max(\overline{\deg}\,\alpha, (k+1)(2d_1)^{\exp O(r)}), \quad \mathcal{T}(\mathbf{u}) \neq 0,$$

$$\overline{h}(\alpha) \leq (2d_1)^{\exp O(r)}((k+1+\overline{\deg}\,\alpha)^{q+5}(h_1 + h(\alpha_j(\mathbf{u})))).$$

Proof This is a modification of Lemmas 5.6 and 5.7 from Evertse and Győry (2013); see also Lemmas 7.4.6 and 7.4.7 in Chapter 7. □

3.2 Illustration of the Application of the Effective Specialization Method to Diophantine Equations

We briefly illustrate how to apply the specialization method to Diophantine equations over finitely generated domains. As an example, consider the *Thue equation*

$$F(x, y) = \delta \quad \text{in } x, y \in A, \tag{3.2.1}$$

where F is a binary form of degree ≥ 3 in $A[X, Y]$ with a nonzero discriminant and where $\delta \in A \setminus \{0\}$.

Step 1. Let $x, y \in A$ be a solution of (3.2.1). Having upper bounds for the degrees and heights of representatives of δ and the coefficients of F, Lemma 3.1.2 gives effective upper bounds for the $\overline{\deg}$-values and $\overline{h}$-values of δ and the coefficients of F. Then, by means of Lemma 3.1.3, one gets effective upper bounds for the H_{L_i}-values of the conjugates of δ and the coefficients of F over K_0. Applying effective results of Schmidt (1978), Mason (1984), or Theorem 5.4.1 from Chapter 5 on Thue equations over function fields, one can derive effective upper bounds for $H_{L_i}(x^{(j)})$ and $H_{L_i}(y^{(j)})$ for all i, j and, subsequently, effective upper bounds for $\overline{\deg}\,x$ and $\overline{\deg}\,y$ from Lemma 3.1.3.

Step 2. Next, let the set $\mathcal{A}$ from Lemma 3.1.1 consist of δ and the discriminant of F. Choose $\mathbf{u} \in \mathbb{Z}^q$ such that $|\mathbf{u}| \leq \max(\overline{d}, (2d_1)^{\exp O(r)})$, $\mathcal{T}(\mathbf{u}) \neq 0$, and subject to these conditions, $H := \max(h(x_j(\mathbf{u})), h(y_j(\mathbf{u})))$ is maximal; here $\overline{d}$ denotes the maximum of the $\overline{\deg}$-values of x, y, δ, and the coefficients of F. Let $F_{\mathbf{u},j}$ be the binary form obtained by applying $\varphi_{\mathbf{u},j}$ to the coefficients of F. It follows from (3.1.3) and from our choice of $\mathcal{A}$ that $\delta_j(\mathbf{u}) \neq 0$, and the discriminant of $F_{\mathbf{u},j}$ is also different from zero.

Step 3. Clearly,

$$F_{\mathbf{u},j}(x_j(\mathbf{u}), y_j(\mathbf{u})) = \delta_j(\mathbf{u}).$$

We can now apply the explicit result of Győry and Yu (2006) about Thue equations; see also Theorem 4.4.1 in Chapter 4 to obtain an effective upper bound for H. Then Lemma 3.1.4 (ii) gives an effective upper bound for $\overline{h}(x)$ and $\overline{h}(y)$. Finally, Lemma 3.1.2 yields an effective upper bound C for the degrees and heights, i.e., for the sizes of certain representatives $\tilde{x}, \tilde{y}$ of x, y.

Step 4. This last step makes it possible to effectively determine all the solutions of equation (3.2.1). Adapting the proof of Proposition 2.1.1 from Chapter 2, we can enumerate all pairs $\tilde{x}, \tilde{y}$ from $\mathbb{Z}[X_1, \ldots, X_r]$ of size at most C. Using an ideal membership algorithm for $\mathbb{Z}[X_1, \ldots, X_r]$, see Section 6.1, we can check for each of these pairs $\tilde{x}, \tilde{y}$ whether $\tilde{F}(\tilde{x}, \tilde{y}) - \tilde{\delta} \in \mathcal{I}$, where $\tilde{F}$ denotes a binary form with coefficients in $\mathbb{Z}[X_1, \ldots, X_r]$ that represent the corresponding coefficients of F. Then we can make a list of all pairs passing this test. This list contains at least one representative for each solution of (3.2.1). Subsequently, we can check, for any two pairs $\tilde{x}_1, \tilde{y}_1$ and $\tilde{x}_2, \tilde{y}_2$ from this list, whether they represent the same solution of (3.2.1) by checking if $\tilde{x}_1 - \tilde{x}_2, \tilde{y}_1 - \tilde{y}_2 \in \mathcal{I}$. If these two pairs represent the same solution of (3.2.1), then we remove one of them from our list. This finally results in a list with precisely one representative for each solution.

Remark The above procedure applies also to unit equations, hyper- and superelliptic equations (see Chapter 9), as well as to discriminant equations and decomposable form equations, including discriminant form equations, index form equations, and some norm form equations; cf. Győry (1983, 1984b), because there are effective function field and number field results for these equations. As for unit equations $ax + by = c$ in $x, y \in A^*$, one may apply the above procedure to systems of equations $ax + by = c$, $x \cdot x' = 1$, $y \cdot y' = 1$ in $x, y, x', y' \in A$. Then, in Step 4, the general version of Proposition 2.1.1 concerning systems of equations must be used.

3.3 Sketch of the Method Reducing Equations to Unit Equations

Lang (1960) was the first to emphasize the importance of *unit equations of the form*

$$ax + by = 1 \text{ in } x, y \in A^*, \tag{3.3.1}$$

where A is a finitely generated integral domain over $\mathbb{Z}$, and a, b are nonzero elements of the quotient field K of A. Generalizing the results of Siegel (1921) and others obtained over number fields, he proved that equation (3.3.1) has only finitely many solutions. These results imply the finiteness of the number of solutions of some other classical equations in two unknowns; see, e.g., Lang (1962).

In the number field case when K is a number field and A is the ring of integers or a ring of S-integers of K, Győry (1974, 1979) gave explicit bounds for the solutions of (3.3.1). He applied his results to obtain the first effective bounds for the solutions of polynomial Diophantine equations in an arbitrary number of unknowns, including discriminant equations and a wide class of decomposable form equations; see, e.g., Győry (1974, 1980a,b). For ineffective generalizations for the case of arbitrary finitely generated domains over $\mathbb{Z}$, see Győry (1982).

Lang's ineffective finiteness theorem concerning equation (3.3.1) was made effective in a quantitative form in Evertse and Győry (2013); see also Theorem 2.2.1 and Corollary 2.2.2. It is applied in an effective way to discriminant equations in Evertse and Győry (2017a, 2017b), and to a wider class of decomposable form equations in a quantitative form, in Chapter 10 of the present book.

In this section, we briefly outline the method of reducing the above-mentioned equations to unit equations. In fact, the equations are reduced to the so-called connected systems of unit equations. We illustrate in some special cases and in a simplified form how to apply the effective theorem of Evertse and Győry (2013) to equation (3.3.1) to decomposable form equations and discriminant equations via systems of unit equations. The general theorems concerning decomposable form equations and discriminant equations and their proofs can be found in Chapter 2 and Chapter 10, respectively. As will be pointed out in Subsection 3.3.3, the proofs of the general, quantitative versions concerning decomposable form equations are more complicated and need several further tools from Chapter 8.

3.3.1 Effective Finiteness Result for Systems of Unit Equations

Let $A = \mathbb{Z}[z_1, \ldots, z_r]$ be a finitely generated integral domain with $A \supseteq \mathbb{Z}$, $r > 0$, and with quotient field K. Let $n \geq 3$ be an integer and $I_1, \ldots, I_k$ subsets of $\{1, 2, \ldots, n\}$ with

$$2 \leq |I_j| \leq 3 \text{ for } j = 1, \ldots, k, \quad I_1 \cup \cdots \cup I_k = \{1, \ldots, n\},$$

where $|S|$ denotes the cardinality of a set S. Many Diophantine problems can be reduced to systems of unit equations of the form

$$\sum_{i \in I_1} \lambda_{1,i} \delta_i = 0, \ldots, \sum_{i \in I_k} \lambda_{k,i} \delta_i = 0 \text{ in } (\delta_1, \ldots, \delta_n) \in (A^*)^n, \qquad (3.3.2)$$

where the coefficients $\lambda_{j,i}$ are nonzero elements of K. For $k = 1$, this is a homogeneous unit equation in at most three unknowns.

Denote by $\mathcal{G}$ the graph whose vertex set is $\{1, \ldots, n\}$ and whose edges are the pairs $\{i, i'\}$ belonging to the same set I_j, for some j with $1 \le j \le k$. The system of unit equations (3.3.2) is said to be *connected* if the graph $\mathcal{G}$ is connected.

Various versions and special cases of Theorem 3.3.1 below were explicitly or implicitly used by Győry (1976, 1980b, 1982, 1983, 1984b, 1990) and Győry and Papp (1978). These versions were all over number fields and over a restricted class of finitely generated domains, and they were mostly in quantitative form. Theorem 3.3.1 is in fact the core of Győry's approach, reducing certain important classes of equations to systems of unit equations in two unknowns and then, over number fields, applying effective results concerning unit equations.

From Corollary 2.2.2, due to Evertse and Győry (2013), we deduce the following.

Theorem 3.3.1 *Suppose that the system of equations* (3.3.2) *is connected. Then up to a proportional factor from A^*,* (3.3.2) *has only finitely many solutions. Further, if A and the coefficients $\lambda_{j,i}$ in* (3.3.2) *are given effectively, then all solutions can be determined effectively.*

The finiteness assertion is a special case of Theorem 4 of Győry (1990) which holds in the more general form when in (3.3.2) $2 \le |I_j| \le n$ holds for $j = 1, \ldots, k$, and only those solutions $(\delta_1, \ldots, \delta_n)$ are considered for which none of the equations in (3.3.2) has a proper vanishing subsum. Obviously, this assumption is necessary for the finiteness. Over finitely generated domains, the second assertion of Theorem 3.3.1 is new.

We note that a quantitative variant of Theorem 3.3.1 can be obtained by using Theorem 2.2.1 instead of Corollary 2.2.2.

Proof of Theorem 3.3.1 (sketch) Let $(\delta_1, \ldots, \delta_n)$ be a solution of equation (3.3.2). We show that for each $i, i' \in \{1, 2, \ldots, n\}$, $\delta_i / \delta_{i'}$ can take only finitely many values from A^*, and if A and the $\lambda_{i,j}$ are effectively given, all these values can be effectively determined. This immediately implies Theorem 3.3.1.

By assumption, the system of equations (3.3.2) is connected and $I_1 \cup \cdots \cup I_k = \{1, \ldots, n\}$. Hence, it is easy to see that there are $j_1, \ldots, j_\ell$ in $\{1, \ldots, k\}$ with the following three properties:

- $I_{j_1} \cup \cdots \cup I_{j_\ell} = \{1, \ldots, n\}$;
- for $t = 1, \ldots, \ell$, the system of equations

$$\sum_{i \in I_{j_1}} \lambda_{j_1 i} \delta_i = 0, \dots, \sum_{i \in I_{j_t}} \lambda_{j_t i} \delta_i = 0 \qquad (3.3.3)$$

to be solved in $\delta_i \in A^*$, for $i \in I_{j_1} \cup \dots \cup I_{j_t}$, is connected;
– for $t = 1, \dots, \ell - 1$, $I_{j_{t+1}}$ has at least one element not contained in $I_{j_1} \cup \dots \cup I_{j_t}$.

Then for $t = 1, \dots, \ell - 1$, system (3.3.3) and the (j_{t+1})th equation have a common unknown.

For $t = 1$, our claim is a consequence of Corollary 2.2.2. Then we can proceed by induction on t, and our theorem follows.   □

In the following two subsections, we illustrate how to apply Theorem 3.3.1 to decomposable form equations and discriminant equations. For convenience, we prove our effective finiteness results in a simplified, qualitative form. The precise general, quantitative statements and their proofs can be found in Chapters 2 and 10, respectively.

3.3.2 Reduction of Decomposable Form Equations to Unit Equations

Consider now the *decomposable form equation*

$$F(x_1, \dots, x_m) = \delta \text{ in } x_1, \dots, x_m \in A, \qquad (3.3.4)$$

where $\delta \in A \setminus \{0\}$, and F is a decomposable form of degree $n \geq 3$ with coefficients in A that factorizes into linear forms, say

$$\ell_i = \alpha_{i,1} X_1 + \dots + \alpha_{i,m} X_m \quad (i = 1, \dots, n).$$

Put $\mathcal{L} = \{\ell_1, \dots, \ell_n\}$. Suppose that rank $\mathcal{L} = m$ and that $\mathcal{L}$ contains at least three pairwise linearly independent linear forms.

For simplicity, we assume that

$$\delta \in A^*, \text{ the coefficients of } \ell_1, \dots, \ell_n \text{ all belong to } A. \qquad (3.3.5)$$

Denote by $\mathcal{G}(\mathcal{L})$ the graph with the vertex system $\mathcal{L}$ in which ℓ_i and ℓ_j are *connected* by an edge if ℓ_i and ℓ_j are linearly dependent on K, or they are linearly independent and there is a $q \notin \{i, j\}$ such that $\lambda_i \ell_i + \lambda_j \ell_j + \lambda_q \ell_q = 0$ for some nonzero $\lambda_i, \lambda_j, \lambda_q \in K$.

The following proposition can be deduced from Theorem 3.3.1.

Proposition 3.3.2 *Suppose that $\mathcal{G}(\mathcal{L})$ is connected. Then, under the assumptions (3.3.5), equation (3.3.4) has only finitely many solutions. Moreover, if A, δ, and the coefficients $\alpha_{i,j}$ are effectively given, all solutions of (3.3.4) can be effectively determined.*

This is a special case of Corollary 2.6.2 on decomposable form equations.

In the proof sketch of Proposition 3.3.2 that we give below, we combine the main arguments of Győry and Papp (1978) over number fields with some effective results from Chapter 6 over finitely generated domains.

Proof (sketch) Let $\mathbf{x} = (x_1, \ldots, x_m) \in A^m$ be a solution of equation (3.3.4). It follows from (3.3.5) that $\ell_i(\mathbf{x})$ is a unit of A, say $\ell_i(\mathbf{x}) = \delta_i$, $i = 1, \ldots, n$. By assumption, the graph $\mathcal{G}(\mathcal{L})$ is connected. Consider all pairs i, j for which ℓ_i, ℓ_j are connected by an edge in $\mathcal{G}(\mathcal{L})$. Then for each such i, j, we have

$$\lambda_i \delta_i + \lambda_j \delta_j = 0 \text{ or } \lambda_i \delta_i + \lambda_j \delta_j + \lambda_q \delta_q = 0, \qquad (3.3.6)$$

with the nonzero elements $\lambda_i, \lambda_j, \lambda_q$ of K. Here we may assume that λ_i, λ_j, $\lambda_q \in A \backslash \{0\}$.

Since $\mathcal{G}(\mathcal{L})$ is connected, the system of unit equations (3.3.6) is also connected. By applying Theorem 3.3.1 to this system of equations, we obtain

$$\delta_i = \varepsilon \delta_i' \text{ for } i = 1, \ldots, n,$$

where $\varepsilon \in A^*$ is still an unknown and δ_i' can take only finitely many values for $i = 1, \ldots, n$. But it follows from (3.3.4) that $\varepsilon^n = \delta / \delta_1', \ldots, \delta_n'$, whence ε and hence δ_i can take only finitely many values for each i. Finally, in view of the assumption rank $\mathcal{L} = m$, from the systems of equations

$$\ell_i(\mathbf{x}) = \delta_i, \ i = 1, \ldots, n, \qquad (3.3.7)$$

we obtain the finiteness of the number of solutions $\mathbf{x}$.

Now assume that A, δ and the coefficients α_{ij} are effectively given. Then appropriate values for λ_i, λ_j and λ_q can be determined effectively from the coefficients of ℓ_i, ℓ_j and ℓ_q. By Theorem 3.3.1, $\delta_1', \ldots, \delta_n'$ can be computed effectively. Just above (3.3.7) we observed that $\varepsilon^n = \delta / \delta_1' \cdots \delta_n'$, that is, ε is a zero of the polynomial $X^n - \delta / \delta_1' \cdots \delta_n'$. Hence, it can be determined by using Theorem 6.2.3. Further, from (3.3.7), we can determine $\mathbf{x} \in K^m$ for each possible value of $\delta_1, \ldots, \delta_n$. Using Theorem 6.3.3 it can be checked whether the $\mathbf{x}$ so obtained is an element of A^n. Subsequently, it can be checked whether $\mathbf{x}$ satisfies (3.3.4). $\qquad \square$

3.3.3 Quantitative Versions

As before, let $A = \mathbb{Z}[z_1, \ldots, z_r]$ be an integral domain of characteristic 0, $\mathcal{I}$ the ideal of polynomials in $\mathbb{Z}[X_1, \ldots, X_r]$ vanishing at $(z_1, \ldots, z_r)$, K the quotient field of A, and $\overline{K}$ an algebraic closure of K.

In the previous subsection, we considered the decomposable form equation (3.3.4) over A. For simplicity, we assumed that (3.3.5) holds, i.e., that $\delta \in A^*$

and the coefficients $\alpha_{i,j}$ of the linear factors of F are elements of A. Then
(3.3.4) leads to a system of unit equations over A. However, in our general
Theorem 2.6.1 and Corollary 2.6.2, this is not the case; the coefficients $\alpha_{i,j}$
of the linear factors belong to a finite extension G of K. In this case, equation
(3.3.4) can be reduced to a finite system of unit equations in two unknowns, but
with units from a subring $A' \supset A$ of G that is finitely generated over $\mathbb{Z}$. Then
Theorem 2.6.1 can be deduced by using Theorem 2.2.1 with A' instead of A.
To do so, in the proof of Theorem 2.6.1, we use the so-called "degree–height
estimates" for the elements of $\overline{K}$.

As an analogue of the naive height (height of the minimal polynomial over
$\mathbb{Z}$) of an algebraic number, we introduce in Chapter 8 the notion of the "degree–
height estimate" for the elements of $\overline{K}$. Given a monic polynomial $P \in K[X]$,
we call $(p_0, \ldots, p_n)$ a *tuple of representatives* for P if $p_0, \ldots, p_n \in \mathbb{Z}[X_1, \ldots, X_r]$,
$p_0 \notin \mathcal{I}$ and

$$P(X) = X^n + \frac{p_1(z_1, \ldots, z_r)}{p_0(z_1, \ldots, z_r)} X^{n-1} + \cdots + \frac{p_n(z_1, \ldots, z_r)}{p_0(z_1, \ldots, z_r)}.$$

We write

$$P \prec (d^*, h^*)$$

if P has a tuple of representatives $(p_0, \ldots, p_n)$ with total degree $\deg p_i \leq d^*$
and logarithmic height $h(p_i) \leq h^*$ for $i = 0, \ldots, n$, and call (d^*, h^*) a *degree–
height estimate* for P. If $\alpha \in \overline{K}$ and P_α denotes the monic minimal polynomial
of α over K, we define a tuple of representatives for α to be a tuple of representatives for P_α. We write

$$\alpha \prec (d^*, h^*) \qquad \text{if} \quad P_\alpha \prec (d^*, h^*)$$

and call (d^*, h^*) a *degree–height estimate* for α.

In Chapter 8, whose results are new, we give a degree–height estimate for
$\beta \in \overline{K}$ in terms of degree–height estimates for $\alpha_1, \ldots, \alpha_m \in \overline{K}$, if β is related
to the α_i by $P(\beta, \alpha_1, \ldots, \alpha_m) = 0$ for some given $P \in \mathbb{Z}[X_0, X_1, \ldots, X_m]$. Such
estimates can be used in the proof of Theorem 2.6.1 to construct in an effective
way a finitely generated domain $A' \supset A$ in G and certain scalar multiples
ℓ_i' of ℓ_i, for $i = 1, \ldots, n$, such that $\ell_1'(\mathbf{x}), \ldots, \ell_n'(\mathbf{x})$ are the units of A' for
any solution $\mathbf{x}$ of (3.3.4). Then one can follow a quantitative version of the
arguments of the proof of Proposition 3.3.2 and can use Theorem 2.2.1 with
A' instead of A, as well as some estimates from Chapter 8 to prove Theorem
2.6.1 in the case when $\mathcal{G}(\mathcal{L})$ is connected. In the general case, when $\mathcal{G}(\mathcal{L})$ is
not connected, some further argument is needed from Step 4 of the proof of
Theorem 2.6.1.

3.3.4 Reduction of Discriminant Equations to Unit Equations

In the remaining part of this chapter, let again A be a finitely generated integral domain over $\mathbb{Z}$. Let $n \geq 2$ be an integer, $\delta \in A\backslash\{0\}$, and consider the *discriminant equation*

$$D(f) = \delta \text{ in monic polynomials } f \in A[X] \text{ of degree } n. \qquad (3.3.8)$$

We recall that the monic polynomials $f, f' \in A[X]$ are called *strongly A-equivalent* if $f'(X) = f(X + a)$ for some $a \in A$. Then f and f' have the same degree and same discriminant.

For simplicity, here we restrict ourselves to the special situation when

$$\left. \begin{array}{l} \delta \in A^*, \; A \text{ is integrally closed and all zeros of } f \text{ belong to} \\ \text{the quotient field } K \text{ of } A \text{ (and hence to } A\text{).} \end{array} \right\} \qquad (3.3.9)$$

We deduce from Theorem 3.3.1 the following proposition.

Proposition 3.3.3 *Under the assumptions* (3.3.9), *the solutions of equations* (3.3.8) *lie in finitely many strong A-equivalence classes of solutions. Moreover, if A and δ are effectively given, a full set of representatives of these equivalence classes can be effectively determined.*

This is a special case of Corollary 2.8.5.

The first proof of Proposition 3.3.3 (reducing directly to unit equations; sketch). Let A and δ be as described in Proposition 3.3 and $f \in A[X]$ be a monic polynomial of degree $n \geq 2$ with zeros $\alpha_1, \dots, \alpha_n$ and with the properties (3.3.8) and (3.3.9). Then we have

$$D(f) = \prod_{1 \leq i < j \leq n} (\alpha_i - \alpha_j)^2 = \delta \in A^*, \qquad (3.3.10)$$

where $\alpha_1, \dots, \alpha_n$ are the zeros of f in A. This implies that

$$\delta_{i,j} := \alpha_i - \alpha_j \in A^* \text{ for each } i, j, \text{ with } 1 \leq i < j \leq n.$$

First, suppose that $n \geq 3$. Consider the system of unit equations

$$\delta_{i,j} + \delta_{j,q} + \delta_{q,i} = 0 \text{ in } \delta_{i,j}, \delta_{j,q}, \delta_{q,i} \in A^*$$

for distinct $i, j, q \in \{1, \dots, n\}$. We show that this system of equations is connected. Indeed, for $\delta_{i,j}, \delta_{i',j'}$, with $(i, j) \neq (i', j')$ and $i \neq j$, $i' \neq j'$, we have

$$\delta_{i,j} + \delta_{j,i'} + \delta_{i',i} = 0, \quad \delta_{j,i'} + \delta_{i',j'} + \delta_{j',j} = 0$$

if $i' \neq j$, and

$$\delta_{i,j} + \delta_{i',j'} + \delta_{j',i} = 0$$

if $i' = j$. Hence, by Theorem 3.3.1, we obtain

$$\delta_{ij} = \varepsilon \delta'_{ij} \text{ for any distinct } i, j, \tag{3.3.11}$$

with a common factor $\varepsilon \in A^*$ and with $\delta'_{i,j} \in A^*$ which may take only finitely many values. This is obviously true for $n = 2$ as well. Furthermore, if A and δ are effectively given, by Theorem 3.3.1, the $\delta'_{i,j}$ can also be effectively determined.

Now (3.3.10) and (3.3.11) give

$$\varepsilon^{n(n-1)} = \delta \prod_{1 \le i < j \le n} (1/\delta'_{i,j})^2.$$

This implies that there are only finitely many $\varepsilon, \delta_{i,j} \in A^*$ and polynomials $f'(X) = \prod_{i=1}^{n}(X - \delta_{i,1}) \in A[X]$ under consideration. Further, if A and δ are effectively given, then using Theorem 6.2.3, we can effectively determine the zeros in A of the polynomials $X^{n(n-1)} - \theta$ for all $\theta := \delta \prod_{1 \le i < j \le n} (1/\delta'_{i,j})^2$ in question. Consequently, all $\varepsilon, \delta_{i,j}$ and $f'(X)$ can be effectively determined. But $f'(X) = f(X + \alpha_1)$, i.e., f is strongly A-equivalent to f'. Finally, from among the f', one can easily select a maximal set of pairwise strongly A-inequivalent polynomials f, satisfying (3.3.8) and (3.3.9).   □

Proposition 3.3.3 can also be deduced from Proposition 3.3.2 on decomposable form equations, following the arguments of the proof of Theorem 4.8.1. However, we recall that the proof of Proposition 3.3.2 is also based on effective results on unit equations.

The second proof of Proposition 3.3.3 (reducing to decomposable form equations; sketch). Let again A and δ be as above and $f \in A[X]$ be a monic polynomial of degree $n \ge 2$ with zeros $\alpha_1, \ldots, \alpha_n$ and with the properties (3.3.8) and (3.3.9). Then we have again (3.3.10). Writing now $x_i := \alpha_i - \alpha_1$ for $i = 2, \ldots, n$, we have $x_i \in A$ for each i. Putting

$$F(X_2, \ldots, X_n) = X_2 \cdots X_n \prod_{2 \le i < j \le n} (X_i - X_j),$$

(3.3.10) implies

$$F(x_2, \ldots, x_n) = \pm \delta_0 \text{ in } x_2, \ldots, x_n \in A, \tag{3.3.12}$$

where $\delta_0^2 = \delta$, and if (3.3.12) is solvable, $\delta_0 \in A^*$ must hold. The decomposable form F is of degree $n(n-1)/2$, and it is easily seen that for $n \ge 3$, it satisfies the assumptions of Proposition 3.3.2. Hence, by Proposition 3.3.2 equation (3.3.12) has only finitely many solutions $(x_2, \ldots, x_n)$. Further, if A and δ are effectively given, then by Theorem 6.2.3, the quantity δ_0 can also be effectively determined and Proposition 3.3.2 gives that all solutions $(x_2, \ldots, x_n)$ can be effectively found. For $n = 2$, the same assertion holds because in this case $x_2 = \pm\delta_0$. Now we can argue, as in the first proof of Proposition 3.3.3, to show that $f'(X) = f(X + \alpha_1)$ is strongly A-equivalent to f and is effectively computable. Finally, our proof can be completed as in the first proof.   □

3.4 Comparison of Our Two Effective Methods

Comparing our effective methods over finitely generated domains over $\mathbb{Z}$, it is easy to observe that the "unit equation" method, reducing equations to appropriate systems of unit equations, is technically less complicated, at least in the qualitative case. The other "effective specialization" method, involving effective specializations, is more complicated to apply. To some classes of equations, for example, to Thue equations, discriminant equations, and decomposable form equations, both methods can be applied, whereas in the case of unit equations, hyper- and superelliptic equations, the Schinzel–Tijdeman equation, and the Catalan equation, only the "effective specialization" method applies.

It is interesting to note that over number fields, the superelliptic equations can be reduced to systems of unit equations via Thue equations. However, this reduction uses Lemma 4.5.4 from Chapter 4 (which in turn uses estimates for class numbers, regulators, and fundamental units) for which there is no analogue over arbitrary finitely generated domains. Hence, the reduction of superelliptic equations to unit equations cannot be extended to arbitrary finitely generated domains.

To illustrate both methods, we used the "effective specialization" method for Thue equations and the "unit equation" method for decomposable form equations and discriminant equations.

4

Effective Results over Number Fields

In our *first general effective method*, equations over finitely generated domains are reduced to equations of the same type over number fields and over function fields. Then the best known or best applicable effective results over number fields/function fields can be applied to estimate from above the sizes of the solutions of the initial equations over finitely generated domains.

Our *second effective method* reduces, if possible, equations to unit equations in two unknowns. Such equations are e.g. Thue equations, discriminant equations, and certain other decomposable form equations. Then using explicit bounds for unit equations, one can derive explicit bounds for the solutions of the initial equations, as well.

In Chapter 9, we apply our first method to unit equations in two unknowns, Thue equations, hyper- and superelliptic equations, the Schinzel–Tijdeman equation, and the Catalan equation over finitely generated domains. The second method will be extended in Chapter 10 from the number field case to the finitely generated situation, reducing decomposable form equations and discriminant equations to unit equations in two unknowns over finitely generated domains. We note that in the general case, the results concerning decomposable form equations are new.

In Sections 4.3–4.6 of *the present chapter*, we present the best applicable explicit bounds for the solutions of those equations over number fields, which are considered in Chapter 9. Although in Chapter 10 we shall not need equations over number fields, for the convenience of the reader, in Sections 4.7 and 4.8 we have included the best explicit results for decomposable form equations and discriminant form equations as well over number fields.

To avoid long and complicated computations but emphasizing the role of the ingredients, we shall sketch the proofs of less precise versions of the presented results over number fields.

4.1 Notation and Preliminaries

We start with introducing some notation and recall some basic facts on number fields. For further details, we refer e.g. to Evertse and Győry (2015, Chapter 1).

Let L be an algebraic number field. Denote by $d, \mathcal{O}_L, \mathcal{M}_L, D_L, h_L, r$, and R_L its degree, ring of integers, set of places, discriminant, class number, unit rank, and regulator, respectively. The set $\mathcal{M}_L$ consists of *real infinite places*, these are the embeddings $\sigma : L \hookrightarrow \mathbb{R}$, *complex infinite places*, these are the pairs of conjugate complex embeddings $\{\sigma, \overline{\sigma} : L \hookrightarrow \mathbb{C}\}$, and *finite places*, these are the prime ideals of $\mathcal{O}_L$. We denote by S_∞ the set of all infinite places, i.e., both real and complex, of L. To every $v \in \mathcal{M}_L$, we associate a *normalized absolute value* $|\cdot|_v$ such that for $\alpha \in L$ we have

$$
\begin{aligned}
|\alpha|_v &:= |\sigma(\alpha)| && \text{if } v = \sigma \text{ is real,} \\
|\alpha|_v &:= |\sigma(\alpha)|^2 = |\overline{\sigma}(\alpha)|^2 && \text{if } v = \{\sigma, \overline{\sigma}\} \text{ is complex,} \\
|\alpha|_v &:= N(\mathfrak{p})^{-\operatorname{ord}_\mathfrak{p}(\alpha)} && \text{if } v = \mathfrak{p} \text{ is a prime ideal of } \mathcal{O}_L.
\end{aligned}
$$

Here, $N(\mathfrak{p}) := |\mathcal{O}_L/\mathfrak{p}|$ denotes the absolute norm of $\mathfrak{p}$, and $\operatorname{ord}_\mathfrak{p}(\alpha)$ denotes the exponent of $\mathfrak{p}$ in the unique prime ideal factorization of $[\alpha]$ (i.e., the fractional ideal generated by α), with $\operatorname{ord}_\mathfrak{p}(0) = \infty$. The absolute values just defined satisfy the *product formula*

$$
\prod_{v \in \mathcal{M}_L} |\alpha|_v = 1 \quad \text{for} \quad \alpha \in L^*.
$$

If M is a finite extension of L, and V and v are places of M and L, respectively, then we say that V *lies above* v or v *below* V, notation $V|v$, if the restriction of $|\cdot|_V$ to L is a power of $|\cdot|_v$. In case of finite places of V and v, this means that $V = \mathfrak{P}$, and $v = \mathfrak{p}$ are prime ideals in M and L, respectively with $\mathfrak{P} \supset \mathfrak{p}$.

Given $v \in \mathcal{M}_L$, we denote by L_v the *completion* of L at v and by $\overline{L_v}$ an algebraic closure of L_v. The absolute value $|\cdot|_v$ has a unique extension to $\overline{L_v}$ that we denote also by $|\cdot|_v$. If v is real, then $\overline{L_v} = \mathbb{C}$, and $|\cdot|_v$ is the ordinary absolute value on $\mathbb{C}$; thus, it satisfies the triangle inequality. If v is complex, then $|\cdot|_v$ is the square of the ordinary absolute value on $\mathbb{C}$, and thus, $|\cdot|_v^{1/2}$ satisfies the triangle inequality. If v is finite, then $|\cdot|_v$ satisfies the ultrametric inequality. Summarizing, for $\alpha, \beta \in \overline{L_v}$ we have

$$
\left.
\begin{aligned}
|\alpha + \beta|_v &\leq |\alpha|_v + |\beta|_v && \text{if } v \text{ is real,} \\
|\alpha + \beta|_v^{1/2} &\leq |\alpha|_v^{1/2} + |\beta|_v^{1/2} && \text{if } v \text{ is complex,} \\
|\alpha + \beta|_v &\leq \max(|\alpha|_v, |\beta|_v) && \text{if } v \text{ is finite.}
\end{aligned}
\right\} \qquad (4.1.1)
$$

From this general fact, we deduce the following useful inequality. We define the quantities $s(v)$ ($v \in \mathcal{M}_L$) as follows:

$$s(v) := \begin{cases} 1 & \text{if } v \text{ is real,} \\ 2 & \text{if } v \text{ is complex,} \\ 0 & \text{if } v \text{ is finite.} \end{cases} \qquad (4.1.2)$$

Lemma 4.1.1 *Let v be a place of L, m a positive integer, and $\alpha \in \overline{L_v}$ such that $|\alpha|_v \leq (2m)^{-s(v)}$. Then*

$$|(1+\alpha)^m - 1|_v \leq (2m)^{s(v)}|\alpha|_v.$$

Proof Assume without loss of generality that $\alpha \neq 0$. Then

$$A := \frac{(1+\alpha)^m - 1}{\alpha} = \sum_{k=1}^{m} \binom{m}{k}\alpha^{k-1}.$$

If v is finite, then by the ultrametric inequality, $|A|_v \leq 1$. Assume that v is infinite. Then $| \cdot |^{1/s(v)}$ is the ordinary absolute value on $\mathbb{C}$, and thus, by the triangle inequality,

$$|A|_v^{1/s(v)} \leq \sum_{k=1}^{m} \binom{m}{k}|\alpha|_v^{(k-1)/s(v)} \leq \sum_{k=1}^{m} m^k |\alpha|_v^{(k-1)/s(v)} \leq 2m.$$

The lemma follows. □

Let $\alpha \in \overline{\mathbb{Q}}$ and choose a number field L such that $\alpha \in L$. The *absolute multiplicative height* of α is defined by

$$H(\alpha) := \prod_{v \in \mathcal{M}_L} \max(1, |\alpha|_v)^{1/[L:\mathbb{Q}]},$$

while its *absolute logarithmic height* or briefly *height* is given by

$$h(\alpha) := \log H(\alpha) = \frac{1}{[L:\mathbb{Q}]} \sum_{v \in \mathcal{M}_L} \log \max(1, |\alpha|_v).$$

These notions are independent of the choice of L.

The *denominator* of $\alpha \in \overline{\mathbb{Q}}^*$, denoted by den α, is defined as the smallest positive rational integer d_0 for which $d_0\alpha$ is an algebraic integer.

The logarithmic height has the following important properties:

$$\begin{cases} h(\alpha_1 \cdots \alpha_k) \leq \sum_{i=1}^{k} h(\alpha_i) & \text{for } \alpha_1, \ldots, \alpha_k \in \overline{\mathbb{Q}}; \\ h(\alpha_1 + \cdots + \alpha_k) \leq \log k + \sum_{i=1}^{k} h(\alpha_i) & \text{for } \alpha_1, \ldots, \alpha_k \in \overline{\mathbb{Q}}; \\ h(\alpha^m) = |m|h(\alpha) & \text{for } \alpha \in \overline{\mathbb{Q}}^*, m \in \mathbb{Z}; \\ h(\zeta\alpha) = h(\alpha) & \text{for } \alpha \in \overline{\mathbb{Q}} \text{ and } \zeta \text{ a root of unity;} \\ h(\alpha) \geq \dfrac{\log \text{den } \alpha}{\deg \alpha} & \text{for } \alpha \in \overline{\mathbb{Q}}^*. \end{cases} \qquad (4.1.3)$$

Further, we need the following more advanced result.

Lemma 4.1.2 *Let α be an algebraic number of degree $d \geq 1$, which is not equal to 0 or to a root of unity. Then*

$$h(\alpha) \geq m(d) := \begin{cases} \log 2 & \text{if } d = 1, \\ 2/d(\log 3d)^3 & \text{if } d \geq 2. \end{cases}$$

Proof See Voutier (1996). Asymptotically, this lower bound is not the most optimal, but it is most convenient for our purposes. See, e.g. Dobrowolski (1979) for an estimate, which is still the best in terms of d. □

For a number field L and $v \in \mathcal{M}_L$, the *v-adic norm* of a vector $\mathbf{x} = (x_1, \ldots, x_n) \in \overline{L_v}$ is defined by

$$|\mathbf{x}|_v := \max(|x_1|_v, \ldots, |x_n|_v).$$

Let $\mathbf{x} = (x_1, \ldots, x_n) \in \overline{\mathbb{Q}}^n$ and choose an algebraic number field L such that $\mathbf{x} \in L^n$. Then the *multiplicative height* and *homogeneous multiplicative height* of $\mathbf{x}$ are defined by

$$H(\mathbf{x}) := \left(\prod_{v \in \mathcal{M}_L} \max(1, |\mathbf{x}|_v) \right)^{1/d}, \quad H^{\text{hom}}(\mathbf{x}) := \left(\prod_{v \in \mathcal{M}_L} |\mathbf{x}|_v \right)^{1/d},$$

respectively, where $d = [L : \mathbb{Q}]$. As in the preceding case $n = 1$, these *heights* are independent of the choice of L. We define the corresponding logarithmic heights by

$$h(\mathbf{x}) := \log H(\mathbf{x}), \quad h^{\text{hom}}(\mathbf{x}) := \log H^{\text{hom}}(\mathbf{x}),$$

respectively. For instance,

$$h(\mathbf{x}) = \log \max(|x_1|, \ldots, |x_n|), \quad h^{\text{hom}}(\mathbf{x}) = \log \left(\frac{\max(|x_1|, \ldots, |x_n|)}{\gcd(x_1, \ldots, x_n)} \right)$$

$$\text{for } \mathbf{x} = (x_1, \ldots, x_n) \in \mathbb{Z}^n \setminus \{\mathbf{0}\}. \tag{4.1.4}$$

From the definitions, it is clear that

$$\max_{1 \leq i \leq n} h(x_i) \leq h(\mathbf{x}) \leq \sum_{i=1}^{n} h(x_i) \quad \text{for } \mathbf{x} = (x_1, \ldots, x_n) \in \overline{\mathbb{Q}}^n. \tag{4.1.5}$$

Further,

$$h^{\text{hom}}(\lambda \mathbf{x}) = h^{\text{hom}}(\mathbf{x}) \quad \text{for } \mathbf{x} \in \overline{\mathbb{Q}}^n, \lambda \in \overline{\mathbb{Q}}^*. \tag{4.1.6}$$

This is shown by applying the product formula with any number field L containing λ and the coordinates of $\mathbf{x}$.

Let again L be a number field. For a polynomial $P \in L[X_1, \ldots, X_n]$ and for $v \in \mathcal{M}_L$, we define

$$|P|_v := |\mathbf{x}_P|_v,$$

where $\mathbf{x}_P$ is a vector consisting of the nonzero coefficients of P. We will need the following estimate.

Lemma 4.1.3 *Let $P_1, \ldots, P_q \in L[X_1, \ldots, X_n]$ be polynomials in n variables and $P := P_1, \ldots, P_q$ their product. Suppose that the partial degrees of P have a sum at most D. If $v \in \mathcal{M}_L$, we have*

$$2^{-nDs(v)} \prod_{j=1}^{q} |P_j|_v \le |P|_v \le 2^{nDs(v)} \prod_{j=1}^{q} |P_j|_v \quad \text{if } v \text{ is infinite,}$$

where $s(v) = 1$ if v is real and $s(v) = 2$ if v is complex, while

$$|P|_v = \prod_{j=1}^{q} |P_j|_v \quad \text{if } v \text{ is finite.}$$

Proof See Bombieri and Gubler (2006) and Lemmas 1.6.11 and 1.6.3. □

We deduce some consequences. For a polynomial $P \in \overline{\mathbb{Q}}[X_1, \ldots, X_n]$, we define

$$h(P) := h(\mathbf{x}_P), \quad h^{\text{hom}}(P) := h^{\text{hom}}(\mathbf{x}_P).$$

Corollary 4.1.4 *Let $P_1, \ldots, P_q \in \overline{\mathbb{Q}}[X_1, \ldots, X_n]$ be nonzero polynomials and $P := P_1, \ldots, P_q$ their product. Suppose that the partial degrees of P have a sum at most D. Then*

$$\left| h^{\text{hom}}(P) - \sum_{j=1}^{q} h^{\text{hom}}(P_j) \right| \le D \log 2.$$

Proof Pick a number field L containing the coefficients of $P_1, \ldots, P_q$, take the logarithms of the inequalities, and identity from Lemma 4.1.3 and sum over $v \in \mathcal{M}_L$. □

Corollary 4.1.5 *Let $P(X) = (X - \alpha_1) \cdots (X - \alpha_n) \in \overline{\mathbb{Q}}[X]$. Then*

$$\left| h(P) - \sum_{i=1}^{n} h(\alpha_i) \right| \le n \log 2.$$

Proof Immediate consequence of Corollary 4.1.4. □

We have the following estimate for the inhomogeneous heights of polynomials with rational integer coefficients.

Corollary 4.1.6 *Let $P_1, \ldots, P_q \in \mathbb{Z}[X_1, \ldots, X_n]$ and $P := P_1, \ldots, P_q$ their product. Suppose that the partial degrees of P have a sum at most D. Then*

$$|h(P) - \sum_{i=1}^{q} h(P_i)| \le D \log 2.$$

Proof Use that $h(Q) = \log |Q|_\infty$ if Q is a polynomial with coefficients in $\mathbb{Z}$ and apply Lemma 4.1.3. □

In addition to the preceding information, we will frequently use the following simple estimates for heights and lengths of polynomials with integer coefficients. Given a polynomial Q with integer coefficients, its height $H(Q)$ is the maximum of the absolute values of its coefficients, while its *length* $L(Q)$ is the sum of the absolute values of its coefficients. Denote by $n(Q)$ the number of nonzero coefficients of Q. Then for $Q \in \mathbb{Z}[X_1, \ldots, X_m]$, we have

$$H(Q) \le L(Q) \le n(Q)H(Q) \le \binom{\deg Q + m}{m} H(Q) \le 2^{\deg Q + m} H(Q), \quad (4.1.7)$$

where as usual we denote by $\deg Q$ the total degree of Q. Further, for $Q_1, Q_2 \in$ and $\mathbb{Z}[X_1, \ldots, X_m]$, we have

$$L(Q_1 + Q_2) \le L(Q_1) + L(Q_2), \quad L(Q_1 Q_2) \le L(Q_1) L(Q_2) \quad (4.1.8)$$

(see e.g., Waldschmidt [2000, p. 76]). From these inequalities, we deduce the following:

Lemma 4.1.7 *Let $P_1, \ldots, P_q \in \mathbb{Z}[X_1, \ldots, X_n]$ and $F \in \mathbb{Z}[X_1, \ldots, X_q]$ be nonzero polynomials. Then for the composed polynomial $G := F(P_1, \ldots, P_q)$, we have*

$$\deg G \le \deg F \cdot \max_{1 \le i \le q} \deg P_i,$$

$$L(G) \le L(F)\Big(\max_{1 \le i \le q} L(P_i)\Big)^{\deg F},$$

$$H(G) \le n(F)H(F)\Big(\max_{1 \le i \le q} n(P_i)H(P_i)\Big)^{\deg F}.$$

Proof The estimate for $\deg G$ is obvious. The estimate for $L(G)$ follows directly from (4.1.8), and then the estimate for $H(G)$ follows from (4.1.7). □

As in the preceding L is an algebraic number field of degree d. Let S be a finite set of places of L, which contains the set S_∞ of infinite places. Denote by s the cardinality of S. We have $d \le 2s$. Recall that the ring of S-integers $\mathcal{O}_S$ is defined as

$$\mathcal{O}_S = \{\alpha \in L : |\alpha|_v \le 1 \quad \text{for} \quad v \in \mathcal{M}_L \backslash S\}.$$

For $S = S_\infty$, this is just the ring of integers $\mathcal{O}_L$ in L. Denote by $\mathcal{O}_L^*$ and, more generally, by $\mathcal{O}_S^*$ the unit groups of $\mathcal{O}_L$ and $\mathcal{O}_S$, respectively. By the

Dirichlet–Chevalley–Weil S-unit theorem, that is, the extension to S-units of Dirichlet's unit theorem, the group $\mathcal{O}_S^*$ has rank $s - 1$. This means that there are $\varepsilon_1, \ldots, \varepsilon_{s-1} \in \mathcal{O}_S^*$ such that every $\varepsilon \in \mathcal{O}_S^*$ can be expressed uniquely as

$$\varepsilon = \zeta \varepsilon_1^{b_1} \cdots \varepsilon_{s-1}^{b_{s-1}},$$

where ζ is a root of unity in L, and $b_1, \ldots, b_{s-1}$ are rational integers. Such a system $\{\varepsilon_1, \ldots, \varepsilon_{s-1}\}$ is called a *fundamental system of S-units* and for $S = S_\infty$, a *fundamental system of units* in L. Notice that rank $\mathcal{O}_L^* = r_1 + r_2 - 1 =: r$, where r_1 is the number of real places, and r_2 is the number of complex places of L.

Pick $s-1$ places $v_1, \ldots, v_{s-1}$ from S, i.e., we omit one place. The *S-regulator* is defined by

$$R_S := |\det(\log |\varepsilon_i|_{v_j})_{i,j=1,\ldots,s-1}|.$$

This quantity is nonzero and independent of the choice of $\varepsilon_1, \ldots, \varepsilon_{s-1}$ and of the choice $v_1, \ldots, v_{s-1}$ from S. In the case that $S = S_\infty$, the S-regulator R_S is equal to the regulator R_L of L.

If S contains finite places, i.e., prime ideals of $\mathcal{O}_L$, we denote by $\mathfrak{p}_1, \ldots, \mathfrak{p}_t$ as the prime ideals in S, and we put

$$P_S := \max(N(\mathfrak{p}_1), \ldots, N(\mathfrak{p}_t)), \quad Q_S := N(\mathfrak{p}_1 \cdots \mathfrak{p}_t) \quad \text{if } S \supsetneq S_\infty,$$

$$P_{S_\infty} := 1, \qquad\qquad\qquad Q_{S_\infty} := 1.$$

$$(4.1.9)$$

The S-regulator R_S and the regulator R_L are related by

$$R_S = h_S R_L \prod_{i=1}^{t} \log N(\mathfrak{p}_i), \qquad (4.1.10)$$

where h_S is a (positive) divisor of the class number h_L. We have

$$h_L R_L \leq |D_L|^{1/2} (\log^* |D_L|)^{d-1}, \qquad (4.1.11)$$

see Louboutin (2000) and Győry and Yu (2006), while on the other hand,

$$R_L > 0.2052, \qquad (4.1.12)$$

see Friedman (1989). From (4.1.10) and (4.1.11) we infer

$$R_S \leq |D_L|^{1/2} (\log^* |D_L|)^{d-1} (\log P_S)^t, \qquad (4.1.13)$$

while in the opposite direction, we have, by (4.1.10) and (4.1.12),

$$R_S \geq \begin{cases} (\log 2)(\log 3)^{s-2} & \text{if } K = \mathbb{Q}, \, s = |S| \geq 3, \\ 0.2052(\log 2)^{s-2} & \text{if } K \neq \mathbb{Q}, \, s \geq 3. \end{cases}$$

For $\alpha \in L^*$, the fractional ideal $[\alpha]$ generated by α can be written uniquely as a product of two fractional ideals $\mathfrak{a}_1 \cdot \mathfrak{a}_2$, where $\mathfrak{a}_1$ is composed of $\mathfrak{p}_1, \ldots, \mathfrak{p}_t$, and $\mathfrak{a}_2$ is composed solely of prime ideals different from $\mathfrak{p}_1, \ldots, \mathfrak{p}_t$. The *S-norm* of α is now defined as $N_S(\alpha) := N(\mathfrak{a}_2)$. Another expression for the S-norm is

$$N_S(\alpha) = \prod_{v \in S} |\alpha|_v.$$

Combining this with $h(\alpha) = d^{-1} \sum_{v \in S} \log \max(1, |\alpha|_v)$ for $\alpha \in \mathcal{O}_S$, we derive the very useful inequality

$$\log N_S(\alpha) \le dh(\alpha) \le s \max_{v \in S} \log |\alpha|_v \quad \text{for } \alpha \in \mathcal{O}_S \backslash \{0\}. \qquad (4.1.14)$$

In case that α is an S-unit we obtain, by applying (4.1.14) to α^{-1},

$$\min_{v \in S} \log |\alpha|_v \le -\frac{d}{s} h(\alpha) \quad \text{for } \alpha \in \mathcal{O}_S^*. \qquad (4.1.15)$$

In many of our estimates, we need an effective version of the Dirichlet–Chevalley–Weil S-unit theorem. We subsequently state a suitable version. Let again L be a number field of degree d, and denote by r the rank of $\mathcal{O}_L^*$. Further, let as before S be a finite set of places of L containing the infinite places and s the cardinality of S. Define the constants

$$c_1 := \frac{((s-1)!)^2}{2^{s-2}d^{s-1}} \quad \text{for } s \ge 2,$$

$$c_2 := 116e \cdot \frac{((s-1)!)^2 \sqrt{s-2}}{2^s} \cdot \log^* d \quad \text{for } s \ge 3,$$

and

$$c_3 := \begin{cases} \dfrac{(s-1)!)^2}{2^{s-2} \log 2} & \text{for } s \ge 2, \, d = 1, \\[2ex] \dfrac{((s-1)!)^2}{2^{s-1}} \cdot (\log(3d))^3 & \text{for } s \ge 2, \, d \ge 2, \end{cases}$$

$$c_4 := \begin{cases} 0 & \text{for } r = 0, \\ 1/d & \text{for } r = 1, \\ 29e \cdot r! r \sqrt{r-1} \cdot \log d & \text{for } r \ge 2. \end{cases}$$

Proposition 4.1.8 *There exists a fundamental system $\{\varepsilon_1, \ldots, \varepsilon_{s-1}\}$ of S-units in L such that*

(i) $\displaystyle\prod_{i=1}^{s-1} h(\varepsilon_i) \le c_1 R_S,$

(ii) $\displaystyle\max_{1 \le i \le s-1} h(\varepsilon_i) \le c_2 R_S \quad \text{if } s \ge 3,$

(iii) if $v_1, \ldots, v_{s-1}$ are any distinct places from S, then the absolute values of the entries of the inverse matrix of $(\log |\varepsilon_i|_{v_j})_{i,j=1,\ldots,s-1}$ *do not exceed* c_3.

Proof (i) and (iii) were proved in Bugeaud and Győry (1996a, 1996b). The inequality (ii) is an improvement, at least in terms of s, of the corresponding statements of Bugeaud and Győry (1996a, 1996b) and Bugeaud (1998). The main tool in the proof of (i) and (iii) is Minkowski's theorem on the successive minima of symmetric convex bodies from the geometry of numbers. Assertion (ii) was proved by combining (i) with a similar type of result as Lemma 4.1.2. See also Evertse and Győry (2015, Prop. 4.3.9) for a proof of (i), (ii), and (iii). □

We shall also need the following lemma.

Proposition 4.1.9 *For every nonzero $\alpha \in \mathcal{O}_S$ and for every integer $n \geq 1$, there exists an $\varepsilon \in \mathcal{O}_S^*$ such that*

$$h(\varepsilon^n \alpha) \leq \frac{1}{d} \log N_S(\alpha) + n \left(c_4 R_L + \frac{h_L}{d} \log Q_S \right).$$

Proof See e.g. Győry and Yu (2006, Lemma 3) or Evertse and Győry (2015, Proposition 4.3.12). The basic idea is as follows: The vectors $(\log |\varepsilon^n|_v)_{v \in S}$ ($\varepsilon \in \mathcal{O}_S^*$) form a full lattice in the vector space of $(x_v)_{v \in S} \in \mathbb{R}^s$ with $\sum_{v \in S} x_v = 0$; hence, every point from this vector space is within bounded distance from a point from this lattice. The lemma follows from an explicit bound for this distance. □

We finish this section with some estimates for discriminants. As before, we denote by D_L the discriminant of a number field L.

Lemma 4.1.10 *Suppose that L is the compositum of the algebraic number fields $L_1, \ldots, L_k$. Then D_L divides $D_{L_1}^{[L:L_1]} \cdots D_{L_k}^{[L:L_k]}$ in $\mathbb{Z}$.*

Proof See Stark (1974). □

Lemma 4.1.11 *Let L be a number field of degree d; let $G \in L[X]$ be a polynomial without multiple zeros, which factorizes over an extension of L as $a_0 (X - \vartheta_1) \cdots (X - \vartheta_n)$; and let $M := L(\vartheta_1, \ldots, \vartheta_k)$ with $1 \leq k \leq n$. Then for the discriminant of M, we have the estimate*

$$|D_M| \leq (n e^{h(G)})^{2kn^k d} |D_L|^{n^k}.$$

In the case that $k = 1$, we have the sharper estimate

$$|D_M| \leq n^{(2n-1)d} e^{(2n^2-2)h(G)} |D_L|^n.$$

Proof This is Lemma 4.1 of Bérczes, Evertse, and Győry (2013). □

Throughout our work, we shall use $A \ll_{a,b,...} B$ or $B \gg_{a,b,...} A$ to denote that $|A|$ is less than c times B, where c is an effectively computable positive number, depending only on $a, b, \ldots$, which may be different at each occurrence of the symbol $\ll$. Further, we define

$$\log^* u := \max\{1, \log u\} \quad \text{for} \quad u > 0, \log^* 0 := 1.$$

4.2 Effective Estimates for Linear Forms in Logarithms

In this section, we present some results from **Baker's effective theory of logarithmic forms** that are used in Section 4.3–4.6. We formulate, without proof, some consequences of the best-known effective estimates for linear forms in logarithms, due to Matveev (2000) in the complex case and Yu (2007) in the p-adic case.

We first give a brief introduction. For the moment, $\overline{\mathbb{Q}}$ denotes the algebraic closure of $\mathbb{Q}$ in $\mathbb{C}$, and algebraic numbers are supposed to belong to $\overline{\mathbb{Q}}$. Here and in the next paragraph, log denotes any fixed determination of the logarithm.

Gel'fond (1934) and Schneider (1934) proved independently of each other that if α and β are algebraic numbers such that $\alpha \neq 0$ or 1 and β is not rational, then $\alpha^\beta := \exp(\beta \log \alpha)$ is transcendental for any choice of $\log \alpha$. An equivalent formulation of the Gel'fond–Schneider theorem is that if α_1 and α_2 are nonzero algebraic numbers such that $\log \alpha_1$ and $\log \alpha_2$ are linearly independent over $\mathbb{Q}$ for any choice of the logarithms, then they are linearly independent over $\overline{\mathbb{Q}}$. Gel'fond (1935) gave a nontrivial effective lower bound for $|\beta_1 \log \alpha_1 + \beta_2 \log \alpha_2|$, where β_1 and β_2 denote algebraic numbers, not both 0, and α_1 and α_2 denote algebraic numbers different from 0 and 1 such that $\log \alpha_1 / \log \alpha_2$ is not rational.

In his celebrated series of papers, Baker (1966, 1967a, 1967b, 1968a) made a significant breakthrough by generalizing the Gel'fond–Schneider theorem to arbitrarily many logarithms. Baker (1966, 1967b) proved that if $\alpha_1, \ldots, \alpha_n$ denote nonzero algebraic numbers such that $\log \alpha_1, \ldots, \log \alpha_n$ are linearly independent over $\mathbb{Q}$, then 1 and $\log \alpha_1, \ldots, \log \alpha_n$ are linearly independent over $\overline{\mathbb{Q}}$. Further, Baker (1967a, 1967b, 1968a) gave nontrivial lower bounds for $|\beta_1 \log \alpha_1 + \cdots + \beta_n \log \alpha_n|$, where $\alpha_1, \ldots, \alpha_n$ are nonzero algebraic numbers such that $\log \alpha_1, \ldots, \log \alpha_n$ are linearly independent of $\mathbb{Q}$, and $\beta_1, \ldots, \beta_n$ are algebraic numbers, not all 0.

Baker's general effective estimates led to significant applications in number theory. Later, many improvements, generalizations, and applications were established by Baker and others. For comprehensive accounts of Baker's theory, analogues for p-adic and elliptic logarithms and algebraic groups and extensive bibliographies, the reader can consult Baker (1975, 1988), Baker and Masser

(1977), Lang (1978), Feldman and Nesterenko (1998), Waldschmidt (2000), Wüstholz (2002), Baker and Wüstholz (2007), and Bugeaud (2018).

For applications to Diophantine equations, the effective estimates of Baker (1968a, 1968b) in which $\beta_1, \ldots, \beta_n$ are rational integers proved to be particularly useful. Using his estimates in this special case, Baker (1968b, 1968c, 1969) gave the first explicit upper bounds for the solutions of Thue equations, Mordell equations, and super- and hyperelliptic equations over $\mathbb{Z}$. For further applications of Baker's theory to Diophantine problems, we refer to Baker (1975), Győry (1980b, 2002), Sprindžuk (1982, 1993), Shorey and Tijdeman (1986), Bilu Yu (1995), Smart (1998), Evertse and Győry (2015, 2017a), Bugeaud (2018), and to the references given there.

Since the 1990s, some **other effective methods** have also been developed for Diophantine equations by various authors, including Bombieri (1993), Bombieri and Cohen (1997, 2003), Bugeaud (1998), Bennett and Skinner (2004), Siksek (2013), Murty and Pasten (2013), Pasten (2017), von Känel (2014), von Känel and Matschke (2016), Kim (2017), Poonen (2019), Le Fourn (2020), Győry (2019), Triantafillou (2020), and Freitas, Kraus and Siksek (2020a, 2020b). See also Evertse and Győry (2015, Section 4.5). However, at present, Baker's theory is the most suitable to derive effective bounds for the solutions of our equations over number fields.

We now state some consequences of the best known results, due to Matveev (2000) and Yu (2007), from Baker's theory. These are the main tools in the next sections, in the proofs for unit equations, hyper- and superelliptic equations, and the Catalan equation. We note that in the complex case Matveev (2000) gives lower bounds for linear forms in logarithms. For applications, it will be more convenient to consider some consequences concerning

$$\Lambda = \alpha_1^{b_1} \cdots \alpha_n^{b_n} - 1,$$

where $\alpha_1, \ldots, \alpha_n$ ($n \geq 2$) are nonzero algebraic numbers and where $b_1, \ldots, b_n$ are rational integers, not all zero.

Throughout this section, L is a number field containing $\alpha_1, \ldots, \alpha_n$, and d denotes the degree of L. Set

$$B^* := \max\{|b_1|, \ldots, |b_n|\}$$

and let $A_1, \ldots, A_n$ be reals with

$$A_i \geq \max\{dh(\alpha_i), \pi\}, \quad i = 1, \ldots, n.$$

We state two propositions that are consequences of results of Matveev (2000).

Proposition 4.2.1 *Suppose $\Lambda \neq 0$, $b_n = \pm 1$, and B satisfies*

$$B \geq \max\{B^*, 2eA_n \max(n\pi/\sqrt{2}, A_1, \ldots, A_{n-1})\}.$$

Then

$$\log|\Lambda| > -c_5(n,d)A_1 \cdots A_n \log(B/(\sqrt{2}A_n))$$

where

$$c_5(n,d) = \min\{1.451(30\sqrt{2})^{n+4}(n+1)^{5.5}, \pi 2^{6.5n+27}\}d^2 \log(ed).$$

Proof This is Proposition 4 in Győry and Yu (2006). It is an easy consequence of Corollary 2.3 of Matveev (2000). □

We put $\chi := 1$ if L is real and $\chi := 2$ otherwise. Let now

$$A_i' := dh(\alpha_i) + \pi, \quad i = 1, \ldots, n.$$

Proposition 4.2.2 *Suppose that $\Lambda \neq 0$, and that B' satisfies*

$$B' \geq \max\{1, \max(|b_i|A_i'/A_n'; 1 \leq i \leq n)\}.$$

Then we have

$$\log|\Lambda| > -c_6(n,d)A_1' \cdots A_n' \log(e(n+1)B'),$$

where

$$c_6(n,d) = 2\pi \min\left\{\frac{1}{\chi}\left(\frac{1}{2}e(n+1)\right)^{\chi} 30^{n+4}(n+1)^{3.5}, 2^{6n+26}\right\} d^2 \log(ed).$$

Proof This is Lemma 3.6 in Koymans (2017). It is easily deduced from Corollary 2.3 of Matveev (2000). □

Consider again Λ defined as above. Let B and B_n be real numbers satisfying

$$B \geq \max\{|b_1|, \ldots, |b_n|\}, \quad B \geq B_n \geq |b_n|.$$

Denote by $\mathfrak{p}$ a prime ideal in $\mathcal{O}_L$ lying above the prime number p, and by $e_\mathfrak{p}$ and $f_\mathfrak{p}$ the ramification index and residue class degree of $\mathfrak{p}$, respectively. Thus, $N(\mathfrak{p}) = p^{f_\mathfrak{p}}$.

The following result is due to Yu (2007).

Proposition 4.2.3 *Assume that $\mathrm{ord}_\mathfrak{p}b_n \leq \mathrm{ord}_\mathfrak{p}b_i$ for $i = 1, \ldots, n$, and set*

$$A_i'' := \max\{h(\alpha_i), 1/(16e^2d^2)\}, \quad i = 1, \ldots, n.$$

If $\Lambda \neq 0$, then for any real δ with $0 < \delta \leq 1/2$, we have

$$\mathrm{ord}_\mathfrak{p}\Lambda < c_7(n,d)e_\mathfrak{p}^n \frac{N(\mathfrak{p})}{(\log N(\mathfrak{p}))^2} \max\left\{A_1'' \cdots A_n'' \log M, \frac{\delta B}{B_n c_8(n,d)}\right\},$$

where

$$c_7(n,d) = (16ed)^{2(n+1)}n^{3/2}\log(2nd)\log(2d),$$
$$c_8(n,d) = (2d)^{2n+1}\log(2d)\log^3(3d)$$

and

$$M = (B_n/\delta)c_9(n,d)N(\mathfrak{p})^{n+1}A_1'' \cdots A_{n-1}''$$

with

$$c_9(n,d) = 2e^{(n+1)(6n+5)}d^{3n}\log(2d).$$

Proof This is the second consequence of the Main Theorem in Yu (2007). □

As before, L denotes a number field of degree d and $\mathcal{M}_L$ its set of places, $\alpha_1,\ldots,\alpha_n$ are $n(\geq 2)$ nonzero elements of L, and $b_1,\ldots,b_n$ are rational integers, not all zero. Let Λ be defined again as above, and put

$$\Omega := \prod_{i=1}^{n}\max\{h(\alpha_i),m(d)\},$$

$$B := \max\{3,|b_1|,\ldots,|b_n|\}$$

where $m(d)$ is the lower bound from Lemma 4.1.2. For a place $v \in \mathcal{M}_L$, we write

$$N(v) := \begin{cases} 2 & \text{if } v \text{ is infinite,} \\ N(\mathfrak{p}) & \text{if } v = \mathfrak{p} \text{ is finite.} \end{cases}$$

The following proposition is in fact a combination of some inequalities of Matveev (2000) and Yu (2007).

Proposition 4.2.4 *Suppose that $\Lambda \neq 0$. Then for $v \in \mathcal{M}_L$, we have*

$$\log|\Lambda|_v > -c_{10}(n,d)\frac{N(v)}{\log N(v)}\Omega\log B,$$

where $c_{10}(n,d) = 12(16ed)^{3n+2}(\log^* d)^2$.

Proof This is Proposition 3.10 in Bérczes, Evertse, and Győry (2013). For v infinite, it is deduced from Corollary 2.3 of Matveev (2000), while for v finite, from the first consequence of the Main Theorem of Yu (2007, p. 190). We have used Lemma 4.1.2 to incorporate the small values of B. □

4.3 S-Unit Equations

Keeping the notation introduced in Section 4.1, L is a number field; S is a finite set of places of L containing the infinite places; and $\mathcal{O}_S$ is the ring of S-integers of L and $\mathcal{O}_S^*$ the unit group of $\mathcal{O}_S$, that is, the group of S-units in L.

Let α and β be nonzero elements of L with

$$\max\{h(\alpha),h(\beta)\} \leq H,$$

where, for technical reasons, we assume that $H \geq \max(1, \pi/d)$. Consider the *S-unit equation*

$$\alpha x + \beta y = 1 \quad \text{in} \quad x, y \in \mathcal{O}_S^*. \tag{4.3.1}$$

For $S = S_\infty$, this is called a(n ordinary) *unit equation*.

The first effective finiteness theorems for equation (4.3.1) were proved for $S = S_\infty$ by Győry (1973, 1974); and for general S by Győry (1979); and independently, in a less precise form, by Kotov and Trelina (1979). In the proofs, Baker's theory of logarithmic forms was used. The results in Győry (1974, 1979) and Kotov and Trelina (1979) are quantitative, providing explicit upper bounds for the heights of the solutions x and y of (4.3.1).

In Chapter 9, we shall use the following theorem, due to Győry and Yu (2006). Here, d denotes the degree of L, s denotes the cardinality of S, $P_S = \max(1, N(\mathfrak{p}_1), \ldots, N(\mathfrak{p}_t))$ where $\mathfrak{p}_1, \ldots, \mathfrak{p}_t$ are the prime ideals in S, and R_S denotes the *S*-regulator.

Theorem 4.3.1 *All solutions of x and y of equation* (4.3.1) *satisfy*

$$\max\{h(x), h(y)\} \leq c_{11} P_S R_S (1 + (\log^* R_S)/\log^* P_S) H,$$

where

$$c_{11} = s^{2s+3.5} \cdot 2^{7s+27} (\log 2s) d^{2(s+1)} (\log^*(2d))^3.$$

In the proof of Theorem 4.3.1, the main tools are Propositions 4.2.1 and 4.2.3 (logarithmic forms estimates) and a variation on Proposition 4.1.8 (the effective *S*-unit theorem).

Combining the method of proof with his new approach, Le Fourn (2019) has recently improved Theorem 4.3.1 with P_S replaced by P_S', which denotes the third largest norm of prime ideals from S. For a further improvement, see Győry (2019). However, these improvements would not give better bounds in Chapter 9.

To avoid long and complicated computations but emphasize the role of the main tools, we shall sketch the proof of the following less precise version of Theorem 4.3.1.

Let x and y be a solution of equation (4.3.1). *Then*

$$\max\{h(x), h(y)\} \ll_{L,S} H. \tag{4.3.2}$$

Here and below, the positive constants implied by $\ll_{L,S}$ depend only on L and S and are effectively computable.

For a complete proof of Theorem 4.3.1, the reader can consult Győry and Yu (2006).

Sketch of the proof of (4.3.2) Let x and y be a solution of (4.3.1). We may assume that $h(x) \geq h(y)$. Further, the case $s = 1$ being trivial, we assume that $s \geq 2$.

Let $\{\varepsilon_1, \ldots, \varepsilon_{s-1}\}$ be a fundamental system of S-units with the properties specified in Proposition 4.1.8. Then y can be written in the form

$$y = \zeta \varepsilon_1^{b_1} \cdots \varepsilon_{s-1}^{b_{s-1}}, \tag{4.3.3}$$

where ζ is a root of unity in L, and $b_1, \ldots, b_{s-1}$ are rational integers. Set

$$B := \max\{3, |b_1|, \ldots, |b_s|\}$$

and let $v_1, \ldots, v_s$ be distinct places from S. Then it follows from (4.3.3) that

$$\log |y|_{v_j} = \sum_{i=1}^{s-1} b_i \log |\varepsilon_i|_{v_j}, \quad j = 1, \ldots, s - 1.$$

Together with Cramer's rule and (iii) of Proposition 4.1.8, this implies that

$$B \ll_{L,S} h(y) \ll_{L,S} B. \tag{4.3.4}$$

Set $\alpha_s = \zeta\beta, b_s = 1$ and

$$\Lambda = \varepsilon_1^{b_1} \cdots \varepsilon_{s-1}^{b_{s-1}} \alpha_s^{b_s} - 1.$$

Let $v \in S$ for which $|x|_v$ is minimal. Then, using (4.3.1) and (4.1.15), we deduce that

$$\log |\Lambda|_v = \log |\alpha x|_v \leq -\frac{d}{s} h(x) + dH. \tag{4.3.5}$$

First, assume that v is infinite. We may assume that

$$B \geq c(L, S)H$$

with an appropriate, effectively computable positive number $c(L, S)$ depending only on L and S, since otherwise (4.3.1) and (4.3.4) would give immediately (4.3.2). Applying Proposition 4.2.1 to $\log |\Lambda|_v$ and using Lemma 4.1.8, (4.3.4), and $h(x) \geq h(y)$, we infer that

$$\log |\Lambda|_v \gg_{L,S} \left(-H \log \left(\frac{h(x)}{H} \right) \right).$$

Together with (4.3.5) this gives (4.3.2).

Second, assume that v is finite and corresponds to the prime ideal $\mathfrak{p}$. Then it follows that

$$\log |\alpha x|_v = -(\text{ord}_\mathfrak{p}\Lambda) \log N(\mathfrak{p}). \tag{4.3.6}$$

We apply now Proposition 4.2.3 to $\mathrm{ord}_{\mathfrak{p}}\Lambda$ with the choice $\delta = \frac{c(L,S)}{2} \cdot \frac{H}{B} \leq \frac{1}{2}$. Then, using Proposition 4.1.8, inequality (4.3.4), and $h(y) \leq h(x)$, we get

$$\mathrm{ord}_{\mathfrak{p}}\Lambda \ll_{L,S} H \log\left(\frac{h(x)}{H}\right).$$

Together with (4.3.5) and (4.3.6) this gives (4.3.2).   □

The following consequence of Theorem 4.3.1, due to Győry and Yu (2006), will be very useful.

Corollary 4.3.2 *Let α_1, α_2, and α_3 be nonzero elements in L with logarithmic heights at most $H(\geq 2)$. Then, for every solution x_1, x_2, and x_3 of*

$$\alpha_1 x_1 + \alpha_2 x_2 + \alpha_3 x_3 = 0$$

$$in \; x_k \in \mathcal{O}_S \backslash \{0\} \; with \; N_S(x_k) \leq N \quad for \; k = 1, 2, 3 \tag{4.3.7}$$

there is an $\varepsilon \in \mathcal{O}_S^*$ such that

$$\max_{1 \leq k \leq 3} h(\varepsilon x_k) \leq 2.001 c_{11} P_S R_S (1 + (\log^* R_S)/(\log^* P_S))\mathcal{N}$$

where

$$\mathcal{N} = c_4 R_L + \frac{h_L}{d} \log Q_S + H + \frac{1}{d} \log N$$

with the constants c_{11} from Theorem 4.3.1 and c_4 from Proposition 4.1.9.

We prove a weaker version of Corollary 4.3.2; that is, we show that *for every solution (x_1, x_2, x_3) of* (4.3.7), *there is $\varepsilon \in \mathcal{O}_S^*$ such that*

$$\max_{1 \leq k \leq 3} h(\varepsilon x_k) \ll_{L,S} \max(H, \log N). \tag{4.3.8}$$

Proof of (4.3.8) Put $H^* := \max(H, \log N)$. Pick a solution $(x_1, x_2,$ and $x_3)$ of (4.3.7). By Proposition 4.1.9, for $k = 1, 2, 3$, there are $\mu_k \in \mathcal{O}_S \backslash \{0\}$ with $h(\mu_k) \ll_{L,S} \log N$ and $\varepsilon_k \in \mathcal{O}_S^*$, such that $x_k = \mu_k \varepsilon_k$. This gives

$$\alpha_1 \mu_1 \varepsilon_1 + \alpha_2 \mu_2 \varepsilon_2 + \alpha_3 \mu_3 \varepsilon_3 = 0,$$

or equivalently,

$$\beta_1 \cdot \frac{\varepsilon_1}{\varepsilon_3} + \beta_2 \cdot \frac{\varepsilon_2}{\varepsilon_3} = 1 \quad \text{where} \quad \beta_1 = -\frac{\alpha_1 \mu_1}{\alpha_3 \mu_3}, \; \beta_2 = -\frac{\alpha_1 \mu_1}{\alpha_3 \mu_3}.$$

By the height properties (4.1.3) we have $h(\beta_1)$ and $h(\beta_2) \ll_{L,S} H^*$, and so, by (4.3.2),

$$h(\varepsilon_1/\varepsilon_3), \; h(\varepsilon_2/\varepsilon_3) \ll_{L,S} H^*.$$

Now taking $\varepsilon := \varepsilon_1^{-1}$ and invoking the height properties (4.1.3), we readily obtain (4.3.8).   □

4.4 Thue Equations

As before, L is a number field and S a finite set of places of L, containing all infinite places. Let $F(X, Y) \in L[X, Y]$ be a binary form of degree $n \geq 3$ with at least three pairwise nonproportional linear factors over $\overline{L}$, a fixed algebraic closure of L. Further, let δ be a nonzero element of L, and consider the *Thue equation* over $\mathcal{O}_S$,

$$F(x, y) = \delta \quad \text{in} \quad x, y \in \mathcal{O}_S. \tag{4.4.1}$$

For a polynomial P with algebraic coefficients, we denote by $h(P)$ the maximum of the absolute logarithmic heights of its coefficients.

In the classical case $L = \mathbb{Q}$ and $\mathcal{O}_S = \mathbb{Z}$, the first effective upper bound for the heights of the solutions of equation (4.4.1) was given by Baker (1968b), using one of his effective estimates concerning linear forms in logarithms. Later, Baker's effective result has been improved and generalized by several people; for references, see e.g. Shorey and Tijdeman (1986), Sprindžuk (1993), Győry (2002), Evertse and Győry (2015), and Bugeaud (2018).

For convenience, choose L such that F factors into linear forms over L. Then the best-known bound to date for the solutions of equation (4.4.1) is due to Győry and Yu (2006). As before, d denotes the degree of L, while s denotes the cardinality of S, the quantities P_S and Q_S are defined by (4.1.9), and R_S denotes the S-regulator.

Theorem 4.4.1 *Assume that F splits into linear factors over L. Then all solutions of x and y of equation (4.4.1) satisfy*

$$\max\{h(x), h(y)\}$$
$$\leq c_{12} P_S R_S \left(1 + \frac{\log^* R_S}{\log^* P_S}\right) \cdot \left(c_4 R_L + \frac{h_L}{d} \log Q_S + 2ndH_1 + H_2\right),$$

where

$$H_1 := \max(1, h(F)), \quad H_2 = \max(1, h(\delta)),$$
$$c_{12} := 250 n^6 s^{2s+3.5} \cdot 2^{7s+29} (\log 2s) d^{2s+4} (\log^*(2d))^3$$

and c_4 is the constant from Proposition 4.1.9, i.e.,

$$c_4 = 0 \text{ if } r = 0, \quad 1/d \text{ if } r = 1, \quad 29e \cdot r! r \sqrt{r-1} \cdot \log d \text{ if } r \geq 2.$$

This is Corollary 3 of Győry and Yu (2006). It is a special case of Theorem 3 of Győry and Yu (2006) concerning decomposable form equations.

In terms of S, a better bound has been recently obtained in Győry (2019), replacing P_S by P_S'; see the remark after Theorem 4.3.1. However, this improvement would not lead to a better bound in Chapter 9.

We shall outline a proof of the following less precise version of Theorem 4.4.1, under the same assumptions as in Theorem 4.4.1:

All solutions of x and y of equation (4.4.1) *satisfy*

$$\max\{h(x), h(y)\} \ll_{n,L,S} \max(H_1, H_2). \tag{4.4.2}$$

We recall that here and below, constants implied by $\ll_{a_1,\ldots,a_r}$ are positive and effectively computable and depend only on the parameters $a_1, \ldots, a_r$ in the subscript.

Sketch of the proof of (4.4.2) We start with some remarks. There is an $a \in \mathbb{Z}$ with $1 \le a \le n$ such that $F(1, a) \ne 0$. Consider the binary form $G(X, Y) := F(X, aX + Y)$ in which the coefficient of X^n is $F(1, 0) \ne 0$ and the heights of the coefficients of G are at most $\ll_n H_1$. Denote by g_0 the product of the denominators of the coefficients of G. Then we can write

$$g_0 G(X, Y) = a_0 X^n + a_1 X^{n-1} Y + \cdots + a_n Y^n$$
$$= a_0 (X - \alpha_1 Y) \cdots (X - \alpha_n Y),$$

where $a_0, \ldots, a_n$ are already integers in L with heights $\ll_{n,d} H_1$. Further, at least three from among $\alpha_1, \ldots, \alpha_n$, say, α_1, α_2, and α_3, are pairwise distinct.

Let x and y be a solution of (4.4.1). Then

$$x' = a_0 x, \quad y' = -ax + y \tag{4.4.3}$$

is a solution of the equation

$$(x' - \alpha_1' y') \cdots (x' - \alpha_n' y') = \delta', \tag{4.4.4}$$

where $\delta' = g_0 a_0^{n-1} \delta \in \mathcal{O}_S$ and $\alpha_i' = a_0 \alpha_i$ for $i = 1, \ldots, n$. We have

$$(\alpha_i')^n + a_1 a_0 (\alpha_i')^{n-1} + \cdots + a_n a_0^{n-1} = 0,$$

which implies that α_i' is an integer in L, and $h(\alpha_i') \ll_{n,d} H_1$ for each i. Further,

$$h(\delta') \ll_{n,d} \max(H_1, H_2). \tag{4.4.5}$$

Putting $\delta_i' = x' - \alpha_i' y'$, we have $\delta_i' \in \mathcal{O}_S \setminus \{0\}$. It follows from (4.4.4) that δ_i' divides δ' in $\mathcal{O}_S$. Therefore,

$$\log N_S(\delta_i') \le \log N_S(\delta') \ll_{n,d} \max(H_1, H_2).$$

The linear forms $X - \alpha_i' Y$ are pairwise nonproportional for $i = 1, 2, 3$. Hence, there are nonzero integers α_1, α_2, and α_3 in L such that

$$\alpha_1 \delta_1' + \alpha_2 \delta_2' + \alpha_3 \delta_3' = 0$$

and $h(\alpha_i) \ll_{n,d} H_1$ for $i = 1, 2, 3$. Then, by (4.3.8), which was proved in the previous section, there is an $\varepsilon \in \mathcal{O}_S^*$ such that

$$h(\varepsilon \delta_i') \ll_{n,L,S} \max(H_1, H_2) \quad \text{for} \quad i = 1, 2, 3.$$

But then from

$$(\varepsilon x') - \alpha_i'(\varepsilon y') = \varepsilon \delta_i', \quad i = 1, 2,$$

it follows that $h(\varepsilon x')$ and $h(\varepsilon y') \ll_{n,L,S} \max(H_1 \text{ and } H_2)$ and thus $h(x'/y') \ll_{n,L,S} \max(H_1 \text{ and } H_2)$. Using this together with (4.4.4) and (4.4.5), we obtain $h(x')$ and $h(y') \ll_{n,L,S} \max(H_1 \text{ and } H_2)$, and finally, a combination with (4.4.3) gives (4.4.2). $\qquad\square$

4.5 Hyper- and Superelliptic Equations, the Schinzel–Tijdeman Equation

Again, S is a finite set of places of a number field L, containing all infinite places. Let

$$f(X) = a_0 X^n + a_1 X^{n-1} + \cdots + a_n \in \mathcal{O}_S[X]$$

be a polynomial of degree $n \geq 2$ with nonzero discriminant, $\delta \in \mathcal{O}_S \backslash \{0\}, m \geq 2$ an integer, and consider the *hyperelliptic equation*

$$f(x) = \delta y^2 \quad \text{in} \quad x, y \in \mathcal{O}_S, \tag{4.5.1}$$

where $n \geq 3$, and the *superelliptic equation*

$$f(x) = \delta y^m \quad \text{in} \quad x, y \in \mathcal{O}_S, \tag{4.5.2}$$

where $n \geq 2$ and $m \geq 3$.

For $L = \mathbb{Q}, S = S_\infty$, the first explicit upper bounds for the solutions of (4.5.1) and (4.5.2) were obtained by Baker (1968b, 1968c, 1969). Over $\mathbb{Z}$, Schinzel and Tijdeman (1976) were the first to consider equation (4.5.2) in the more general situation when also the exponent m is unknown, and they gave an effective upper bound for m. Quantitative improvements and generalizations were later obtained by many authors, including Brindza (1984), who gave an effective upper bound for the solutions x and y of (4.5.1) and (4.5.2) under the most general condition, where f is allowed to have multiple roots. For further references, see Shorey and Tijdeman (1986), Sprindžuk (1993), Győry (2002), Evertse and Győry (2015), and Bugeaud (2018).

The following best-known explicit results are due to Bérczes, Evertse, and Győry (2013). As before, d denotes the degree of L, s the cardinality of S, and the quantities P_S and Q_S are defined by (4.1.9). Further, we put

$$\widehat{h} := \frac{1}{d} \sum_{v \in \mathcal{M}_L} \log \max(1, |\delta|_v, |a_0|_v, \ldots, |a_n|_v). \tag{4.5.3}$$

Theorem 4.5.1 *All solutions of x and y of equation* (4.5.1) *satisfy*

$$h(x) \text{ and } h(y) \leq c_{13}|D_L|^{8n^3} Q_S^{20n^3} e^{50n^4 d\widehat{h}},$$

where $c_{13} = (4ns)^{212n^4 s}$.

Theorem 4.5.2 *All solutions of x and y of equation* (4.5.2) *satisfy*

$$h(x) \text{ and } h(y) \leq c_{14}|D_L|^{2m^2 n^2} Q_S^{3m^2 n^2} e^{8m^2 n^3 d\widehat{h}},$$

where $c_{14} = (6ns)^{14m^3 n^3 s}$.

Finally, consider the *Schinzel–Tijdeman equation*

$$f(x) = \delta y^m \quad \text{in} \quad x, y \in \mathcal{O}_S, \quad m \in \mathbb{Z}_{\geq 2} \tag{4.5.4}$$

where $m \geq 2$ is also an unknown.

Theorem 4.5.3 *All solutions of* (4.5.4) *such that $y \neq 0$ and y is not a root of unity satisfy*

$$m \leq c_{15}|D_L|^{6n} P_S^{n^2} e^{11nd\widehat{h}},$$

where $c_{15} = (10n^2 s)^{40ns}$.

The preceding theorems are Theorems 2.1, 2.2, and 2.3 in Bérczes, Evertse, and Győry (2013).

Let $f(X)$ be as in Theorem 4.5.1 (when $n \geq 3$) or in Theorem 4.5.2 (when $n \geq 2$) with nonzero discriminant, and G the splitting field of f over L. Then

$$f(X) = a_0(X - \alpha_1) \cdots (X - \alpha_n) \quad \text{with} \quad \alpha_1, \ldots, \alpha_n \in G.$$

Let $M_i = L(\alpha_i)$, and denote by T_i the set of places of M_i lying above the places of S and by $\mathcal{O}_{T_i}$ the ring of T_i-integers in M_i, $i = 1, \ldots, n$.

We state without proof the following crucial lemma.

Lemma 4.5.4 *Let x and $y \in \mathcal{O}_S$ be a solution of* (4.5.1) *(when $m = 2$) or* (4.5.2) *(when $m \geq 3$) with $y \neq 0$. Then for $i = 1, \ldots, n$ there are γ_i, ξ_i with*

$$x - \alpha_i = \gamma_i \xi_i^m, \qquad \xi_i \in \mathcal{O}_{T_i}, \quad \gamma_i \in M_i^*$$

and

$$h(\gamma_i) \leq c_{16} e^{2nd\widehat{h}} |D_L|^n \left(80(dn)^{dn+2} + \frac{1}{d}\log Q_S\right),$$

where $c_{16} = m(n^3 d)^{nd}$.

For a proof, see (ii) in Lemma 4.2 of Bérczes, Evertse, and Győry (2013). The first version of this lemma in a less general and ineffective form was proved by Siegel (1926), and in a less general but effective form by Baker (1969). The proof uses the ideal factorization of $[x - \alpha_i]$ in $\mathcal{O}_{T_i}$, estimates for the class number of M_i, and an analogue of Proposition 4.1.9 for $\mathcal{O}_{T_i}$.

Subsequently, we sketch proofs of less-explicit versions of Theorems 4.5.1, 4.5.2, and 4.5.3. We first prove the following weaker version of Lemma 4.5.4:

Let x and $y \in \mathcal{O}_S$ be as in the statement of Lemma 4.5.4. Then for $i = 1,\ldots,n$ there are γ_i, ξ_i with

$$x - \alpha_i = \gamma_i \xi_i^m, \quad \xi_i \in \mathcal{O}_{T_i}. \quad \gamma_i \in M_i^*, \quad h(\gamma_i) \ll_{n,L,S} m \cdot c(n,d)^{\widehat{h}}. \quad (4.5.5)$$

Here, and in the remainder of this section, $c(n, d)$ will denote effectively computable numbers exceeding 1 that depend only on n and d, and that may be different at each occurrence.

Proof It suffices to prove (4.5.5) for $i = 1$. We write M and T for M_1 and T_1. We denote by $[\beta_1,\ldots,\beta_r]$ the fractional ideal of $\mathcal{O}_T$ generated by $\beta_1,\ldots,\beta_r \in M$. Let $g(X) = (X - a_0\alpha_2)\cdots(X - a_0\alpha_n) = a_0^n f(X/a_0)/(X - a_0\alpha_1)$. Then $g \in \mathcal{O}_T[X]$. We first prove that there are ideals $\mathfrak{C}$ and $\mathfrak{A}$ of $\mathcal{O}_T$ such that

$$[a_0(x - \alpha_1)] = \mathfrak{C} \cdot \mathfrak{A}^m, \quad \mathfrak{C} \supseteq [(\delta a_0 g(a_0\alpha_1))^{m-1}]. \quad (4.5.6)$$

Observe that there is $h \in \mathcal{O}_T[X]$ such that

$$g(X) - g(a_0\alpha_1) = (X - a_0\alpha_1)h(X).$$

Substituting $a_0 x$ for X, we infer

$$g(a_0\alpha_1) \in [a_0(x - \alpha_1), g(a_0x)]. \quad (4.5.7)$$

Further,

$$a_0^n \delta y^m = a_0(x - \alpha_1)g(a_0x). \quad (4.5.8)$$

For a prime ideal $\mathfrak{P}$ of $\mathcal{O}_T$ (or equivalently, a prime ideal of $\mathcal{O}_K$ outside T), denote by $\mathrm{ord}_{\mathfrak{P}}(\beta)$ the exponent of $\mathfrak{P}$ in the prime ideal factorization of $[\beta]$. For every prime ideal of $\mathcal{O}_T$, write $\mathrm{ord}_{\mathfrak{P}}(a_0(x - \alpha_1)) = v_{\mathfrak{P}} + mw_{\mathfrak{P}}$ with $v_{\mathfrak{P}}$, $w_{\mathfrak{P}}$ nonnegative integers and $0 \leq v_{\mathfrak{P}} \leq m - 1$. The preceding two relations imply that $v_{\mathfrak{P}} = 0$ for those prime ideals $\mathfrak{P}$ that do not divide $\delta a_0 g(a_0 \alpha_1)$. Now clearly, (4.5.6) holds with

$$\mathfrak{C} = \prod_{\mathfrak{P} | \delta a_0 g(a_0 \alpha_1)} \mathfrak{P}^{v_{\mathfrak{P}}}, \quad \mathfrak{A} = \prod_{\mathfrak{P}} \mathfrak{P}^{w_{\mathfrak{P}}}.$$

We now proceed to prove (4.5.5). An ideal $\mathfrak{B}$ of $\mathcal{O}_T$ can be expressed uniquely as $\mathfrak{B}^* \mathcal{O}_T$, where $\mathfrak{B}^*$ is an ideal of $\mathcal{O}_K$ composed of prime ideals outside T. We define $N_T \mathfrak{B} := |\mathcal{O}_K / \mathfrak{B}^*|$. For instance, by Lang (1994, pp. 119, 120), there is nonzero $\xi \in \mathfrak{A}^*$ with $|N_{L/\mathbb{Q}}(\xi)| \le |D_M|^{1/2} |\mathcal{O}_K / \mathfrak{A}^*|$. This translates into $N_T(\xi) \le |D_M|^{1/2} N_T \mathfrak{A}$, i.e., $[\xi] = \mathfrak{B}\mathfrak{A}$ where $\mathfrak{B}$ is an ideal of $\mathcal{O}_T$ with $N_T \mathfrak{B} \le |D_M|^{1/2}$. Similarly, there exists $\gamma \in L$ with $[\gamma] = \mathfrak{D}\mathfrak{C}$, where $\mathfrak{D}$ is an ideal of $\mathcal{O}_T$ with $N_T \mathfrak{D} \le |D_M|^{1/2}$. As a consequence, we have

$$a_0(x - \alpha) = \frac{\gamma}{\gamma'} \xi^m,$$

where $\gamma, \gamma' \in \mathcal{O}_T$, and

$$[\gamma'] = \mathfrak{D}\mathfrak{B}^m.$$

Using (4.5.6) and the choice of $\mathfrak{B}$ and $\mathfrak{D}$, we get

$$N_T(\gamma) \le |D_M|^{1/2} N_T(\delta a_0 g(a_0 \alpha_1))^{m-1}, \quad N_T(\gamma') \le |D_M|^{(m+1)/2}.$$

By Lemma 4.1.11, we have $|D_M| \ll_L c(n, d)^{\widehat{h}}$. Combined with (4.1.3) and (4.1.14), this shows

$$N_T(\gamma), N_T(\gamma') \ll_{L,n,S} c(n, d)^{m\widehat{h}}.$$

By (4.1.11) we have $R_M \ll_L c(n, d)^{\widehat{h}}$. Using this together with Proposition 4.1.9, we infer that there are T-units $\eta, \eta' \in \mathcal{O}_T^*$ such that

$$h(\gamma \eta^m), h(\gamma' \eta'^m) \ll_{n,L,S} m \cdot c(n, d)^{\widehat{h}}.$$

Putting

$$\gamma_1 := a_0^{-1} \gamma {\gamma'}^{-1} (\eta {\eta'}^{-1})^m, \quad \xi_1 = \eta' \eta^{-1} \xi,$$

we obtain $x - \alpha_1 = \gamma_1 \xi_1^m$, where $\gamma_1 \in M^*$ with $h(\gamma_1) \ll_{n,L,S} m \cdot c(n, d)^{\widehat{h}}$ and $\xi_1 \in \mathcal{O}_T$. This proves (4.5.5). $\qquad \square$

We now sketch the proof of the following weaker version of Theorem 4.5.2:

All solutions x and $y \in \mathcal{O}_S$ of (4.5.2) with $y \ne 0$ satisfy

$$\max\{h(x), h(y)\} \ll_{m,n,L,S} c(n, d)^{m^2 \widehat{h}}. \tag{4.5.9}$$

Sketch of the proof of (4.5.9) Let $m \geq 3$, and let x and $y \in \mathcal{O}_S$ be a solution of $f(x) = \delta y^m$ with $y \neq 0$. Then we have $x - \alpha_i = \gamma_i \xi_i^m$ with γ_i, ξ_i as in (4.5.5), $i = 1, \ldots, n$. Let $N = L(\alpha_1, \alpha_2, \sqrt[m]{\gamma_1/\gamma_2}, \varrho)$, where ϱ is a primitive mth root of unity. Let T be the set of places of N, lying above the places from S, and $\mathcal{O}_T$ the ring of T-integers in N.

Then

$$\gamma_1 \xi_1^m - \gamma_2 \xi_2^m = \alpha_2 - \alpha_1, \quad \xi_1, \xi_2 \in \mathcal{O}_T. \tag{4.5.10}$$

The left-hand side is a binary form in ξ_1, ξ_2 with nonzero discriminant, which splits into linear factors over N. We are going to apply Theorem 4.4.1 with N and T instead of L and S. Notice that $[N : L] \leq m^2 n^2$. A repeated application of Lemma 4.1.11 gives a discriminant estimate

$$|D_N| \ll_{m,n,L,S} c(n,d)^{m^2 \widehat{h}}$$

and together with (4.1.11) and (4.1.13), this implies the estimates

$$h_N, \ R_T \ll_{m,n,L,S} c(n,d)^{m^2 \widehat{h}}.$$

By applying Theorem 4.4.1 to (4.5.10) and inserting these bounds, we get

$$h(\xi_1) \ll_{m,n,L,S} c(n,d)^{m^2 \widehat{h}} \max(H_1, H_2), \tag{4.5.11}$$

where

$$H_1 := \max\{1, h(\gamma_1), h(\gamma_2)\}, \quad H_2 := \max\{1, h(\alpha_2 - \alpha_1)\}.$$

Further, we have $H_2 \ll_n \widehat{h}$ and by (4.5.5),

$$H_1 \ll_{m,n,L,S} c(n,d)^{\widehat{h}}. \tag{4.5.12}$$

Using

$$h(x) \leq \log 2 + h(\alpha_1) + h(\gamma_1) + mh(\xi_1), \quad h(y) \leq m^{-1}(h(\delta) + h(f) + nh(x))$$

and combining these with (4.5.11) and (4.5.12), inequality (4.5.9) follows. □

In the proof of Theorem 4.5.1, we shall use the following.

Lemma 4.5.5 *Let* $\gamma_1, \gamma_2, \gamma_3, \beta_{12},$ *and* β_{13} *be nonzero elements of L such that*

$$\beta_{12} \neq \beta_{13}, \quad \sqrt{\gamma_1/\gamma_2}, \quad \sqrt{\gamma_1/\gamma_3} \in L,$$
$$h(\gamma_i) \leq H_1 \quad for \quad i = 1, 2, 3, \quad h(\beta_{12}), h(\beta_{13}) \leq H_2.$$

Then for the solutions of $x_1, x_2,$ and x_3 of the system of equations

$$\gamma_1 x_1^2 - \gamma_2 x_2^2 = \beta_{12}, \quad \gamma_1 x_1^2 - \gamma_3 x_3^2 = \beta_{13} \quad in \quad x_1, x_2, x_3 \in \mathcal{O}_S, \tag{4.5.13}$$

we have

$$\max_{1 \le i \le 3} h(x_i)$$

$$\le c_{17} P_S R_S \left(1 + \frac{\log^* R_S}{\log^* P_S}\right)\left(R_L + \frac{h_L}{d} \log Q_S + dH_1 + H_2\right) \qquad (4.5.14)$$

where $c_{17} = s^{2s+4} 2^{7s+60} d^{2s+d+2}$.

Proof This is Proposition 3.12 of Bérczes, Evertse, and Győry (2013). The idea of the proof is to reduce (4.5.13) to the decomposable form equation

$$F(x_1, x_2, x_3) = \delta \quad \text{in} \quad x_1, x_2, x_3 \in \mathcal{O}_S,$$

where

$$\delta = \beta_{12}\beta_{13}\beta_{23} \quad \text{with} \quad \beta_{23} = \beta_{13} - \beta_{12}$$

and

$$F(X_1, X_2, X_3) = \prod_{1 \le i < j \le 3} (\gamma_i X_i^2 - \gamma_j X_j^2).$$

Here, F is a decomposable form of degree 6 with splitting field L, whose linear factors form a triangularly connected system; see Section 2.6. To this equation, one can apply Theorem 3 of Győry and Yu (2006), which is a quantitative number field version of Theorem 2.6.1, to obtain (4.5.14). In Section 4.7, we have stated a special case of this result of Győry and Yu (Theorem 4.7.1) and at the end of Section 4.7, we have included a sketch of its proof. $\qquad \square$

We sketch now the proof of the following less explicit version of Theorem 4.5.1.

All solutions of x and $y \in \mathcal{O}_S$ of equation (4.5.1) with $y \ne 0$ satisfy

$$\max\{h(x), h(y)\} \ll_{n,L,S} c(n,d)^{\widehat{h}}. \qquad (4.5.15)$$

Sketch of the proof of (4.5.15) Let x and $y \in \mathcal{O}_S$ be a solution of the equation $f(x) = \delta y^2$ with $y \ne 0$. Then we have $x - \alpha_i = \gamma_i \xi_i^2$ $(i = 1, \ldots, n)$ with the γ_i, ξ_i as in (4.5.5). Let

$$N = L(\alpha_1, \alpha_2, \alpha_3, \sqrt{\gamma_1/\gamma_3}, \sqrt{\gamma_2/\gamma_3}),$$

let T be the set of places of N lying above the places from S, and let $\mathcal{O}_T$ be the ring of T-integers in N. Then

$$\gamma_1 \xi_1^2 - \gamma_2 \xi_2^2 = \alpha_2 - \alpha_1, \quad \gamma_1 \xi_1^2 - \gamma_3 \xi_3^2 = \alpha_3 - \alpha_1, \quad \xi_1, \xi_2 \in \mathcal{O}_T. \quad (4.5.16)$$

Notice that $[N : L] \leq 4n^3$. Further, a repeated application of Lemma 4.1.11 gives $|D_N| \ll_{n,L,S} c(n,d)^{\widehat{h}}$ and together with (4.1.11) and (4.1.13), this implies the estimates

$$h_N, \; R_T \ll_{n,L,S} c(n,d)^{\widehat{h}}.$$

By applying Lemma 4.5.5 to (4.5.16) with N and T instead of L and S, inserting the estimates for h_N and R_T and following the same computations as above, (4.5.15) follows. □

Finally, we shall sketch the proof of a less explicit version of Theorem 4.5.3.

Let L and S be as above; let $f(X) \in \mathcal{O}_S[X]$ be a polynomial of degree $n \geq 2$ with nonzero discriminant; and consider the Schinzel–Tijdeman equation (4.5.4), where both of x and $y \in \mathcal{O}_S$ and $m \geq 2$ are unknowns.

All solutions of x and $y \in \mathcal{O}_S$ and $m \in \mathbb{Z}_{\geq 3}$ of (4.5.4) such that $y \neq 0$ and y is not a root of unity satisfy

$$m \ll_{n,L,S} c(n,d)^{\widehat{h}}. \tag{4.5.17}$$

We start with some preliminaries and a lemma. Let again $f(X) = a_0(X - \alpha_1) \cdots (X - \alpha_n)$. For $i = 1, \ldots, n$, let $M_i = L(\alpha_i)$, and denote by $d_{M_i}, h_{M_i}, R_{M_i}$ the degree, class number, and regulator of M_i. Further, let T_i be the set of places of M_i lying above the places in S, and denote by t_i the cardinality of T_i and by R_{T_i} the T_i-regulator of M_i. By Lemma 4.1.11 and (4.1.11) and (4.1.13), we have the estimates

$$h_{M_i}, \; R_{T_i} \ll_{n,L,S} c(n,d)^{\widehat{h}}. \tag{4.5.18}$$

The group of T_i-units is finitely generated and, by Proposition 4.1.8, we can choose a fundamental system of T_i-units $\eta_{i,1}, \ldots, \eta_{i,t_i-1}$ such that

$$\prod_{j=1}^{t_i-1} h(\eta_{i,j}), \; \max_{1 \leq j \leq t_i-1} h(\eta_{i,j}) \ll_{n,L,S} c(n,d)^{\widehat{h}}. \tag{4.5.19}$$

Lemma 4.5.6 *Let x and $y \in \mathcal{O}_S$ and $m \geq 3$ be a solution of equation (4.5.4) such that $y \neq 0$, and y is not a root of unity. Then for $i = 1,2$ there are $\gamma_i, \xi_i \in M_i^*$, and integers $b_{i,1}, \ldots, b_{i,t_i-1}$ of absolute value at most $m/2$ such that*

$$(x - \alpha_i)^{h_{M_1} h_{M_2}} = \eta_{i,1}^{b_{i,1}} \cdots \eta_{i,t_i-1}^{b_{i,t_i-1}} \gamma_i \xi_i^m, \tag{4.5.20}$$

$$h(\gamma_i) \leq (2n^3 s)^{6ns} |D_L|^{2n} e^{4nd\widehat{h}} (\widehat{h} + \log^* P_S). \tag{4.5.21}$$

This is Lemma 5.1 in Bérczes, Evertse, and Győry (2013). We prove the following less precise result with instead of (4.5.21) the estimate

$$h(\gamma_i) \ll_{n,L,S} c(n,d)^{\widehat{h}} \text{ for } i = 1,2. \tag{4.5.22}$$

Proof We prove this only for $i = 1$. We write M and T for M_1 and T_1 and use again $[\cdot]$ to denote ideals in $\mathcal{O}_T$. Let again $g(X) = (X - a_0\alpha_1) \cdots (X - a_0\alpha_n)$. By (4.5.8) and (4.5.7), we have for every prime ideal $\mathfrak{P}$ of $\mathcal{O}_T$,

$$\mathrm{ord}_{\mathfrak{P}}(a_0^n \delta) + m\,\mathrm{ord}_{\mathfrak{P}}(y) = \mathrm{ord}_{\mathfrak{P}}(a_0(x - \alpha_1)) + \mathrm{ord}_{\mathfrak{P}}(g(a_0 x)),$$
$$\mathrm{ord}_{\mathfrak{P}}(a_0(x - \alpha_1)) \le \mathrm{ord}_{\mathfrak{P}}(g(a_0\alpha_1)), \text{ or } \mathrm{ord}_{\mathfrak{P}}(g(a_0 x)) \le \mathrm{ord}_{\mathfrak{P}}(g(a_0\alpha_1)),$$

hence,

$$\mathrm{ord}_{\mathfrak{P}}(a_0(x - \alpha_1)) = v + mw \text{ with } v, w \in \mathbb{Z},\ |v| \le \mathrm{ord}_{\mathfrak{P}}(\delta a_0^n g(a_0\alpha_1)), w \ge 0.$$

It follows that there are ideals $\mathfrak{C}_1, \mathfrak{C}_2$, and $\mathfrak{A}$ of $\mathcal{O}_T$ such that

$$[a_0(x - \alpha_1)] = \mathfrak{C}_1\mathfrak{C}_2^{-1}\mathfrak{A}^m, \quad \mathfrak{C}_1, \mathfrak{C}_2 \supseteq [a_0^n \delta g(a_0\alpha_1)].$$

Raising to the $(h_{M_1} h_{M_2})$th power to make all ideals principal, and invoking (4.5.18), we infer that there are $\theta_1, \theta_2, \xi \in \mathcal{O}_T$ such that

$$[(a_0(x - \alpha_1))^{h_{M_1} h_{M_2}}] = [\theta_1 \theta_2^{-1} \xi^m],$$
$$\log N_T(\theta_j) \le h_{M_1} h_{M_2} \log N_T(a_0^n \delta g(a_0\alpha_1)) \ll_{n,L,S} c(n,d)^{\widehat{h}} \text{ for } j = 1, 2. \tag{4.5.23}$$

Proposition 4.1.9 and again (4.5.18) imply that for $j = 1, 2$ there is $\varepsilon_j \in \mathcal{O}_T^*$ such that $\theta_j' := \varepsilon_j \theta_j$ satisfies

$$h(\theta_j') \ll_{n,L,S} c(n,d)^{\widehat{h}}.$$

Combined with (4.5.23), this gives

$$(a_0(x - \alpha_1))^{h_{M_1} h_{M_2}} = \eta \theta_1' \theta_2'^{-1} \xi^m$$

with $\eta \in \mathcal{O}_T^*$. By the S-unit Theorem, we can express η as

$$\eta = \zeta \eta_{1,1}^{b_{1,1}} \cdots \eta_{1,t_1-1}^{b_{1,t_1-1}} \cdot \varepsilon^m$$

with ζ a root of unity, $b_{1,k}$ $(k = 1, \ldots, t_1 - 1)$ integers of absolute value at most $m/2$, and $\varepsilon \in \mathcal{O}_T^*$. (4.5.20) and (4.5.22) hold with $\gamma_1 := \zeta \theta_1' \theta_2'^{-1}, \xi_1 := \varepsilon\xi$.

$\square$

Sketch of the proof of (4.5.17) Let x and $y \in \mathcal{O}_S$ and $m \ge 3$ be a solution of equation (4.5.4) such that $y \ne 0$, and y is not a root of unity. We assume that

$$h(x) \gg_{n,L,S} c(n,d)^{\widehat{h}}. \tag{4.5.24}$$

This is no loss of generality for if $h(x) \ll_{n,L,S} c(n,d)^{\widehat{h}}$, then (4.5.4) and Lemma 4.1.2 imply (4.5.17). In the course of our proof, the constant $c(n,d)$ in (4.5.24) will be chosen sufficiently large to make all our estimates work.

Let $M = L(\alpha_1, \alpha_2)$, $d_M = [M : \mathbb{Q}]$, T the set of places of M lying above the places from S, and t the cardinality of T. Put

$$\Lambda := 1 - \left(\frac{x - \alpha_1}{x - \alpha_2}\right)^{h_{M_1} h_{M_2}}.$$

By (4.1.14), there is $w \in T$ such that $\log |x - \alpha_2|_w \geq \frac{d_M}{t} h(x - \alpha_2)$. Now a combination of Lemma 4.1.1 (4.5.18) and (4.5.24) gives

$$\log |\Lambda|_w \ll_{n,L,S} (h_{K_1} h_{K_2})^2 + \log \left| 1 - \frac{x - \alpha_1}{x - \alpha_2} \right|_w$$

$$\ll_{n,L,S} c(n,d)^{\widehat{h}} + \log \frac{|\alpha_1 - \alpha_2|_w}{|x - \alpha_2|_w} \ll_{n,L,S} (-h(x)). \qquad (4.5.25)$$

To obtain a lower bound for $\log |\Lambda|_w$, we substitute the identity

$$\left(\frac{x - \alpha_1}{x - \alpha_2}\right)^{h_{K_1} h_{K_2}} = \frac{\gamma_1}{\gamma_2} \eta_{1,1}^{b_{1,1}} \cdots \eta_{1,t_1-1}^{b_{1,t_1-1}} \cdot \eta_{2,1}^{-b_{2,1}} \cdots \eta_{2,t_2-1}^{-b_{2,t_2-1}} \left(\frac{\xi_1}{\xi_2}\right)^m,$$

from (4.5.20) and then apply Proposition 4.2.4. Notice that by (4.5.19) and (4.5.22) we have

$$h(\xi_1/\xi_2) \ll_{n,L,S} m^{-1} \left(h(\gamma_1/\gamma_2) + c(n,d)^{\widehat{h}} m + h\left(\frac{x - \alpha_1}{x - \alpha_2}\right) \right)$$

$$\ll_{n,L,S} c(n,d)^{\widehat{h}} (1 + m^{-1} h(x)).$$

By inserting this, as well as the bounds from (4.5.22) and (4.5.19) into the lower bound from Proposition 4.2.4, we obtain

$$\log |\Lambda|_w \gg_{n,L,S} (-c(n,d)^{\widehat{h}} (1 + m^{-1} h(x)) \log m). \qquad (4.5.26)$$

Comparing (4.5.25) and (4.5.26) and combining this with (4.5.24), the estimate of (4.5.17) easily follows. $\qquad \square$

4.6 The Catalan Equation

Consider now the *Catalan equation* in the following generalized form

$$x^m \pm y^n = 1 \quad \text{in } x, y \in \mathcal{O}_S, \ m, n \in \mathbb{Z}$$
$$\text{with } x \text{ and } y \text{ not roots of unity and } m \text{ and } n > 1, mn > 4, \qquad (4.6.1)$$

where again S is a finite set of places, containing all infinite places, of a number field L. As was mentioned in Section 2.5, in the classical case of $L = \mathbb{Q}$ and $\mathcal{O} = \mathbb{Z}$, Tijdeman (1976) proved that the solutions of x, y and m, n are bounded above by an effectively computable absolute constant. His proof relies on Baker's theory of logarithmic forms.

Brindza, Győry, and Tijdeman (1986) generalized Tijdeman's proof for equations (4.6.1) where $\mathcal{O}_S = \mathcal{O}_L$ is the ring of integers of L. They showed that in this case, the heights of the solutions of (4.6.1) can be effectively bounded above by a number, which depends only on L. Brindza (1987) further generalized this to equations (4.6.1) where S is an arbitrary finite set of places. The following theorem, due to Koymans (2016), is a more explicit version of Brindza's result. Again, d denotes the degree of L, s denotes the cardinality of S, the quantities P_S and Q_S are defined by (4.1.9), and R_S denotes the S-regulator.

Theorem 4.6.1 *Suppose that in* (4.6.1) *m and n are primes. Then all solutions of* (4.6.1) *satisfy*

$$\max\{m,n\} < (sP_S^2)^{c_1^* sP_S}|D_L|^{6P_S}P_S^{P_S^2} =: C,$$

where c_1^* is an effectively computable positive absolute constant, and

$$\max\{h(x),h(y)\} < (C \cdot s)^{C^6}(|D_L|Q_S)^{C^4}.$$

Furthermore, if in (4.6.1) *m and n are arbitrary positive integers, we have*

$$\max\{m,n\} < (C \cdot s)^{C^6}(|D_L|Q_S)^{C^4}.$$

The proof of Theorem 4.6.1 is a generalization of the proof given for ordinary rings of integers in Brindza, Győry, and Tijdeman (1986). The main tools are Propositions 4.2.2 and 4.2.3 from Baker's theory of logarithmic forms, while also Theorems 4.3.1, 4.5.1, and 4.5.2, and Proposition 4.1.8 play an important role.

Following the proof of Koymans, we shall sketch the proof of the following less precise version of Theorem 4.6.1. In all subsequent statements, constants implied by the Vinogradov symbols $\ll_{L,S}$ and $\gg_{L,S}$ are effectively computable and depend only on L and S.

Suppose that in (4.6.1) *m and n are prime. Then all solutions of* (4.6.1) *satisfy*

$$\max\{m,n\} \ll_{L,S} 1, \tag{4.6.2}$$

and

$$\max\{h(x),h(y)\} \ll_{L,S} 1. \tag{4.6.3}$$

Furthermore, if in (4.6.1) *m and n are arbitrary positive integers, we have*

$$\max\{m,n\} \ll_{L,S} 1. \tag{4.6.4}$$

We mention here that inequality (4.6.3) follows at once from (4.6.2) and Theorems 4.5.1 and 4.5.2. Further, if m and n are arbitrary positive integers instead of just primes, we pick prime divisors m' and n' of m and n, respectively, estimate m' and n' by means of (4.6.2), then estimate $h(x')$ and $h(y')$ where $x' := x^{m/m'}$ and $y' := y^{n/n'}$ by means of (4.6.3), and subsequently obtain (4.6.4) by applying Voutier's inequality Lemma 4.1.2. So it suffices to prove (4.6.2) under the assumption that m and n are primes. This will be assumed henceforth.

Another important ingredient of the proof of Theorem 4.6.1 is the following result of Koymans (2017), which is of interest in itself.

Lemma 4.6.2 *If $p, x_1, x_2,$ and y are such that p is a prime, $x_1, x_2 \in \mathcal{O}_S^*$, and $y \in \mathcal{O}_S$ with and $y \neq 0$ and $y \notin \mathcal{O}_S^*$ and*

$$x_1 + x_2 = y^p \qquad (4.6.5)$$

then

$$p \leq (2s)^{c_2^* s} P_S^2 R_S^4 \qquad (4.6.6)$$

with an effectively computable positive absolute contant c_2^*.

Proof This is Theorem 4.2 in Koymans (2017). It is a generalization of Lemma 6 in Brindza, Győry, and Tijdeman (1986). Koymans' proof is a more modern and simplified version of that of Theorem 9.3 in Shorey and Tijdeman (1986). We briefly sketch a proof of the weaker inequality

$$p \ll_{L,S} 1, \qquad (4.6.7)$$

which is in fact sufficient for (4.6.2).

Below, we write $\ll, \gg$ for $\ll_{L,S}, \gg_{L,S}$. Choose a fundamental system of S-units $\{\eta_1, \ldots, \eta_{s-1}\}$ of $\mathcal{O}_S^*$, where $s := |S|$. Let $x_1, x_2, y,$ and p be as in Lemma 4.6.2. Then

$$x_1 = \zeta_1 \eta_1^{a_1} \cdots \eta_{s-1}^{a_{s-1}}, \quad x_2 = \zeta_2 \eta_1^{b_1} \cdots \eta_{s-1}^{b_{s-1}},$$

where ζ_1 and ζ_2 are roots of unity, and the a_i and b_i are integers. We may and shall assume that

$$0 \leq b_i < p \text{ for } i = 1, \ldots, s-1; \qquad (4.6.8)$$

indeed, these inequalities can be achieved after multiplying $x_1, x_2,$ and y^p with a suitable pth power of an S-unit. From this assumption, together with the lower bound for $h(y)$ arising from Lemma 4.1.2, we deduce that for $v \in S$,

$$\left| \sum_{i=1}^{s-1} a_i \log |\eta_i|_v \right| = |\log |x_1|_v| \ll p \max(1, |\log |y|_v|) \ll ph(y),$$

and then, on multiplying the vector $(a_1, \ldots, a_{s-1})$ with the inverse of any $(s-1) \times (s-1)$-submatrix of $(\log |\eta_i|_v)_{i=1,\ldots,s-1,\, v \in S}$,

$$\max(|a_1|, \ldots, |a_{s-1}|) \ll ph(y).$$

The next step is to derive upper and lower bounds for $|x_2 y^{-p}|_v = |1 - x_1 y^{-p}|_v$ for $v \in S$. Here, we assume without loss of generality that $p > P_S$ (the maximum of the norms of the finite places in S). First, suppose that v is an infinite place. Second, by Proposition 4.2.2 with $n = s$, $A_s' \ll h(y)$, $b_s = p$, $A_i' \ll 1$, $b_i \ll ph(y)$ for $i < s$,

$$\log |x_2 y^{-p}|_v = \log |1 - x_1 y^{-p}|_v = \log |1 - \zeta_1 \eta_1^{a_1} \cdots \eta_{s-1}^{a_{s-1}} y^{-p}|_v$$
$$\gg -h(y) \log p. \tag{4.6.9}$$

If v is finite, then $|p|_v = 1$ since p is a prime exceeding P_S. Now we can apply Proposition 4.2.3 with $A_n'' \ll h(y)$, $\delta = \frac{1}{2}$, $B_n = p$, $B \ll ph(y)$, and obtain again (4.6.9). By (4.1.14), there is $v \in S$ such that $\log |y|_v \geq \frac{d}{s} h(y)$. Then by (4.6.9) we have

$$\log |x_2|_v - \frac{d}{s} ph(y) \gg -h(y) \log p,$$

while $\log |x_2|_v \ll p$ by (4.6.8). Assuming $p \gg 1$ as we may, we infer $h(y) \ll 1$. Since $x_2 \in \mathcal{O}_S^*$, we have $\prod_{v \in S} |x_2|_v = 1$, and since $y \in \mathcal{O}_S$, $y \neq 0$, and $y \notin \mathcal{O}_S^*$, we have $\prod_{v \in S} |y|_v \geq 2$. Using these inequalities and summing (4.6.9) over $v \in S$, we obtain

$$-p \log 2 \gg -\log p,$$

which implies (4.6.7). $\qquad\qquad\qquad\qquad\qquad\qquad\qquad\qquad\square$

Sketch of the proof of (4.6.2) The proof will be carried out in several steps. Again, constants implied by Vinogradov symbols $\ll$ and $\gg$ will be effectively computable and depend only on L and S. We fix a solution $(x, y, m, \text{and } n)$ of (4.6.1), with m and n primes.

Step 1. *Simplifications.*

In view of Theorem 4.5.3, we may assume that m and n are primes exceeding P_S (the maximum of the norms of the prime ideals in S). If x and y are S-units, then we obtain $h(x^m)$ and $h(y^n) \ll 1$ from Theorem 4.3.1 and subsequently m and $n \ll 1$ from Lemma 4.1.2. If exactly one of x and y, say, x, is an S-unit, then by (4.6.7), we have $n \ll 1$ and subsequently also $m \ll 1$ by Theorem 4.5.3. So we may assume that neither of x nor y is an S-unit. This implies also that in the course of the proof, we may assume that $h(x)$ and $h(y) \gg 1$. By enlarging S with a finite number of places, we can say that all nonzero elements z of L with $h(z) \ll 1$ are S-units. So by the preceding arguments, if

$h(x) \ll 1$ or $h(y) \ll 1$, then m and $n \ll 1$ follows. Lastly, we may assume that $m \neq n$. Suppose that $m = n$ is a prime. Then $u = x^m$ and $v = -xy$ satisfy $u(u \pm 1) = v^m$, v is nonzero and not an S-unit, and so $m \ll 1$ by Theorem 4.5.3.

So summarizing, it suffices to deal with the equation

$$x^m + y^n = 1 \quad \text{in } x \text{ and } y \in \mathcal{O}_S \text{ with } x \text{ and } y \neq 0, \, x \text{ and } y \notin \mathcal{O}_S^*, \quad (4.6.10)$$
$$\text{and primes } m \text{ and } n \text{ with } m > n > P_S.$$

Henceforth, $(x, y, m, \text{and } n)$ will be a fixed solution of (4.6.10), and we may assume that $h(x)$ and $h(y) > C$ for any effectively computable constant depending on L and S of our choice.

Step 2. *A special case.*

Assume that

$$(x - 1)^m + (y - 1)^n = 0. \qquad (4.6.11)$$

The exponent n is not an S-unit, since $n > P_S$. Let $\mathfrak{q}$ be a prime ideal of $\mathcal{O}_S$ dividing n. Then (4.6.11) together with (4.6.10) implies $(x-1)^m \equiv x^m \pmod{\mathfrak{q}}$. This shows that x and $x - 1$ are coprime with n. Moreover, m divides the order of the unit group of $\mathcal{O}_S/\mathfrak{q}$, which is smaller than n^d.

Let a and b be integers with $am + bn = 1$. Then by (4.6.11), we have $x = 1 + \varepsilon^n$ and $y = 1 - \varepsilon^m$ where $\varepsilon = (x - 1)^b (1 - y)^a$. Clearly, $\varepsilon \in \mathcal{O}_S$, a substitution into (4.6.10), gives

$$(1 + \varepsilon^n)^m + (1 - \varepsilon^m)^n - 1 = 0,$$

implying

$$1 + \sum_{k=1}^{m-1} \binom{m}{k} \varepsilon^{nk} + \sum_{k=1}^{n-1} \binom{n}{k} (-1)^k \varepsilon^{mk} = 0. \qquad (4.6.12)$$

As observed we may assume that, say, $h(x)$ and $h(y) > 3$, which implies that ε is not a root of unity. Hence, there is a place v of L with $|\varepsilon|_v < 1$. This place v cannot be finite, since otherwise, the left-hand side of (4.6.12) would have v-adic absolute value 1. So ε is an S-unit, and (4.1.15) implies that there is an infinite place v of K with $\log |\varepsilon|_v \leq -\frac{d}{s} h(\varepsilon)$. There is an embedding $\sigma : L \hookrightarrow \mathbb{C}$ such that $|\cdot|_v = |\sigma(\cdot)|^{s(v)}$ with $s(v) = 1$ or 2. So by Lemma 4.1.2, we have $|\sigma(\varepsilon)| \leq c(d)^{-1}$, where $c(d) > 1$ is effectively computable and depends only on d. Using $m < n^d$, we obtain the following lower bound for the absolute value of the quantity obtained by applying σ to the left-hand side of (4.6.12):

$$1 - \sum_{k=1}^{m-1} n^{dk} c(d)^{-nk} - \sum_{k=1}^{n-1} n^k c(d)^{-mk}.$$

If $n \gg 1$, then $n^d c(d)^{-n} < \frac{1}{4}$, say, and this lower bound is strictly positive, contradicting (4.6.12). So under the hypothesis (4.6.11), we conclude $n \ll 1$, and then also $m \ll 1$, since $m < n^d$.

Having disposed of the case (4.6.11), we assume henceforth that

$$(x-1)^m + (y-1)^n \neq 0. \tag{4.6.13}$$

Step 3. *Ideal arithmetic.*

We denote by $[\alpha_1, \ldots, \alpha_k]$ the fractional ideal of $\mathcal{O}_S$ generated by $\alpha_1, \ldots, \alpha_k \in L$. Fix a system of fundamental S-units $\eta_1, \ldots, \eta_{s-1}$ of $\mathcal{O}_S^*$. Since

$$[x-1] \cdot \left[\tfrac{x^m - 1}{x-1} \right] = [y]^n, \quad \left[x-1, \tfrac{x^m - 1}{x-1} \right] \supseteq [m],$$

we have an ideal factorization $[x-1] = \mathfrak{a}_1 \mathfrak{a}_2^{-1} \cdot \mathfrak{b}^n$, where $\mathfrak{a}_1, \mathfrak{a}_2$ and $\mathfrak{b}$ are ideals of $\mathcal{O}_S$ with $\mathfrak{a}_1$ and $\mathfrak{a}_2 \supseteq [m]$. By raising to the (h_L)th power, we get a factorization $[x-1]^{h_L} = [\theta_0] \cdot [\omega]^n$ with $\omega \in \mathcal{O}_S \backslash \{0\}$ and $\theta_0 \in L^*$ with numerator and denominator of θ_0 dividing m^{h_L}. Using Propositions 4.1.8 and 4.1.9, one infers that θ_0 and ω can be chosen in such a way, that

$$(x-1)^{h_L} = \eta_1^{u_1} \cdots \eta_{s-1}^{u_{s-1}} \theta_0 \omega^n, \tag{4.6.14}$$

where the u_i are rational integers with $0 \leq u_i < n$ for $i = 1, \ldots s-1$, and where $\omega \in \mathcal{O}_S \backslash \{0\}$ and $\theta_0 \in K^*$ with

$$h(\theta_0) \ll \log m. \tag{4.6.15}$$

Similarly, we can write

$$(1-y)^{h_L} = \eta_1^{v_1} \cdots \eta_{s-1}^{v_{s-1}} \tau_0 \sigma^m \tag{4.6.16}$$

with rational integers v_i such that $0 \leq v_i < m$ for $i = 1, \ldots, s-1$, and with $\sigma \in \mathcal{O}_S \backslash \{0\}$ and $\tau_0 \in K^*$ such that

$$h(\tau_0) \ll \log n. \tag{4.6.17}$$

Step 4. *First bounds for m and n.*

We show that

$$m \ll h(y) \log m, \tag{4.6.18}$$

$$n \ll h(x) \log m. \tag{4.6.19}$$

Notice that (4.6.19) follows from (4.6.18) via $|mh(x) - nh(y)| \ll 1$. We prove (4.6.18). Let $\Lambda_1 := 1 - \frac{(-y)^n}{x^m} = \frac{1}{x^m}$. By (4.1.14), there is $v \in S$ such that

$$\log |\Lambda_1|_v = -m \log |x|_v \leq -\frac{md}{s} h(x),$$

while Proposition 4.2.4 produces a lower bound

$$\log |\Lambda_1|_v \gg -h(x)h(y)\log m.$$

By combining the two bounds, we get (4.6.18).

Step 5. *A bound for n.*

We now show that

$$n \ll (\log m)^4. \tag{4.6.20}$$

We assume without loss of generality that

$$n > (\log m)^4, \quad m \gg 1. \tag{4.6.21}$$

Then by (4.6.19), we have

$$h(x) \gg (\log m)^3. \tag{4.6.22}$$

We deal with the expression

$$\Lambda_2 := \frac{(1-y)^{nh_L}}{(x-1)^{mh_L}} - 1,$$

and derive an upper and lower bound for $|\Lambda_2|_v$ for appropriate v. We assume for the moment that $\Lambda_2 \neq 0$. By (4.1.14), there is $v \in S$ such that

$$\log |x|_v \geq \frac{d}{s}h(x).$$

For this v, we have $\log |x|_v \gg (\log m)^3$. Now by Lemma 4.1.1 (4.6.10) and (4.6.22), we have

$$\log \left| \frac{(1-y)^{nh_L}}{y^{nh_L}} - 1 \right|_v \ll \log n - \log |y|_v \ll \log n - \log |x|_v$$

$$\ll \log m - \frac{d}{s}h(x) \ll -h(x).$$

Likewise,

$$\log \left| \frac{x^{mh_L}}{(x-1)^{mh_L}} - 1 \right|_v \ll \log m - \log |x-1|_v \ll -h(x),$$

and lastly, $\log |y^{nh_L} x^{-mh_L} - 1|_v \ll -m \log |x|_v \ll -mh(x)$. So

$$\log |\Lambda_2|_v = \log \left| \frac{(1-y)^{nh_L}}{y^{nh_L}} \cdot \frac{x^{mh_L}}{(x-1)^{mh_L}} \cdot y^{nh_L} x^{-mh_L} - 1 \right|_v \ll -h(x).$$

$$\tag{4.6.23}$$

By inserting (4.6.14) and (4.6.16), using the inequalities (4.6.15) and (4.6.17) and applying Proposition 4.2.4, we get

$$\log |\Lambda_2|_v = \log |\eta_1^{nv_1 - mu_1} \cdots \eta_{s-1}^{nv_{s-1} - mu_{s-1}} \tau_0^n \theta_0^{-m} (\sigma/\omega)^{mn}|_v \gg -(\log m)^3 H_0,$$

(4.6.24)

where $H_0 := \max(h(\sigma), h(\omega))$. From (4.6.14), (4.6.15), (4.6.21), and (4.6.19), we infer

$$h(\omega) \ll n^{-1}(n + h(\theta_0) + h(x - 1)) \ll n^{-1}(n + \log m + h(x))$$

$$\ll 1 + \frac{h(x)}{n} \ll \frac{\log m}{n} \cdot h(x)$$

while (4.6.16), (4.6.17) and (4.6.18) give $h(\sigma) \ll h(y) \cdot (\log m)/m \ll h(x) \cdot (\log m)/n$. So altogether, $H_0 \ll h(x) \cdot (\log m)/n$. Now a combination with (4.6.23) and (4.6.24) gives $n \ll (\log m)^4$.

We still have to deal with the case $\Lambda_2 = 0$. We now consider instead

$$\Lambda_3 := \frac{(1 - y)^n}{(x - 1)^m} - 1.$$

By (4.6.13), we have $\Lambda_3 \neq 0$. By precisely the same argument as above, we get (4.6.23) with Λ_3 instead of Λ_2. Further, $\Lambda_3 = \zeta - 1$ for a root of unity $\zeta \neq 1$ of L, so we certainly have the analogue of (4.6.24). Again, we obtain $n \ll (\log m)^4$.

Step 6. *Finishing the proof.*

Let

$$\Lambda_4 := \frac{x^{mh_L}}{(1 - y)^{nh_L}} - 1.$$

Assume for the moment that $\Lambda_4 \neq 0$. By (4.1.14), we can choose $v \in S$ such that $\log |y|_v \geq \frac{d}{s} h(y)$. By an argument similar as in Step 5, we get

$$\log |\Lambda_4|_v = \log \left| \frac{x^{mh_L}}{y^{nh_L}} \cdot \frac{y^{nh_L}}{(y - 1)^{nh_L}} - 1 \right|_v \ll \log n - \log |y|_v \ll -h(y)$$

$$\ll -\frac{m}{n} h(x).$$

(4.6.25)

By virtue of (4.6.16), we can write

$$\Lambda_4 = \eta_1^{d_1} \cdots \eta_{s-1}^{d_{s-1}} \tau_0^{-n} \left(\frac{x^{h_L}}{\sigma^n} \right)^m,$$

with rational integers d_i such that $|d_i| < mn$ for $i = 1, \ldots, s - 1$. Notice that by (4.6.16), (4.6.17), and (4.6.20), we have

$$h(x^{h_L} \sigma^{-n}) \ll h(x) + \frac{n}{m}(h(1 - y) + m + \log n) \ll h(x) + n.$$

By Proposition 4.2.4, inserting this upper bound and (4.6.18), we obtain

$$\log |\Lambda_4|_v \gg -(h(x) + n) \log n \log m.$$

Together with (4.6.25) and $n \ll (\log m)^4$, this implies $m \ll 1$.

We still have to deal with the case $\Lambda_4 = 0$, i.e., $x^{mh_L} = (1 - y)^{nh_L}$. This implies $(1 - y^n)^{h_L} = (1 - y)^{nh_L}$, whence

$$1 - y^n = \xi(1 - y)^n,$$

where ξ is a (h_L)th root of unity in L. We show that $n \ll 1$. Suppose without loss of generality that $n > h_L$. Then in fact, n is coprime with h_L, so $\xi = \zeta^n$ for some root of unity ζ of L, and thus,

$$(\zeta(1 - y))^n + y^n = 1.$$

As observed in Step 1, this implies $n \ll 1$. Then, using the fact $g(y) = 0$ for the polynomial $g(X) := (1 - X)^{nh_L} - (1 - X^n)^{h_L}$, one can deduce that $h(y) \ll 1$. Combined with (4.6.10), this gives $h(x^m) \ll 1$, and then an application of Lemma 4.1.2 finally leads to $m \ll 1$. $\qquad\square$

4.7 Decomposable Form Equations

Keeping the preceding notation, let again L be a number field, and S a finite set of places of L containing the set S_∞ of infinite places. Consider now the *decomposable form equation*

$$F(\mathbf{x}) = \delta \quad \text{in} \quad \mathbf{x} = (x_1, \ldots, x_m) \in \mathcal{O}_S^m, \tag{4.7.1}$$

where $\delta \in L \backslash \{0\}$ and $F(\mathbf{X}) = F(X_1, \ldots, X_m)$ is a decomposable form of degree $n \geq 3$ in $m \geq 2$ variables, which factorizes into linear forms over L. As in Chapter 2, we write

$$F = \ell_1, \ldots, \ell_n \tag{4.7.2}$$

where $\ell_1, \ldots, \ell_n$ are linear forms in the variables $X_1, \ldots, X_m$ with coefficients in L, and denote by $\mathcal{L}_F$ the system $(\ell_1, \ldots, \ell_n)$. Suppose that $\mathcal{L}_F$ has at least three pairwise nonproportional linear forms. Let $\mathcal{G}(\mathcal{L}_F)$ denote the *triangular graph* of $\mathcal{L}_F$ as defined by (2.6.4), i.e., $\mathcal{G}(\mathcal{L}_F)$ has vertex system $\mathcal{L}_F$, and ℓ_i and ℓ_j with $i \neq j$ are connected by an edge if ℓ_i and ℓ_j are linearly dependent on L, or they are linearly independent, but there is $q \notin \{i, j\}$ such that $\lambda_i \ell_i + \lambda_j \ell_j + \lambda_q \ell_q = 0$ for some nonzero λ_i, λ_j, and $\lambda_q \in L$. Let $\mathcal{L}_1, \ldots, \mathcal{L}_k$ be the vertex systems of the connected components of $\mathcal{G}(\mathcal{L}_F)$, and denote by $[\mathcal{L}_j]$ the L-vector space spanned by the linear forms from $\mathcal{L}_j$. To be in accordance with earlier work to which we will refer, we suppose that F in (4.7.1) satisfies conditions slightly stronger than (2.6.5), i.e.,

$$\mathcal{L}_F \text{ has rank } m, \tag{4.7.3}$$

$$X_m \in [\mathcal{L}_1] \cap \cdots \cap [\mathcal{L}_k]; \tag{4.7.4}$$

of course, (4.7.4) is automatically satisfied if (4.7.3) holds and $k = 1$.

In the next theorem, we use the earlier notation, i.e., d, r, h_L, and R_L denote the degree, unit rank, class number, and regulator of L; s is the cardinality of S; R_S is the S-regulator of L, and P_S and Q_S are the largest norm and the product of the norms of the prime ideals corresponding to the finite places in S, with the convention that $P_{S_\infty} = Q_{S_\infty} = 1$. Recall that by the height of an algebraic number, we always meant the absolute logarithmic height.

The following theorem was proved by Győry and Yu (2006) in a slightly more general form.

Theorem 4.7.1 *Let $F \in L[X_1, \ldots, X_m]$ be a decomposable form of degree $n \geq 3$ that factorizes into linear forms over L such that $\mathcal{L}_F$ satisfies the conditions (4.7.3) and (4.7.4). Suppose that the logarithmic heights of the coefficients of the linear forms in $\mathcal{L}_F$ do not exceed $H_1(\geq 1)$. Further, let $\delta \in L \backslash \{0\}$ with logarithmic height at most $H_2(\geq 1)$. With the preceding notation, all solutions $\mathbf{x} = (x_1, \ldots, x_m) \in \mathcal{O}_S^m$ of (4.7.1) with $x_m \neq 0$ if $k > 1$, satisfy*

$$\max_{1 \leq i \leq m} h(x_i) \leq c_{18} P_S R_S \left(1 + \frac{\log^* R_S}{\log^* P_S} \right)$$

$$\times \left(c_4 R_L + \frac{h_L}{d} \log Q_S + mndH_1 + H_2 \right) \qquad (4.7.5)$$

with $c_{18} = 50m(m+1)(n-1)c_{11}$, where c_4, c_{11} denote the constants specified in Proposition 4.1.9 and Theorem 4.3.1.

For $m = 2$, this gives Theorem 4.4.1 on Thue equations with a somewhat different bound.

Recently, in terms of S, the bound in (4.7.5) has been improved in Győry (2019), replacing e.g. P_S by P_S'; see Section 4.3. However, this does not play any role in our work.

Next, suppose that L is a finite normal extension of degree ≥ 3 of a number field K. Let $\alpha_1 = 1$ and $\alpha_2, \ldots, \alpha_m$ be linearly independent elements of L over K such that $K' := K(\alpha_1, \ldots, \alpha_m)$ is of degree $n \geq 3$ over K. Consider the *norm form equation*

$$N_{K'/K}(\alpha_1 x_1 + \cdots + \alpha_m x_m) = \delta \quad \text{in} \quad x_1, \ldots, x_m \in \mathcal{O}_S, \qquad (4.7.6)$$

where $N_{K'/K}(\alpha_1 X_1 + \cdots + \alpha_m X_m) = \prod_{i=1}^n \ell^{(i)}(\mathbf{X})$, with $\ell^{(i)}(\mathbf{X}) = \alpha_1^{(i)} X_1 + \cdots + \alpha_m^{(i)} X_m$, $i = 1, \ldots, n$, the conjugates of $\ell(\mathbf{X})$ with respect to K'/K. With the preceding notation, for $k > 1$, Theorem 4.7.1 implies the following.

Corollary 4.7.2 *Suppose that α_m is of degree ≥ 3 over $K(\alpha_1, \ldots, \alpha_{m-1})$, the heights of $\alpha_2, \ldots, \alpha_m$ do not exceed $H_1(\geq 1)$, and $h(\delta)$ is at most $H_2(\geq 1)$. Then all solutions $(x_1, \ldots, x_m) \in \mathcal{O}_S^m$ of equation (4.7.6) with $x_m \neq 0$ satisfy (4.7.5).*

It is not difficult to show that under the conditions of Corollary 4.7.2, the norm form $N_{K'/K}(\alpha_1 X_1 + \cdots + \alpha_m X_m)$ satisfies the conditions concerning F in Theorem 4.7.1; see e.g. Győry (1981a) or the proof of Corollary 2.7.2. Hence, Corollary 4.7.2 is a consequence of Theorem 4.7.1. Corollary 4.7.2 has a further consequence for equation (4.7.6), corresponding to Corollary 2.7.3.

Let now $1, \alpha_1, \ldots, \alpha_m$ be linearly independent elements of L over K with heights at most $H(\geq 1)$. Assume again that $K' := K(\alpha_1, \ldots, \alpha_m)$ is of degree $n \geq 3$ over K. In the *discriminant form equation*

$$D_{K'/K}(\alpha_1 x_1 + \cdots + \alpha_m x_m) = \delta \quad \text{in} \quad x_1, \ldots, x_m \in \mathcal{O}_S \qquad (4.7.7)$$

the discriminant form $D_{K'/K}(\alpha_1 X_1 + \cdots + \alpha_m X_m)$ satisfies the assumptions concerning F in Theorem 4.7.1 with $k = 1$ and with $n(n-1)$ in place of n. Further, the coefficients of the linear factors of the discriminant form have heights at most $2H_1 + \log 2$. Suppose again that $h(\delta)$ does not exceed $H_2(\geq 1)$.

Corollary 4.7.3 *Under the above assumptions, all solutions of equation (4.7.7) satisfy (4.7.5) with n replaced by $n(n-1)$ and H_1 by $2H_1 + \log 2$.*

Similar to Corollary 2.8.2, our Corollary 4.7.3 has applications to index form equations, integral elements of given discriminant, and simple integral ring extensions of $\mathcal{O}_S$; for details, see e.g. Győry (1981a, 1998).

As was mentioned in Section 1.5, for equation (4.7.7) over $\mathbb{Z}$, the first effective result was obtained by Győry (1976) in quantitative form. This was extended by Győry and Papp (1977) to the case of rings of integers of number fields. For $S = S_\infty$ and $k = 1$ (Győry and Papp [1978]), while for arbitrary S and $k \geq 1$, Győry (1981a) proved the preceding results with weaker effective bounds. Improvements were later obtained among others by Bugeaud and Győry (1996a), Bugeaud (1998) for equation (4.7.6), and by Győry (1998), Győry and Yu (2006), and Győry (2019) for equation (4.7.1).

As was seen in Chapters 1 and 2, equation (4.7.1), (4.7.6), and (4.7.7) can be extended to equations of the type

$$F(x_1, \ldots, x_m) = \eta\delta \quad \text{in} \quad \eta \in \mathcal{O}_S^*, x_1, \ldots, x_m \in \mathcal{O}_S, \qquad (4.7.8)$$

where we may assume that δ and the coefficients of F are contained in $\mathcal{O}_S$. If $\eta, \mathbf{x}_0 = (x_1, \ldots, x_m)$ is a solution of (4.7.8), then so is $\varepsilon^n \eta, \varepsilon \mathbf{x}_0$ for any S-unit ε. However, for each solution $\eta, \mathbf{x}_0$ of (4.7.8), there is an $\varepsilon \in \mathcal{O}_S^*$ such that $\varepsilon \mathbf{x}_0$ is a solution of the equation corresponding to (4.7.1) with δ replaced by $\varepsilon^n \eta\delta$, whose height can be explicitly bounded above by Proposition 4.1.9. Then the preceding results give an explicit upper bound for $\max_{1 \leq i \leq m} h(\varepsilon x_i)$.

We now sketch the proof of the following less explicit version of Theorem 4.7.1.

Under the assumptions of Theorem 4.7.1, all solution $\mathbf{x} = (x_1, \ldots, x_m) \in$
$\mathcal{O}_S^m$ *of* (4.7.1) *with* $x_m \neq 0$ *if* $k > 1$ *satisfy*

$$\max_{1 \leq i \leq m} h(x_i) \ll_{m,n,L,S} \max(H_1, H_2). \tag{4.7.9}$$

Sketch of the proof of (4.7.9) We set $H := \max(H_1, H_2)$. Constants implied by
$\ll$ will be effectively computable and depend on m, n, L, and S only. We frequently use the elementary height properties listed in (4.1.3) without mention.

Let F and δ be as in (4.7.1) and (4.7.2). Let the positive rational integer
a be the product of the denominators of the coefficients of $\ell_1, \ldots, \ell_n$ and put
$\ell_i' := a\ell_i$ $(i = 1, \ldots, n)$, $F' := \ell_1' \cdots \ell_n'$, and $\delta' := a^n \delta$. Then equation (4.7.1)
is equivalent to

$$F'(\mathbf{x}) = \ell_1'(\mathbf{x}) \cdots \ell_n'(\mathbf{x}) = \delta'. \tag{4.7.10}$$

Further,

the coefficients of $\ell_1', \ldots, \ell_n'$ and δ' have logarithmic heights $\ll H$. (4.7.11)

To the system $\mathcal{L}_{F'} = (\ell_1', \ldots, \ell_n')$, we can attach a graph $\mathcal{G}(\mathcal{L}_{F'})$ similar to
$\mathcal{G}(\mathcal{L}_F)$, and we denote the vertex systems of its connected components by
$\mathcal{L}_1', \ldots, \mathcal{L}_k'$. Then it satisfies conditions analogous to (4.7.3) and (4.7.4).

Pick a solution $\mathbf{x} = (x_1, \ldots, x_m) \in \mathcal{O}_S^m$ of (4.7.1) or equivalently, (4.7.10)
with $x_m \neq 0$ if $k > 1$. Let $\{\ell_i', \ell_j'\}$ be an edge of $\mathcal{G}(\mathcal{L}_{F'})$. Assume that ℓ_i', ℓ_j' are
linarly independent. Then there are $q \notin \{i, j\}$, and nonzero $\lambda_i', \lambda_j', \lambda_q' \in L$, such
that $\lambda_i'\ell_i' + \lambda_j\ell_j' + \lambda_q\ell_q' = 0$. This leads to

$$\lambda_i'\ell_i'(\mathbf{x}) + \lambda_j'\ell_j'(\mathbf{x}) + \lambda_q\ell_q'(\mathbf{x}) = 0. \tag{4.7.12}$$

The coefficients λ_i', λ_j', and λ_q' can be chosen as 2×2-determinants with entries
from the coefficients of ℓ_i', ℓ_j', and ℓ_q'. So by (4.7.11),

$$h(\lambda_i'), h(\lambda_j'), h(\lambda_q') \ll H.$$

Further, $\ell_i'(\mathbf{x}), \ell_l'(\mathbf{x})$, and $\ell_j'(\mathbf{x})$ divide δ' in $\mathcal{O}_S$, so

$$\log N_S(\ell_i'(\mathbf{x})), \log N_S(\ell_j'(\mathbf{x})), \log N_S(\ell_q'(\mathbf{x})) \leq \log N_S(\delta') \ll H.$$

Now an application of (4.3.8) yields that there is $\varepsilon \in \mathcal{O}_S^*$ such that

$$h(\varepsilon\ell_i'(\mathbf{x})), h(\varepsilon\ell_j'(\mathbf{x})) \ll H.$$

In case that ℓ_i' and ℓ_j' are linearly dependent, this is trivially true; so this holds
for each edge $\{\ell_i', \ell_j'\}$ of $\mathcal{G}(\mathcal{L}_{F'})$. Thus, we obtain

$$h\left(\frac{\ell_i'(\mathbf{x})}{\ell_j'(\mathbf{x})}\right) \ll H \quad \text{for each edge } \{\ell_i', \ell_j'\} \text{ of } \mathcal{G}(\mathcal{L}_{F'}). \tag{4.7.13}$$

Now let ℓ'_i and ℓ'_j belong to the same connected component of $\mathcal{G}(\mathcal{L}_{F'})$. Then there is a path from ℓ'_i to ℓ'_j, i.e., a sequence of edges $\{\ell'_i, \ell'_{i_1}\}$, $\{\ell'_{i_1}, \ell'_{i_2}\}, \ldots,$ $\{\ell'_{i_t}, \ell'_j\}$ of $\mathcal{G}(\mathcal{L}_{F'})$. Taking t minimal, we have $t \le n$. Writing

$$\frac{\ell'_i(\mathbf{x})}{\ell'_j(\mathbf{x})} = \frac{\ell'_i(\mathbf{x})}{\ell'_{i_1}(\mathbf{x})} \cdot \frac{\ell'_{i_1}(\mathbf{x})}{\ell'_{i_2}(\mathbf{x})} \cdots \frac{\ell'_{i_t}(\mathbf{x})}{\ell'_j(\mathbf{x})}$$

and invoking (4.7.13), we obtain

$$h\left(\frac{\ell'_i(\mathbf{x})}{\ell'_j(\mathbf{x})}\right) \ll H \text{ for each } \ell'_i \text{ and } \ell'_j$$
$$\text{in the same connected component of } \mathcal{G}(\mathcal{L}_{F'}). \qquad (4.7.14)$$

For the moment, let $k > 1$. We want to extend (4.7.14) to all pairs of ℓ'_i and ℓ'_j, not necessarily belonging to the same connected component. Let $i \in \{1, \ldots, n\}$, and let $\mathcal{L}'_j$ be the connected component of ℓ'_i. According to (4.7.4), we have

$$X_m = \sum_{\ell'_u \in \mathcal{L}'_j} \gamma_u \ell'_u$$

with coefficients $\gamma_u \in L$. Taking such an expression with a minimal number of nonzero coefficients, the γ_u are quotients of determinants of order at most m, whose entries are from the coefficients of the $\ell'_u \in \mathcal{L}'_j$. So by (4.7.11), $h(\gamma_u) \ll H$. Now we have a relation

$$\frac{x_m}{\ell'_i(\mathbf{x})} = \sum_{\ell'_u \in \mathcal{L}'_j} \gamma_u \frac{\ell'_u(\mathbf{x})}{\ell'_i(\mathbf{x})}$$

and so by (4.7.14),

$$h\left(\frac{x_m}{\ell'_i(\mathbf{x})}\right) \ll H.$$

This holds for $i = 1, \ldots, n$. Since we assumed $k > 1$, we have $x_m \ne 0$. Thus, we conclude

$$h\left(\frac{\ell'_i(\mathbf{x})}{\ell'_j(\mathbf{x})}\right) \ll H \quad \text{for each } i, j \in \{1, \ldots, n\}.$$

We proved this assuming $k > 1$, but for $k = 1$, this is immediate from (4.7.14). Now from (4.7.10), we infer

$$\ell'_i(\mathbf{x})^n = \delta' \prod_{j=1}^{n} \frac{\ell'_i(\mathbf{x})}{\ell'_j(\mathbf{x})}$$

and thus,

$$h(\ell'_i(\mathbf{x})) \ll H \quad \text{for } i = 1, \ldots, n. \qquad (4.7.15)$$

In view of (4.7.3), we may assume that $\ell'_1, \ldots, \ell'_m$ are linearly independent. By Cramer's rule, we can express each x_i as a quotient Δ_i/Δ, where Δ is the determinant of order m whose jth column consists of the coefficients of ℓ'_j, for $j = 1, \ldots, m$, and where Δ_i is obtained from Δ by replacing the ith row of the latter by $\ell'_1(\mathbf{x}), \ldots, \ell'_m(\mathbf{x})$. Now, (4.7.15) in combination with (4.7.11) gives (4.7.9). This completes our proof. $\qquad\qquad\square$

4.8 Discriminant Equations

Let K be a number field, D_K its discriminant, $\mathcal{M}_K$ the set of places of K, T a finite subset of $\mathcal{M}_K$ containing the set of infinite places T_∞, and L a finite normal extension of K with the parameters d, r, h_L, and R_L previously specified. Let S denote the set of extensions to L of the places in T, with s, P_S, Q_S, and R_S as in Section 4.1.

If f is a monic polynomial with coefficients in $\mathcal{O}_T$, the ring of T-integers of K, and $f'(X) = f(X + a)$ for some $a \in \mathcal{O}_T$, then the discriminants $D(f)$ and $D(f')$ coincide. As before, such polynomials f and f' are called *strongly $\mathcal{O}_T$-equivalent*.

For a polynomial $P(X) = X^n + a_1 X^{n-1} + \cdots + a_n \in K[X]$, we put

$$h(P) := \frac{1}{[K : \mathbb{Q}]} \sum_{v \in \mathcal{M}_K} \log \max(1, |a_1|_v, \ldots, |a_n|_v).$$

From Theorem 4.7.1, one can deduce the following.

Theorem 4.8.1 *Let $f \in \mathcal{O}_T[X]$ be a monic polynomial of degree $n \geq 2$ with zeros in L such that*

$$D(f) = \delta, \tag{4.8.1}$$

where δ is a nonzero element of K with height not exceeding $H(\geq 1)$. Then f is strongly $\mathcal{O}_T$-equivalent to a polynomial $f' \in \mathcal{O}_T[X]$ for which

$$h(f') \leq n((n + 1)C + \log|D_K| + 1). \tag{4.8.2}$$

Here,

$$C := c_{19} P_S R_S (1 + \log^* R_S / \log^* P_S)(c_4 R_L + \frac{h_L}{d} \log Q_S + n^3 + H)$$

with $c_{19} = 50n^3 c_{11}$, where c_4 and c_{11} denote the constants specified in Proposition 4.1.9 and Theorem 4.3.1.

This theorem is an improvement of Theorem 3 of Győry (1998). In the proof of Theorem 4.8.1, one can follow the arguments of the deduction of Theorem 3 from Theorem 1 of Győry (1998). We note that Theorem 4.8.1 could be directly deduced, with a slightly different bound, from Theorem 4.3.1 concerning S-unit equations.

Theorem 4.8.1 has several consequences.

Let again K and L be number fields with the properties and parameters specified in the beginning of this section such that there is a number field K' with $K \subset K' \subseteq L$ and with $n = [K' : K] \geq 2$. Note that if $\xi \in K'$, then every element ξ' of the $\mathcal{O}_T$-coset $\xi + \mathcal{O}_T = \{\xi + a : a \in \mathcal{O}_T\}$ satisfies $D_{K'/K}(\xi') = D_{K'/K}(\xi)$.

Corollary 4.8.2 *Let δ be a nonzero element of K with height at most $H(\geq 1)$. Then for every $\xi \in K'$, such that*

$$D_{K'/K}(\xi) = \delta, \quad \xi \text{ is integral over } \mathcal{O}_T \tag{4.8.3}$$

there are $\xi' \in K'$, $a \in \mathcal{O}_T$ such that

$$h(\xi') \leq (n + 1)C + \log|D_K|, \; \xi = \xi' + a \tag{4.8.4}$$

with the above C.

This is an improvement of Theorem 15 of Győry (1984b). It could be easily deduced from Theorem 4.8.1 but only with a slightly weaker bound. To obtain Corollary 4.8.2 in the present form, it suffices to apply the proof of Theorem 4.8.1 to the minimal polynomial, say, f, of the ξ under consideration. Then $D(f) = \delta$, and following the proof of Theorem 4.8.1, it follows that f is strongly $\mathcal{O}_T$-equivalent to a polynomial f', which has a zero $\xi' \in \xi + \mathcal{O}_T$ such that (4.8.4) holds.

In the classical case where $K = \mathbb{Q}$ and $T = T_\infty$, the first effective results for equations (4.8.1) and (4.8.3) were proved by Győry (1973, 1974) in quantitative forms, without fixing the splitting field of the polynomials f resp. the number field L containing the algebraic numbers ξ. For general K and T, and various applications, see Győry (1976, 1978b, 1981b, 1984b, 1998) and Evertse and Győry (2017a).

As was mentioned in Section 1.6, the following more general versions of equation (4.8.1) in f and equation (4.8.3) in ξ have also important applications:

$$D(f) \in \delta\mathcal{O}_T^* \quad \text{in monic } f \in \mathcal{O}_T[X] \text{ of degree } n \geq 2$$
$$\text{having all its zeros in } \mathcal{O}_S \tag{4.8.5}$$

and

$$D_{K'/K}(\xi) \in \delta\mathcal{O}_T^* \quad \text{in } \xi \in K', \text{ integral over } \mathcal{O}_T, \tag{4.8.6}$$

where $\mathcal{O}_T^*$ is the unit group of $\mathcal{O}_T$.

We recall that the monic polynomials f and $f' \in \mathcal{O}_T[X]$ of degree n are called $\mathcal{O}_T$-*equivalent* if $f'(X) = \varepsilon^n f(\varepsilon^{-1}X + a)$ for some $\varepsilon \in \mathcal{O}_T^*$, $a \in \mathcal{O}_T$. If f satisfies (4.8.5), so does f'. Using Proposition 4.1.9, equation (4.8.5) can be reduced to finitely many equations of the form (4.8.1). Then by means of Theorem 4.8.1, one can prove that each $\mathcal{O}_T$-equivalence class of solutions of (4.8.5) has a representative with explicitly bounded height. A similar effective result can be proved for equation (4.8.6) by using Corollary 4.8.2. In this case, two elements, ξ and ξ' of K' and integral over $\mathcal{O}_T$, are said to be $\mathcal{O}_T$-*equivalent* if $\xi' = \varepsilon\xi + a$ with some $\varepsilon \in \mathcal{O}_T^*$ and $a \in \mathcal{O}_T$. This latter result has an important application to the equation

$$\mathcal{O} = \mathcal{O}_T[\xi] \quad \text{in } \xi \in K', \text{ integral over } \mathcal{O}_T, \qquad (4.8.7)$$

where $\mathcal{O}$ denotes the integral closure of $\mathcal{O}_T$ in K'. If (4.8.7) holds for some $\xi \in \mathcal{O}$, and ξ' is $\mathcal{O}_T$-equivalent to ξ, then $\mathcal{O} = \mathcal{O}_T[\xi']$. The above-mentioned result concerning (4.8.6) implies an effective and quantitative form that every ξ satisfying (4.8.7) is $\mathcal{O}_T$-equivalent to an $\xi' \in \mathcal{O}_T$ whose height can be explicitly bounded. For these and related results, see Győry (1981b, 1984b) and Evertse and Győry (2017a).

We sketch the proof of the following less explicit version of Theorem 4.8.1.

Under the assumption of Theorem 4.8.1, every solution f of equation (4.8.1) is strongly $\mathcal{O}_T$-equivalent to a monic polynomial $f' \in \mathcal{O}_T[X]$ for which

$$h(f') \ll_{n,K,L,S} H. \qquad (4.8.8)$$

In the proof of (4.8.8), we use the next division with remainder lemma. Denote by $\mathcal{O}_K$ the ring of integers of K.

Lemma 4.8.3 *Let $n \geq 2$ be an integer, and let $\beta \in \mathcal{O}_T$. Then there is an $\alpha \in \mathcal{O}_K$ such that*

$$\alpha \equiv \beta \,(\mathrm{mod}\, n)$$

and

$$h(\alpha) \leq \log([K : \mathbb{Q}] \cdot n|D_K|^{1/2}).$$

Proof This is a special case of Lemma 6 of Evertse and Győry (1991).   □

Sketch of the proof of (4.8.8) Assume that $\alpha_1, \ldots, \alpha_n$ are the zeros of f in L. Denote by $\mathcal{O}_S$ the ring of S-integers in L. Writing $x_i = \alpha_i - \alpha_1$ for $i = 1, \ldots, n$, we have $\alpha_i \in \mathcal{O}_S$ and $x_i \in \mathcal{O}_S$. Further, putting

$$F(X_2, \ldots, X_n) = X_2 \cdots X_n \prod_{2 \leq i < j \leq n} (X_i - X_j),$$

(4.8.1) implies

$$F(x_2, \ldots, x_n) = \pm \delta_0 \text{ with } x_2, \ldots, x_n \in \mathcal{O}_S, \qquad (4.8.9)$$

where $\delta_0 \in \mathcal{O}_S \setminus \{0\}$ and $\delta_0^2 = \delta$. We have $h(\delta_0) \leq \frac{1}{2} H$. The decomposable form F is of degree $n(n-1)/2$, and it is easy to verify that for $n \geq 3$ it satisfies the assumptions of Theorem 4.7.1 with $k = 1$. Hence, by the less precise version (4.7.9), we deduce from (4.8.9) that both for $n = 2$ and for $n \geq 3$,

$$\max_{2 \leq i \leq n} h(x_i) \ll_{n,L,S} H = C' \qquad (4.8.10)$$

holds.

The sum $a_0 = \alpha_1 + \cdots + \alpha_n$ is contained in $\mathcal{O}_T$. Setting $\beta = -(x_1 + \cdots + x_n)$, it follows from (4.8.10) that $h(\beta) \leq (n - 1/2)C'$. Further, we have $n\alpha_1 - a_0 = \beta$. By Lemma 4.8.3, there is an $a_1 \in \mathcal{O}_K$ such that $a_1 \equiv a_0 \pmod{n}$ in $\mathcal{O}_T$ and $h(a_1) \leq \log(dn|D_K|^{1/2})$. Set $\alpha_1' = (\beta + a_1)/n$. Then $h(\alpha_1') \leq (n - 1/3)C' + \log|D_K|$. Further, $\alpha_1 = a + \alpha_1'$ with some $a \in \mathcal{O}_T$ and $\alpha_1' \in \mathcal{O}_S$. Finally, with the notation $\alpha_i' = x_i + \alpha_1'$, we get $\alpha_i' = \alpha_i - a$ and

$$h(\alpha_i') \leq (n+1)C' + \log|D_K| \quad \text{for } i = 1, \ldots, n.$$

Put $f'(X) := \prod_{i=1}^{n}(X - \alpha_i)$. Then $f'(X) = f(X + a)$, while Corollary 4.1.5 gives our height estimate (4.8.8). $\qquad \square$

5

Effective Results over Function Fields

As was previously mentioned, S-unit equations, Thue-equations, hyper- and superelliptic equations, and the Catalan equation over finitely generated domains will be reduced in Chapter 9 to equations of the same type over number fields and over function fields. In this chapter, we formulate the best bounds to date for the heights of the integral solutions of the reduced equations over function fields and sketch the main ideas of their proofs. In contrast with the number field case, these bounds in the function field case do not imply the finiteness of the number of solutions.

5.1 Notation and Preliminaries

We start with some notation and definitions and with recalling some preliminary results over function fields. For further details, we refer to Mason (1984) and Evertse and Győry (2015, Chapter 2).

Let $\Bbbk$ be an algebraically closed field of characteristic 0 and K an algebraic function field in one variable over $\Bbbk$, that is, a finitely generated extension of $\Bbbk$ of transcendence degree 1. Recall that K is a finite extension of the rational function field $\Bbbk(z)$ for any $z \in K \backslash \Bbbk$. By a valuation on K over $\Bbbk$, we mean a normalized, discrete valuation on K that is trivial on $\Bbbk$, that is, a surjective map $v : K \mapsto \mathbb{Z} \cup \{\infty\}$ such that

$$v(\alpha) = \infty \Longleftrightarrow \alpha = 0;$$
$$v(\alpha\beta) = v(\alpha) + v(\beta), \quad v(\alpha + \beta) \geq \min(v(\alpha), v(\beta)) \quad \text{for} \quad \alpha, \beta \in K;$$
$$v(\alpha) = 0 \quad \text{for} \quad \alpha \in \Bbbk^*.$$

We denote by $\mathcal{M}_K$ the set of valuations on K over $\Bbbk$.

For a finite extension L of K, we say that a valuation w on L lies above a valuation v on K, notation $w \mid v$, if the restriction of w to K is a multiple of

98

v. In this case, we have $w(\alpha) = e_{w|v}v(\alpha)$ for $\alpha \in K$, where $e_{w|v}$ is a positive integer, called the *ramification index* of w over v.

The valuations over $\Bbbk$ on the field of rational functions $\Bbbk(z)$, with z transcendental over $\Bbbk$, can be described easily as follows. Let $z_a := z - a$ if $a \in \Bbbk$, and $z_\infty := z^{-1}$. For every $a \in \Bbbk \cup \{\infty\}$, we may expand $\alpha \in \Bbbk(z)$ as a formal Laurent series $\sum_{m=n(\alpha)}^{\infty} a_m(\alpha)z_a^m$ with $a_m(\alpha) \in \Bbbk$ for all m and $a_{n(\alpha)}(\alpha) \neq 0$. Then ord_a defined by $\mathrm{ord}_a(\alpha) := n(\alpha)$ defines a valuation on $\Bbbk(z)$. In particular, $\mathrm{ord}_\infty(\alpha) = -\deg(\alpha)$ for $\alpha \in \Bbbk[z]$. The valuations ord_a ($a \in \Bbbk \cup \{\infty\}$) provide all valuations on $\Bbbk(z)$ over $\Bbbk$.

Let K be a function field in one variable and L a finite extension of K. Clearly, every valuation on L lies above a valuation on K. We explain how to construct, for a given valuation v on K, the valuations on L that lie above v. As a special case, we may choose $z \in K\backslash\Bbbk$, take the valuations on $\Bbbk(z)$ just described, and construct from these the valuations on K.

Let v be a valuation on K. Take a local parameter $z_v \in K$ of v, i.e., with $v(z_v) = 1$. Then the completion K_v of K at v is (up to isomorphism) just the field of Laurent series $\Bbbk((z_v))$, and the algebraic closure $\overline{K_v}$ of K_v is the field of Puiseux series in z_v, i.e., $\sum_{m=n}^{\infty} a_m z_v^{m/e}$ with $n \in \mathbb{Z}$, $e \in \mathbb{Z}_{>0}$ and $a_m \in \Bbbk$ for all $m \geq n$, where for every positive integer e we have fixed an eth root of z_v. There are precisely $[L : K]$ K-isomorphic embeddings $L \hookrightarrow \overline{K_v}$, given by

$$\alpha \mapsto \sum_{m=n_i(\alpha)}^{\infty} a_{im}(\alpha)\zeta_i^{jm} z_v^{m/e_i} \quad (i = 1,\ldots,g, \; j = 0,\ldots,e_i - 1), \quad (5.1.1)$$

where $e_1,\ldots,e_g$ are positive integers with $e_1 + \cdots + e_g = [L : K]$, ζ_i is some fixed e_ith root of unity, $a_{im}(\alpha) \in \Bbbk$ for all $m \geq n_i(\alpha)$, and $a_{i,n_i(\alpha)} \neq 0$. We can now define for $i = 1,\ldots,g$ a valuation w_i on L by $w_i(\alpha) := n_i(\alpha)$ for $\alpha \in L^*$, and $w_i(0) = \infty$. These are precisely the valuations on L lying above v. Writing w for w_i, the integer e_i is just the ramification index $e_{w|v}$. Thus, we have

$$\sum_{w|v} e_{w|v} = [L : K] \quad \text{for } v \in \mathcal{M}_K, \quad (5.1.2)$$

where the sum is taken over all valuations on L lying above v. We easily deduce from this the *sum formula* for K,

$$\sum_{v \in \mathcal{M}_K} v(\alpha) = 0 \quad \text{for } \alpha \text{ in } K^*. \quad (5.1.3)$$

Indeed, this is clear if $\alpha \in \Bbbk^*$. Let $\alpha \in K^*\backslash\Bbbk^*$. By using the preceding description of the valuations on $\Bbbk(\alpha)$, with α replacing z, one easily gets $\sum_{u \in \mathcal{M}_{\Bbbk(\alpha)}} u(\alpha) = 0$, and together with (5.1.2), with $\Bbbk(\alpha)$ and K instead of K and L, the identity (5.1.3) follows.

Denote by $g_{K/\Bbbk}$ the *genus* of K over $\Bbbk$. In the case that z is transcendental over $\Bbbk$, one has

$$g_{\Bbbk(z)/\Bbbk} = 0. \tag{5.1.4}$$

We can relate the genus of a finite extension L of K to that of K by means of the *Riemann–Hurwitz formula*

$$2g_{L/\Bbbk} - 2 = [L:K](2g_{K/\Bbbk} - 2) + \sum_{v \in \mathcal{M}_K} \sum_{w|v} (e_{w|v} - 1). \tag{5.1.5}$$

Let S be a finite subset of $\mathcal{M}_K$. We call $\alpha \in K$ an *S-integer* if $v(\alpha) \geq 0$ for all $v \in \mathcal{M}_K \backslash S$, and an *S-unit* if $v(\alpha) = 0$ for all $v \in \mathcal{M}_K \backslash S$. The S-integers form a ring in K, denoted by $\mathcal{O}_S$, and the S-units form a multiplicative group, denoted by $\mathcal{O}_S^*$.

We define the *height* $H_K(\alpha)$ of $\alpha \in K$ relative to $K/\Bbbk$ by

$$H_K(\alpha) := -\sum_{v \in \mathcal{M}_K} \min(0, v(\alpha)).$$

It is clear that $H_K(\alpha) \geq 0$ for $\alpha \in K$ and $H_K(\alpha) = 0$ if and only if $\alpha \in \Bbbk$. If L is a finite extension of K, then by (5.1.2),

$$H_L(\alpha) = [L:K] \cdot H_K(\alpha) \quad \text{for } \alpha \in K.$$

From the description of the valuations on a field of rational functions, one easily deduces that $H_{\Bbbk(\alpha)}(\alpha) = 1$ if $\alpha \notin \Bbbk$. So in particular,

$$H_K(\alpha) = [K : \Bbbk(\alpha)] \quad \text{for } \alpha \in K \backslash \Bbbk. \tag{5.1.6}$$

It is easy to see that

$$H_K(\alpha^m) = |m| H_K(\alpha) \quad \text{for } \alpha \in K^*, \, m \in \mathbb{Z} \tag{5.1.7}$$

(where for negative m one has to employ the sum formula) and

$$H_K(\alpha + \beta) \leq H_K(\alpha) + H_K(\beta), \quad H_K(\alpha\beta) \leq H_K(\alpha) + H_K(\beta) \tag{5.1.8}$$

for all α and $\beta \in K$. Further

$$H_K(\alpha) = \tfrac{1}{2}(H_K(\alpha) + H_K(\alpha^{-1}))$$
$$= \tfrac{1}{2} \sum_{v \in \mathcal{M}_K} |v(\alpha)| \geq \tfrac{1}{2}|T| \quad \text{for } \alpha \in K^*, \tag{5.1.9}$$

where T denotes the set of valuations $v \in \mathcal{M}_K$ for which $v(\alpha) \neq 0$.

We define the height of a vector $\alpha = (\alpha_1, \dots, \alpha_n) \in K^n$ relative to $K/\Bbbk$ by

$$H_K(\alpha) := H_K(\alpha_1, \dots, \alpha_n) = -\sum_{v \in \mathcal{M}_K} \min(0, v(\alpha)),$$

where $v(\alpha) = \min_i(v(\alpha_i))$ is the v-value of α. If L is a finite extension of K, then

$$H_L(\alpha) = [L : K]H_K(\alpha) \quad \text{for } \alpha \in K^n \backslash \{0\}.$$

We note that

$$H_K(\alpha_i) \le H_K(\alpha) \le H_K(\alpha_1) + \cdots + H_K(\alpha_n) \quad (i = 1, \ldots, n). \quad (5.1.10)$$

The *homogeneous height* of $\alpha = (\alpha_1, \ldots, \alpha_n) \in K^n \backslash \{0\}$, relative to $K/\Bbbk$ is defined as

$$H_K^{\text{hom}}(\alpha) := - \sum_{v \in \mathcal{M}_K} v(\alpha).$$

It is clear that

$$H_K^{\text{hom}}(\alpha) \le H_K(\alpha).$$

By the sum formula, we have

$$H_K^{\text{hom}}(\lambda \alpha) = H_K^{\text{hom}}(\alpha) \quad \text{for} \quad \lambda \in K^*. \quad (5.1.11)$$

For instance, let $p_1, \ldots, p_n \in \Bbbk[z]$ with $\gcd(p_1, \ldots, p_n) = 1$. Then

$$H_{\Bbbk(z)}^{\text{hom}}(p_1, \ldots, p_n) = \max(\deg p_1, \ldots, \deg p_n). \quad (5.1.12)$$

Further, if L is a finite extension of K, then, for $\alpha = (\alpha_1, \ldots, \alpha_n) \in K^n \backslash \{0\}$, we have

$$H_L^{\text{hom}}(\alpha) = [L : K]H_K^{\text{hom}}(\alpha). \quad (5.1.13)$$

Since

$$H_K^{\text{hom}}(\alpha_1, \ldots, \alpha_n) = H_K(\alpha_1/\alpha_i, \ldots, \alpha_n/\alpha_i) \quad \text{for all } i \text{ with } \alpha_i \ne 0,$$

we deduce from (5.1.10) that, for $\alpha_1 \ne 0$,

$$H_K(\alpha_i/\alpha_1) \le H_K^{\text{hom}}(\alpha_1, \ldots, \alpha_n) \le \sum_{j=1}^{n} H_K(\alpha_j) + (n-2)H_K(\alpha_i),$$

$$\text{for } i = 1, \ldots, n. \quad (5.1.14)$$

Further, for a polynomial $F \in K[X]$, its height $H_K(F)$ resp. its homogeneous height $H_K^{\text{hom}}(F)$ and its v-value $v(F)$ are defined by the height resp. homogeneous height and the v-value of a vector whose coordinates are the coefficients of F. Clearly, for monic polynomials, the two heights coincide, while in general,

$$H_K^{\text{hom}}(F) \le H_K(F).$$

For any two polynomials F and G in $K[X]$, we have

$$v(FG) = v(F) + v(G) \text{ for } v \in \mathcal{M}_K, \quad H_K^{\mathrm{hom}}(FG) = H_K^{\mathrm{hom}}(F) + H_K^{\mathrm{hom}}(G).$$
$$(5.1.15)$$

If a nonzero polynomial $F(X)$ factorizes in K as $f_0(X - \alpha_1) \cdots (X - \alpha_n)$, then by (5.1.15) and the sum formula, applied to f_0, we obtain

$$H_K(f_0) \leq H_K(F) \quad \text{and} \quad H_K^{\mathrm{hom}}(F) = \sum_{i=1}^n H_K(\alpha_i) \geq \max_{1 \leq i \leq n} H_K(\alpha_i).$$
$$(5.1.16)$$

Lemma 5.1.1 *Let*

$$F = f_0 X^n + f_1 X^{n-1} + \cdots + f_n \in K[X]$$

be a polynomial with $f_0 \neq 0$ and with nonzero discriminant. Let L be the splitting field of F over K. Then

$$g_{L/\Bbbk} \leq [L : K](g_{K/\Bbbk} + nH_K(F)).$$

In particular, if $K = \Bbbk(z)$ and $f_0, \dots, f_n \in \Bbbk[z]$, we have

$$g_{L/\Bbbk} \leq [L : K]n \max(\deg f_0, \dots, \deg f_n).$$

Proof Lemma 4.2 of Bérczes, Evertse, and Győry (2014) gives a slightly better estimate with $H_K^{\mathrm{hom}}(F)$ instead of $H_K(F)$. The second assertion is due to Schmidt (1978). $\qquad\qquad\square$

In the next sections, we present explicit upper bounds for the heights of the solutions of S-unit equations, the Catalan equation, Thue equations, and hyper- and superelliptic equations.

We denote by $|S|$ the cardinality of a set S.

5.2 S-Unit Equations

Let, again, K be a function field in one variable over an algebraically closed field $\Bbbk$ of characteristic 0, and S a finite, nonempty set of valuations on K over $\Bbbk$, of cardinality at least 2. Consider the S-unit equation

$$\alpha x + \beta y = 1 \quad \text{in} \quad x, y \in \mathcal{O}_S^* \backslash \Bbbk^*, \qquad (5.2.1)$$

where α and $\beta \in K^*$. For $\alpha = \beta = 1$, the following theorem is due to Mason (1983). The general case, under the assumption $\alpha x \notin \Bbbk^*$, was established in Evertse and Győry (2015). We prove Theorem 5.2.1 without this assumption.

Theorem 5.2.1 *Every solution of x and y of equation* (5.2.1) *satisfies*

$$\max(H_K(x), H_K(y)) \le 5H + |S| + 2g_{K/\Bbbk} - 2, \tag{5.2.2}$$

where $H = \max(H_K(\alpha), H_K(\beta))$.

Independently of Mason, Győry (1983) proved a version of Theorem 5.2.1 with larger explicit coefficients of $|S|$ and $g_{K/\Bbbk}$. We note that this weaker version would also be sufficient for the applications in Chapter 9. Theorem 5.2.1 is a consequence of the following theorem of Mason (1983, 1984). It is a generalization of an earlier result of Stothers (1981).

Theorem 5.2.2 *Let S be a finite, nonempty subset of* $\mathcal{M}_K$, *and let* x_1, x_2, *and* x_3 *be nonzero elements of K with* $x_1 + x_2 + x_3 = 0$ *such that* $v(x_1) = v(x_2) = v(x_3)$ *for every v in* $\mathcal{M}_K \setminus S$. *Then either* x_1/x_2 *lies in* $\Bbbk$, *or*

$$H_K(x_1/x_2) \le |S| + 2g_{K/\Bbbk} - 2. \tag{5.2.3}$$

Proof We do not give Mason's proof based on derivations but instead another well-known proof based on the Riemann–Hurwitz formula. We assume without loss of generality that $x_1/x_2 \notin \Bbbk$ and that S is precisely the set of all $v \in \mathcal{M}_K$ such that $v(x_1), v(x_2)$, and $v(x_3)$ are distinct. Let $z := x_1/x_2$. Then z and $1 + z$ are S-units. We can write S as a disjoint union $S_0 \cup S_{-1} \cup S_\infty$, where

$$S_0 := \{v \in \mathcal{M}_K : v(z) > 0\}, \quad S_{-1} := \{v \in \mathcal{M}_K : v(z+1) > 0\},$$
$$S_\infty := \{v \in \mathcal{M}_K : v(z) < 0\}.$$

Note that for $a = 0, -1, \infty$, and S_a is precisely the set of valuations on K lying above the valuation ord_a on $\Bbbk(z)$. So by the Riemann–Hurwitz formula (5.1.5), and by (5.1.4) and (5.1.2),

$$2g_{K/\Bbbk} - 2 = [K : \Bbbk(z)](2g_{\Bbbk(z)/\Bbbk} - 2) + \sum_{a \in \Bbbk \cup \{\infty\}} \sum_{v|a}(e_{v|a} - 1)$$

$$\ge -2[K : \Bbbk(z)] + \sum_{a \in \{0, -1, \infty\}} \sum_{v|a}(e_{v|a} - 1)$$

$$= -2[K : \Bbbk(z)] + \sum_{a \in \{0, -1, \infty\}}([K : \Bbbk(z)] - |S_a|) = [K : \Bbbk(z)] - |S|,$$

where we have written $v|a$ for $v|\mathrm{ord}_a$. Using $[K : \Bbbk(z)] = H_K(z)$ (see [5.1.6]), Theorem 5.2.2 follows. □

Proof of Theorem 5.2.1 Let x and y be a solution of equation (5.2.1). Then

$$H_K(x) = H_K(\alpha x \cdot \alpha^{-1}) \le H_K(\alpha x) + H_K(\alpha^{-1}) \le H_K(\alpha x) + H, \tag{5.2.4}$$

and similarly $H_K(y) \leq H_K(\beta y) + H$. If α and $\beta \in \mathbb{k}^*$, then Theorem 5.2.1 follows at once from Theorem 5.2.2. Suppose that α and β are not both in $\mathbb{k}^*$. Then $H \geq 1$.

If $\alpha x \in \mathbb{k}^*$, then $\beta y \in \mathbb{k}^*$; hence, their heights are zero, and (5.2.2) immediately follows.

Now, suppose that αx and $\beta y \notin \mathbb{k}^*$. Let S_α denote the set of valuations $v \in \mathcal{M}_K$ with $v(\alpha) \neq 0$ and define S_β similarly. In view of (5.1.9), we have $|S_\alpha| \leq 2H_K(\alpha) \leq 2H$ and similarly $|S_\beta| \leq 2H$. Then it follows that $v(\alpha x) = v(\beta y) = v(1) = 0$ for $v \in \mathcal{M}_K \backslash (S \cup S_\alpha \cup S_\beta)$. Now using (5.2.4) and applying Theorem 5.2.2 to $\alpha x, \beta y,$ and 1 with $S \cup S_\alpha \cup S_\beta$ instead of S, we obtain

$$\max(H_K(x), H_K(y)) \leq \max(H_K(\alpha x), H_K(\beta y)) + H$$

$$\leq |S| + |S_\alpha| + |S_\beta| + 2g_{K/\mathbb{k}} - 2 + H$$

$$\leq 5H + |S| + 2g_{K/\mathbb{k}} - 2. \qquad \square$$

Remark In contrast with the number field case, for function fields, there are effective results for S-unit equations in the more unknowns case as well; see Mason (1986, 1988), Brownawell and Masser (1986) and, for further references, e.g. Evertse and Győry (2015).

5.3 The Catalan Equation

As before, K is a function field in one variable over an algebraically closed field $\mathbb{k}$ of characteristic 0, and S is a finite set of valuations on K over $\mathbb{k}$. We deduce consequences for a slight generalization of the Catalan equation, i.e.,

$$x^m - y^n = \pm 1, \tag{5.3.1}$$

both in the cases that x and y assume their values in $\mathcal{O}_S$ and that x and y assume their values in K.

Theorem 5.3.1 *(i) Let S be a finite set of valuations of K, let m and n be integers with m and $n \geq 2$, $mn > 4$, and let x and $y \in \mathcal{O}_S \backslash \mathbb{k}$ satisfy (5.3.1). Then*

$$mH_K(x), nH_K(y) \leq 6(|S| + 2g_{K/\mathbb{k}} - 2). \tag{5.3.2}$$

(ii) Let m and n be integers with $n \geq m \geq 2$, $mn \geq 10$ and let x and $y \in K \backslash \mathbb{k}$ satisfy (5.3.1). Then

$$mH_K(x), nH_K(y) \leq 20(g_{K/\mathbb{k}} - 1). \tag{5.3.3}$$

Since $H_K(x)$ and $H_K(y)$ are ≥ 1, (5.3.2) and (5.3.3) imply upper bounds for m and n.

Proof (i) Let S_1 be the set of $v \in \mathcal{M}_K \backslash S$ with $v(x) > 0$ and S_2 the set of $v \in \mathcal{M}_K \backslash S$ with $v(y) > 0$. Then

$$|S_1| \leq H_K(x) = \tfrac{1}{m} H_K(x^m), \quad |S_2| \leq H_K(y) = \tfrac{1}{n} H_K(y^n) \leq \tfrac{1}{n} H_K(x^m).$$

So by Theorem 5.2.2,

$$H_K(x^m) \leq |S| + |S_1| + |S_2| + 2g_{K/\Bbbk} - 2$$

$$\leq |S| + 2g_{K/\Bbbk} - 2 + (\tfrac{1}{m} + \tfrac{1}{n}) H_K(x^m)$$

$$\leq |S| + 2g_{K/\Bbbk} - 2 + \tfrac{5}{6} H_K(x^m),$$

implying (5.3.2).

(ii) Assume without loss of generality that $n \geq m$. Let S_1 be the set of $v \in \mathcal{M}_K$ with $v(x) > 0$, let S_2 be the set of $v \in \mathcal{M}_K$ with $v(y) > 0$, and let S_3 be the set of $v \in \mathcal{M}_K$ with $v(x) < 0$; for these places, we have also $v(y) < 0$. Then

$$|S_1| \leq \tfrac{1}{m} H_K(x^m), \quad |S_2|, |S_3| \leq H_K(y) = \tfrac{1}{n} H_K(x^m),$$

and Theorem 5.2.2 gives

$$H_K(x^m) \leq |S_1| + |S_2| + |S_3| + 2g_{K/\Bbbk} - 2$$

$$\leq 2g_{K/\Bbbk} - 2 + (\tfrac{1}{m} + \tfrac{2}{n}) H_K(x^m)$$

$$\leq 2g_{K/\Bbbk} - 2 + \tfrac{9}{10} H_K(x^m),$$

implying (5.3.3). $\qquad\qquad\qquad\qquad\qquad\qquad\qquad\qquad\qquad\qquad$ $\square$

5.4 Thue Equations

We denote as before by K a function field in one variable over an algebraically closed field $\Bbbk$ of characteristic 0 and by S a finite set of valuations on K over $\Bbbk$. In this section, we consider the Thue equation

$$F(x, y) = 1 \quad \text{in} \quad x, y \in \mathcal{O}_S, \qquad (5.4.1)$$

where F is a binary form of degree $n \geq 3$ with coefficients in K and with nonzero discriminant.

Using a method of Osgood (1973, 1975), Schmidt (1976, 1978) derived bounds for the heights of the solutions of (5.4.1). Later, by means of his Theorem 5.2.2, Mason (1984) gave a better bound for the heights of the integral solutions over $\Bbbk[z]$ in the case when F factorizes into linear factors over K. For a refinement, see Dvornicich and Zannier (1994).

We prove the following version of the theorems of Schmidt and Mason.

Theorem 5.4.1 *Every solution of x and y of equation* (5.4.1) *satisfies*

$$\max(H_K(x), H_K(y)) \leq H_K(x, y)$$
$$\leq (8n + 62)H_K(F) + 4|S| + 8g_{K/k}. \qquad (5.4.2)$$

If the splitting field of F is K, we have the stronger estimate

$$H_K(x, y) \leq 62H_K(F) + 4|S| + 8g_{K/k}. \qquad (5.4.3)$$

Our proof is different from that of Schmidt and Mason. It is based on Theorem 5.2.1. No special importance is attached to the constants in our upper bounds, which could be improved with some extra effort. However, such improvements would be irrelevant for our application in Chapter 9.

Proof of Theorem 5.4.1 Write the binary form F in the form

$$F(X, Y) = a_0 X^n + a_1 X^{n-1} Y + \cdots + a_n Y^n.$$

We may suppose without loss of generality that $a_0 \neq 0$. Indeed, this can be achieved by a linear transformation of the shape

$$X = X', \quad Y = aX' + Y'$$

with $a \in k^*$ such that $F(1, a) \neq 0$. Then putting

$$F'(X', Y') := F(X', aX' + Y') = a_0' X'^m + a_1' X'^{m-1} Y' + \cdots + a_n' Y'^m,$$

we have $a_0' = F(1, a) \neq 0$. Then with $x' = x$ and $y' = -ax + y$, we have $F'(x', y') = 1$. Observe that $\min(v(x'), v(y')) = \min(v(x), v(y))$ for $v \in \mathcal{M}_K$, so $H_K(x', y') = H_K(x, y)$. Further, each coefficient a_i' of F' is a k-linear combination of $a_0, \ldots, a_n$; hence, $v(a_i') \geq \min(v(a_0), \ldots, v(a_n)) = v(F)$ for $v \in \mathcal{M}_K$, $i = 0, \ldots, n$, that is, $v(F') \geq v(F)$ for $v \in \mathcal{M}_K$. But by symmetry, the reverse inequality also holds so we have $v(F') = v(F)$ for $v \in \mathcal{M}_K$. Hence, $H_K(F') = H_K(F)$.

We assume, henceforth, that $a_0 \neq 0$. For the moment, we assume that F has splitting field K. We can write (5.4.1) as

$$a_0(x - \alpha_1 y) \cdots (x - \alpha_n y) = 1 \quad \text{in} \quad x, y \in \mathcal{O}_S, \qquad (5.4.4)$$

where, by our assumption, $\alpha_1, \ldots, \alpha_n$ are distinct elements of K. Denote by S_α the set of valuations $v \in \mathcal{M}_S$ for which $v(a_0) \neq 0$ and $v(\alpha_i) \neq 0$ for $i = 1, \ldots, n$. Notice that in view of (5.1.16),

$$|S_\alpha| \leq 2(H_K(a_0) + \sum_{i=1}^{n} H_K(\alpha_i)) \leq 4H_K(F). \qquad (5.4.5)$$

Putting $S' = S \cup S_\alpha$, the elements $x, y, a_0, \alpha_1, \ldots, \alpha_n$, and $\beta_i = x - \alpha_i y$ for $i = 1, \ldots, n$ are all S'-integers in K. Now it follows from (5.4.4) that β_i is an S'-unit in K for $i = 1, \ldots, n$.

We have

$$-\gamma \frac{\beta_i}{\beta_j} + (\gamma + 1) \frac{\beta_l}{\beta_j} = 1 \tag{5.4.6}$$

for any distinct i, j, and l, where $\gamma = (\alpha_j - \alpha_l)/(\alpha_l - \alpha_i)$ and $\beta_i/\beta_j, \beta_l/\beta_j$ are S'-units. If $\beta_i/\beta_j \in \Bbbk$, then $H_K(\beta_i/\beta_j) = 0$. Otherwise, by Theorem 5.2.1 we have

$$H_K(\beta_i/\beta_j) \le 5H + |S'| + 2g_{K/\Bbbk} - 2$$

for each distinct i and j with $1 \le i$ and $j \le n$, where $H = \max(H_K(\gamma), H_K(\gamma + 1))$. In view of (5.1.16), we have $H \le 2H_K(F)$, while by (5.4.5), we have $|S'| \le |S| + 4H_K(F)$. Thus, we get

$$H_K(\beta_i/\beta_j) \le 14H_K(F) + |S| + 2g_{K/\Bbbk} =: C \tag{5.4.7}$$

for each i, j with $1 \le i, j \le n$, $i \ne j$. Here, we have removed the -2-term to incorporate the case $\beta_i/\beta_j \in \Bbbk$.

We infer from (5.4.4) that

$$\beta_j^{-n} = a_0 \prod_{i=1}^{n} \frac{\beta_i}{\beta_j}, \quad j = 1, \ldots, n,$$

whence, using (5.1.16), we get

$$nH_K(\beta_j) \le H_K(F) + nC. \tag{5.4.8}$$

But we have

$$x = \frac{\beta_i \alpha_j - \alpha_i \beta_j}{\alpha_j - \alpha_i}, \quad y = \frac{\beta_i - \beta_j}{\alpha_j - \alpha_i},$$

and so, using again (5.1.16) and (5.4.8), it follows that

$$H_K(x) \le \tfrac{2n+2}{n} H_K(F) + 2C.$$

For $H_K(y)$, we get the same upper bound, so, in view of $H_K(x, y) \le H_K(x) + H_K(y)$,

$$H_K(x, y) \le \tfrac{4n+4}{n} H_K(F) + 4C.$$

Together with (5.4.7), this gives (5.4.3).

Now assume that F has splitting field L over K. Let $\Delta := [L : K]$, and let T be the set of valuations on L lying above those in S. Then (5.4.3) holds with L and T instead of K and S, i.e.,

$$H_L(x,y) \le 62H_L(F) + 4|T| + 8g_{L/\Bbbk}. \tag{5.4.9}$$

We have $H_L(x,y) = \Delta H_K(x,y)$, $H_L(F) = \Delta H_K(F)$, $|T| \le \Delta|S|$, and lastly, by Lemma 5.1.1,

$$g_{L/\Bbbk} \le \Delta(g_{K/\Bbbk} + nH_K(F)).$$

By inserting this into (5.4.9) and dividing by Δ, we obtain (5.4.2). This completes our proof. $\qquad\square$

5.5 Hyper- and Superelliptic Equations

Let, again, $\Bbbk$ be an algebraically closed field of characteristic 0, K a function field in one variable over $\Bbbk$, and S a finite subset of $\mathcal{M}_K$.

Let $f \in K[X]$ be a polynomial of degree n with nonzero discriminant. Consider the hyperelliptic equation

$$f(x) = y^2 \quad \text{in} \quad x,y \in \mathcal{O}_S \tag{5.5.1}$$

where $n \ge 3$, and the superelliptic equation

$$f(x) = y^m \quad \text{in} \quad x,y \in \mathcal{O}_S \tag{5.5.2}$$

where $m \ge 3$ and $n \ge 2$,

Schmidt (1978) gave an explicit upper bound for the heights of the solutions x and y of (5.5.1) in the case when all the zeros of f lie in K. Using his Theorem 5.2.2, Mason (1983, 1984) derived explicit upper bounds for the heights of the solutions of equation (5.5.1) and (5.5.2) but only under the assumptions that the zeros of f belong to K and S consists of the infinite valuations of K (i.e., those valuations v with $v(z) < 0$, where z is an element of $K\backslash\Bbbk$ that is chosen and fixed in advance). Bérczes, Evertse, and Győry (2014) needed results without these conditions, and so they extended Mason's results to the most general situation when the splitting field of f and the set of valuations S are arbitrary.

Subsequently, we present results similar to Proposition 4.7 and Proposition 4.6 of Bérczes, Evertse, and Győry (2014), with different upper bounds. In our proofs, we will follow Mason. Both proofs will be based on Theorem 5.2.2.

Theorem 5.5.1 *Every solution x,y of equation* (5.5.1) *satisfies*

$$H_K(x) \le (8n + 42)H_K(f) + 8|S| + 8g_{K/\Bbbk}, \tag{5.5.3}$$
$$H_K(y) \le (4n^2 + 21n + 1)H_K(f) + 4n|S| + 4ng_{K/\Bbbk}. \tag{5.5.4}$$

In the case of equation (5.5.2), we can even estimate m from above, provided that $y \notin \Bbbk$.

Theorem 5.5.2 *Every solution of x and y of equation* (5.5.2) *satisfies*

$$H_K(x) \le (6n + 18)H_K(f) + 3|S| + 6g_{K/k}, \tag{5.5.5}$$

$$mH_K(y) \le (6n^2 + 18n + 1)H_K(f) + 3n|S| + 6ng_{K/k}. \tag{5.5.6}$$

It will be more convenient to prove first Theorem 5.5.2 and then Theorem 5.5.1. Similarly as in the case of Thue equations, in both proofs we first consider the case when f splits completely over K and then deduce the general case.

Proof of Theorem 5.5.2 We fix a solution (x and y) of (5.5.2). We start with the special case that $f(X) = a_0 \prod_{i=1}^{n}(X - \alpha_i)$ with $a_0 \in K^*$ and with distinct elements $\alpha_1, \ldots, \alpha_n$ of K. We shall apply Theorem 5.2.2 to the identity

$$(x - \alpha_1) + (\alpha_2 - x) + (\alpha_1 - \alpha_2) = 0. \tag{5.5.7}$$

We assume without loss of generality that $\alpha_1, \ldots, \alpha_n$ are arranged in order of increasing height, and so, in view of (5.1.16),

$$\sum_{i=1}^{s} H_K(\alpha_i) \le sH_K(f)/n \quad \text{for } s = 1, \ldots, n. \tag{5.5.8}$$

Put $\beta_i = \prod_{j \ne i}(\alpha_i - \alpha_j)$, $i = 1, \ldots, n$. Then we have

$$H_K(\beta_i) \le H_K(f) + (n - 2)H_K(\alpha_i) \quad \text{for} \quad i = 1, \ldots, n.$$

Denote by W the set of valuations v on K at which one or more of the following occur:

$$v \in S, \quad v(f) < 0, \quad v(a_0) > 0, \quad v(\beta_1\beta_2) > 0. \tag{5.5.9}$$

The number of v with $v(f) < 0$ is at most $H_K(f)$; the number of v with $v(a_0) > 0$ is at most $H_K(a_0^{-1}) = H_K(a_0) \le H_K(f)$; and lastly, by (5.5.8), the number of v with $v(\beta_1\beta_2) > 0$ is at most

$$H_K((\beta_1\beta_2)^{-1}) = H_K(\beta_1\beta_2) \le H_K(\beta_1) + H_K(\beta_2)$$

$$\le 2H_K(f) + (n - 2)(H_K(\alpha_1) + H_K(\alpha_2)) \le (4 - \tfrac{4}{n})H_K(f).$$

So altogether,

$$|W| \le |S| + (6 - \tfrac{4}{n})H_K(f). \tag{5.5.10}$$

It is easy to check that for $v \in \mathcal{M}_K \backslash W$,

$$v(a_0) = 0, \quad v(f) = 0, \quad v(\alpha_i) = 0 \text{ for } i = 1, \ldots, n,$$
$$v(\alpha_i - \alpha_j) = 0 \text{ for } i = 1, 2, j \ne i, \tag{5.5.11}$$

so that $\min(v(x - \alpha_i), v(x - \alpha_j)) = 0$ for $i = 1, 2, j \ne i$. Hence,

$$v(x - \alpha_i) = mv(y) \equiv 0 \,(\text{mod } m) \quad \text{for } i = 1, 2. \tag{5.5.12}$$

Denoting by S_i, $i = 1, 2$, the set of $v \in \mathcal{M}_K \setminus W$ such that $v(x - \alpha_i) > 0$, and applying Theorem 5.2.2 to (5.5.7), we get

$$H_K\left(\frac{x - \alpha_1}{\alpha_1 - \alpha_2}\right) \leq |W \cup S_1 \cup S_2| + 2g_{K/\mathbb{k}}. \tag{5.5.13}$$

By (5.5.12), we have $m|S_i| \leq H_K(x - \alpha_i) \leq H_K(x) + H_K(\alpha_i)$ for $i = 1, 2$. Therefore, we can deduce from (5.5.13) that

$$H_K(x) - 2H_K(\alpha_1) - H_K(\alpha_2)$$

$$\leq \tfrac{1}{m}(2H_K(x) + H_K(\alpha_1) + H_K(\alpha_2)) + |W| + 2g_{K/\mathbb{k}}.$$

Together with (5.5.10) and (5.5.8) and $m \geq 3$, this gives

$$(1 - \tfrac{2}{m})H_K(x)$$

$$\leq \tfrac{1}{n}H_K(f) + (1 + \tfrac{1}{m}) \cdot \tfrac{2}{n}H_K(f) + |S| + (6 - \tfrac{4}{n})H_K(f)) + 2g_{K/\mathbb{k}}$$

$$\leq 6H_K(f) + |S| + 2g_{K/\mathbb{k}},$$

and thus,

$$H_K(x) \leq 18H_K(f) + 3|S| + 6g_{K/\mathbb{k}}. \tag{5.5.14}$$

Now assume that f has splitting field L over K. Put $\Delta := [L : K]$, and let T be the set of valuations of L lying above the valuations in S. The inequality (5.5.14) holds with L and T instead of K and S. Inserting

$$\left.\begin{array}{l} H_L(x) = \Delta H_K(x), \;\; H_L(y) = \Delta H_K(y), \;\; H_L(f) = \Delta H_K(f), \\ |T| \leq \Delta|S|, \;\; g_{L/\mathbb{k}} \leq \Delta(g_{K/\mathbb{k}} + nH_K(f)) \;\; \text{(from Lemma 5.1.1)} \end{array}\right\} \tag{5.5.15}$$

and then dividing by Δ, inequality (5.5.5) follows.

Writing $f(X) = a_0 X^n + a_1 X^{n-1} + \cdots + a_n$, we get for $v \in \mathcal{M}_K$,

$$mv(y) = v(f(x)) \geq \min_{0 \leq i \leq n}(v(a_i) + iv(x)))$$

$$\geq \min(0, v(f)) + n\min(0, v(x)),$$

and thus,

$$mH_K(y) \leq H_K(f) + nH_K(x), \tag{5.5.16}$$

which combined with (5.5.5) gives (5.5.6). This completes the proof of our theorem. □

Proof of Theorem 5.5.1 Let $(x$ and $y)$ be a solution of (5.5.1) in $\mathcal{O}_S$. We start again with assuming that f has splitting field K, so that $f(X) =$

$a_0(X - \alpha_1) \cdots (X - \alpha_n)$ with $\alpha_1, \ldots, \alpha_n$ distinct elements of K. We assume again that $H_K(\alpha_1) \leq \cdots \leq H_K(\alpha_n)$, so that we have again (5.5.8). Let, again,

$$\beta_i = \prod_{j \neq i} (\alpha_i - \alpha_j) \ (i = 1, \ldots, n).$$

We now take for W the set of valuations v of K satisfying at least one of the conditions

$$v \in S, \quad v(f) < 0, \quad v(a_0) > 0, \quad v(\beta_1 \beta_2 \beta_3) > 0. \tag{5.5.17}$$

Then by a similar computation as in (5.5.10),

$$|W| \leq |S| + (8 - \tfrac{6}{n}) H_K(f). \tag{5.5.18}$$

Moreover, similar to (5.5.11), we have for $v \in \mathcal{M}_K \backslash W$,

$$\begin{aligned} & v(a_0) = 0, \quad v(f) = 0, \quad v(\alpha_i) = 0 \text{ for } i = 1, \ldots, n, \\ & v(\alpha_i - \alpha_j) = 0 \text{ for } i = 1, 2, 3, \ j \neq i, \end{aligned} \tag{5.5.19}$$

and similar to (5.5.12),

$$v(x - \alpha_i) \equiv 0 \,(\mathrm{mod}\, 2) \ \text{ for } v \in \mathcal{M}_K \backslash W, \ i = 1, 2, 3. \tag{5.5.20}$$

Consider the field M generated over K by the square roots of $x - \alpha_1, x - \alpha_2$, and $x - \alpha_3$. Let $\delta := [M : K]$. We first compute an upper bound for the genus of M, to be used in later arguments. By the Riemann–Hurwitz formula (see (5.1.5)), we have

$$2g_{M/\Bbbk} - 2 = \delta(2g_{K/\Bbbk} - 2) + \sum_{v \in \mathcal{M}_K} \sum_{w | v} (e_{w | v} - 1).$$

Let $v \in \mathcal{M}_K \backslash W$ and choose a local parameter z_v of v. By (5.5.20), the square roots of $x - \alpha_i$ ($i = 1, 2, 3$) can be expressed as Laurent series in z_v, so for the embeddings of M in $\overline{K_v}$ as described in (5.1.1), all e_i are equal to 1. In other words, we have $e_{w | v} = 1$ for all $v \in \mathcal{M}_K \backslash W$ and all valuations w of M lying above v. We lastly observe that by (5.1.2), we have $\sum_{w | v} e_{w | v} = \delta$ for $v \in W$. By inserting this into the Riemann–Hurwitz formula, we get

$$2g_{M/\Bbbk} - 2 \leq \delta(2g_{K/\Bbbk} - 2 + |W|). \tag{5.5.21}$$

Let $\mu := (\alpha_2 - \alpha_1)(\alpha_3 - \alpha_2)(\alpha_1 - \alpha_3)$, and let V be the set of valuations of K satisfying at least one of the conditions

$$v \in S, \quad v(a_0) > 0, \quad v(f) < 0, \quad v(\mu) > 0.$$

By (5.5.8), we have

$$\begin{aligned} |V| & \leq |S| + 2H_K(f) + 2(H_K(\alpha_1) + H_K(\alpha_2) + H_K(\alpha_3)) \\ & \leq |S| + (2 + \tfrac{6}{n}) H_K(f). \end{aligned} \tag{5.5.22}$$

For $v \in \mathcal{M}_K \backslash V$, we have, similar to (5.5.19),

$$v(a_0) = 0, \quad v(f) = 0, \quad v(\alpha_i) = 0 \text{ for } i = 1, \ldots, n,$$
$$v(\alpha_i - \alpha_j) = 0 \text{ for } i = 1, 2, 3. \tag{5.5.23}$$

Let U be the set of valuations of M lying above those in V, and denote by $\mathcal{O}_U$ the ring of U-integers, and by $\mathcal{O}_U^*$ the ring of U-units in M. Choose $\xi_1, \xi_2, \xi_3 \in M$ such that

$$\xi_i^2 = x - \alpha_i, \quad i = 1, 2, 3.$$

Then by (5.5.23), we have $\xi_i \in \mathcal{O}_U$ for $i = 1, 2, 3$. Further, let $\gamma_i, \widehat{\gamma_i}$ ($i = 1, 2, 3$) be given by

$$\gamma_1 = \xi_2 - \xi_3, \quad \widehat{\gamma_1} = \xi_2 + \xi_3,$$
$$\gamma_2 = \xi_3 - \xi_1, \quad \widehat{\gamma_2} = \xi_3 + \xi_1,$$
$$\gamma_3 = \xi_1 - \xi_2, \quad \widehat{\gamma_3} = \xi_1 + \xi_2.$$

Then

$$\gamma_1 \widehat{\gamma_1} = \alpha_3 - \alpha_2, \quad \gamma_2 \widehat{\gamma_2} = \alpha_1 - \alpha_3, \quad \gamma_3 \widehat{\gamma_3} = \alpha_2 - \alpha_1,$$

which together with (5.5.23), implies that

$$\gamma_i, \widehat{\gamma_i} \in \mathcal{O}_U^* \text{ for } i = 1, 2, 3.$$

By applying Theorem 5.2.2 to the relations

$$\gamma_1 + \gamma_2 + \gamma_3 = 0, \quad \gamma_1 + \widehat{\gamma_2} - \widehat{\gamma_3} = 0,$$
$$\widehat{\gamma_1} - \gamma_2 - \widehat{\gamma_3} = 0, \quad \widehat{\gamma_1} - \widehat{\gamma_2} + \gamma_3 = 0,$$

and inserting $|U| \leq \delta|V|$ (5.5.21), (5.5.18), and (5.5.22), we infer that the quantities

$$H_M(\gamma_2/\gamma_3), \quad H_M(\widehat{\gamma_2}/\gamma_3), \quad H_M(\gamma_2/\widehat{\gamma_3}), \text{ and } H_M(\widehat{\gamma_2}/\widehat{\gamma_3})$$

are all bounded above by

$$|U| + 2g_{M/k} \leq \delta(|V| + |W| + 2g_{K/k})$$
$$\leq \delta\left(10H_K(f) + 2|S| + 2g_{K/k}\right) =: \delta N.$$

Since $x - \alpha_1 = \xi_1^2 = \frac{1}{4}(\widehat{\gamma_2} - \gamma_2)^2$ and $x - \alpha_3 = \xi_3^2 = \frac{1}{4}(\widehat{\gamma_2} + \gamma_2)^2$, it follows that

$$\frac{2x - \alpha_1 - \alpha_3}{\alpha_2 - \alpha_1} = \frac{1}{2}\left((\widehat{\gamma_2}/\gamma_3)(\widehat{\gamma_2}/\widehat{\gamma_3}) + (\gamma_2/\gamma_3)(\gamma_2/\widehat{\gamma_3})\right),$$

whence

$$H_K\left(\frac{2x - \alpha_1 - \alpha_3}{\alpha_2 - \alpha_1}\right) = \delta^{-1} H_M\left(\frac{2x - \alpha_1 - \alpha_3}{\alpha_2 - \alpha_1}\right) \leq 4N.$$

Together with (5.5.8), this implies

$$H_K(x) \le 4N + 2H_K(\alpha_1) + H_K(\alpha_2) + H_K(\alpha_3) \le 4N + \tfrac{4}{n}H_K(f),$$

which in view of $n \ge 3$ leads to

$$H_K(x) \le 42H_K(f) + 8|S| + 8g_{K/k}. \tag{5.5.24}$$

Now consider the general case that f has arbitrary splitting field L over K, and let $\Delta := [L : K]$. Denote by T the set of valuations of L lying above those in S. Then (5.5.24) holds with L and T instead of K and S. By inserting (5.5.15) and dividing by Δ, we obtain (5.5.3). Together with (5.5.16) with $m = 2$, this implies (5.5.4). This completes our proof.   □

6

Tools from Effective Commutative Algebra

In this chapter, we have collected some algorithmic results for fields finitely generated over $\mathbb{Q}$ and for integral domains of characteristic 0 finitely generated over $\mathbb{Z}$. Our main references are Seidenberg (1974) and Aschenbrenner (2004).

By saying that given any input from a specified set we can determine effectively an output, we mean that there exists an algorithm (i.e., a deterministic Turing machine) which, for any choice of input from the specified set, computes the output in a finite number of steps. We say that an object is *given effectively* if it is given in such a form that it can serve as an input for an algorithm.

We agree once more that uppercase characters such as X, Y denote variables, whereas lowercase characters denote elements of rings or fields. Given a ring R, we denote by $R^{m,n}$ the R-module of $m \times n$-matrices, with elements in R, and by R^n the R-module of n-dimensional column vectors, with coordinates in R.

From matrices A, B with the same number of rows, we form a matrix $[A, B]$ by placing the columns of B after those of A. Likewise, from two matrices A, B with the same number of columns, we form $\left[\begin{smallmatrix} A \\ B \end{smallmatrix} \right]$ by placing the rows of B below those of A.

The logarithmic height $h(\mathcal{A})$ of a finite set $\mathcal{A} = \{a_1, \ldots, a_t\} \subset \mathbb{Z}$ is defined by $h(\mathcal{A}) := \log \max(|a_1|, \ldots, |a_t|)$. The logarithmic height $h(U)$ of a matrix U with entries in $\mathbb{Z}$ is defined by the logarithmic height of the set of entries of U. The logarithmic height $h(P)$ of a polynomial P with coefficients in $\mathbb{Z}$ is the logarithmic height of the set of coefficients of P. By the degree of a polynomial, we always mean its total degree, and the total degree of a polynomial P is denoted by $\deg P$.

As in Section 4.1, we write

$$\log^* u := \max(1, \log u), \text{ for } u > 0, \ \log^* 0 := 1.$$

We use notation $O(\cdot)$ as an abbreviation for $c \times$ the expression between the parentheses, where c is an effectively computable positive absolute constant. At each occurrence of $O(\cdot)$, the value of c may be different.

6.1 Effective Linear Algebra over Polynomial Rings

We have taken some material from Evertse and Győry (2013, 2015) on effective results for systems of linear equations over polynomial rings over a field or over $\mathbb{Z}$, with some small improvements here and there. For the convenience of the reader, we have repeated some details.

Lemma 6.1.1 *Let $U \in \mathbb{Z}^{m,n}$ and $\mathbf{b} \in \mathbb{Z}^m$.*

(i) The $\mathbb{Z}$-module of $\mathbf{y} \in \mathbb{Z}^n$, with $U\mathbf{y} = \mathbf{0}$ is generated by vectors in $\mathbb{Z}^n$ of logarithmic height at most $mh(U) + \frac{1}{2}m \log m$.

(ii) Assume that $U\mathbf{y} = \mathbf{b}$ is solvable in $\mathbb{Z}^n$. Then it has a solution $\mathbf{y} \in \mathbb{Z}^n$, with $h(\mathbf{y}) \leq mh([U, \mathbf{b}]) + \frac{1}{2}m \log m$.

Proof (i) We follow Aschenbrenner (2004, Lemma 4.2 and Section 5). Let $\mathcal{M}$ be the $\mathbb{Z}$-module of $\mathbf{y} \in \mathbb{Z}^n$, with $U\mathbf{y} = \mathbf{0}$. We may assume without loss of generality that $m \leq n$, and U has rank m, so that U has nonsingular submatrices of order m. Let $U_1, \ldots, U_k$ be the nonsingular submatrices of U of order m and put $\delta_j := \det U_j$, for $j = 1, \ldots, k$, and $\delta := \gcd(\delta_1, \ldots, \delta_k)$.

We first prove that for $j = 1, \ldots, k$, there are $\mathbf{y}_{1,j}, \ldots, \mathbf{y}_{n-m,j} \in \mathcal{M}$ such that

$$\text{for every } \mathbf{y} \in \mathcal{M}, \text{ there are } b_{i,j} \in \mathbb{Z} \text{ with } \mathbf{y} = (\delta/\delta_j) \sum_{i=1}^{n-m} b_{i,j}\mathbf{y}_{i,j}, \qquad (6.1.1)$$

$$h(\mathbf{y}_{i,j}) \leq mh(U) + \tfrac{1}{2}m \log m \text{ for } i = 1, \ldots, n - m. \qquad (6.1.2)$$

It suffices to prove this for $j = 1$. After permuting the columns of U, we may assume that U_1 consists of the first m columns of U. Let V_1 consist of the last $n - m$ columns of U. For $\mathbf{y} \in \mathcal{M}$, let $\mathbf{y}^{(1)}$ consist of the first m coordinates of $\mathbf{y}$, and $\mathbf{y}^{(2)}$ consists of the last $n - m$ coordinates of $\mathbf{y}$. Then $U_1\mathbf{y}^{(1)} = -V_1\mathbf{y}^{(2)}$ or equivalently

$$\mathbf{y}^{(1)} = -U_1^{-1}V_1\mathbf{y}^{(2)}.$$

For $i = 1, \ldots, n - m$, let $\mathbf{y}_{i,1}$ be the solution $\mathbf{y}$ of $U\mathbf{y} = \mathbf{0}$ for which $\mathbf{y}^{(2)} = (\delta_1/\delta)\mathbf{e}_i$, where $\mathbf{e}_i$ is the ith standard basis vector of $\mathbb{Z}^{n-m}$. The coordinates of $\mathbf{y}_{i,1}$ are all of the shape $\pm(\delta_1/\delta) \det W/ \det U_1 = \pm\delta^{-1} \det W$, where W is the determinant of some submatrices of U of order m. Hence, $\mathbf{y}_{i,1} \in \mathbb{Z}^n$, implying $\mathbf{y}_{i,1} \in \mathcal{M}$ for $i = 1, \ldots, n - m$. Further, by Hadamard's inequality, we have (6.1.2). If $\mathbf{y} = (b_1, \ldots, b_n)^T \in \mathcal{M}$, then $\mathbf{y} = (\delta/\delta_1^{-1}) \sum_{i=m+1}^n b_i\mathbf{y}_{i-m,1}$. This proves (6.1.1).

There are integers $a_1, \ldots, a_k$ such that $a_1\delta_1 + \cdots + a_k\delta_k = \delta$. Applying (6.1.2), we see that for $\mathbf{y} \in \mathcal{M}$, we have

$$\mathbf{y} = \sum_{j=1}^{k} a_j \Big(\sum_{i=1}^{n-m} b_{i,j} \mathbf{y}_{i,j} \Big).$$

This implies that the $\mathbf{y}_{i,j}$ generate $\mathcal{M}$.

(ii) Assume without loss of generality that U and $[U, \mathbf{b}]$ have rank m. By a result of Borosh et al. (1989), $U\mathbf{y} = \mathbf{b}$ has a solution $\mathbf{y} \in \mathbb{Z}^n$ such that the absolute values of the entries of $\mathbf{y}$ are bounded above by the maximum of the absolute values of the $m \times m$-subdeterminants of $[U, \mathbf{b}]$. The upper bound for $h(\mathbf{y})$ as in the lemma easily follows from Hadamard's inequality.   □

Theorem 6.1.2 *Let F be a field, $r \geq 1$, and $R := F[X_1, \ldots, X_r]$. Further, let V be an $m \times n$-matrix and $\mathbf{b}$ an m-dimensional column vector, both consisting of polynomials from R of degree $\leq d$ where $d \geq 1$.*

(i) The R-module of $\mathbf{x} \in R^n$ with $V\mathbf{x} = \mathbf{0}$ is generated by vectors $\mathbf{x}$ whose coordinates are polynomials of degree at most $(2md)^{2^r}$.

(ii) Suppose that $V\mathbf{x} = \mathbf{b}$ is solvable in $\mathbf{x} \in R^n$. Then it has a solution $\mathbf{x}$ whose coordinates are polynomials of degree at most $(2md)^{2^r}$.

Proof See Aschenbrenner (2004, Theorems 3.2 and 3.4). Results of this type were obtained earlier, but not with a completely correct proof, by Hermann (1926) and Seidenberg (1974).   □

Part (ii) of Theorem 6.1.2 gives an effective method to decide ideal membership in $F[X_1, \ldots, X_r]$, provided that the field F is given effectively (a notion that we are not going to formalize):

Corollary 6.1.3 *Given $b, f_1, \ldots, f_M \in F[X_1, \ldots, X_r]$, it can be decided effectively whether b belongs to the ideal $\mathcal{I} := (f_1, \ldots, f_M)$ of $F[X_1, \ldots, X_r]$.*

Proof Let $d := \max(\deg b, \deg f_1, \ldots, \deg f_M)$. If $b \in \mathcal{I}$, then there are $x_1, \ldots, x_M \in F[X_1, \ldots, X_r]$ of degree at most $(2d)^{2^r}$ such that $b = x_1 f_1 + \cdots + x_M f_M$. By comparing the coefficients of the polynomials on the left- and right-hand side, we get an inhomogeneous system of linear equations over F whose solvability can be checked by standard linear algebra.   □

Corollary 6.1.4 *Let $R := \mathbb{Q}[X_1, \ldots, X_r]$. Further, let V be an $m \times n$-matrix consisting of polynomials in $\mathbb{Z}[X_1, \ldots, X_r]$ of degrees at most d and logarithmic heights at most h where $d \geq 1$, $h \geq 1$. Then the R-module of $\mathbf{x} \in R^n$ with $V\mathbf{x} = \mathbf{0}$ is generated by vectors $\mathbf{x}$, consisting of polynomials in $\mathbb{Z}[X_1, \ldots, X_r]$ of degree at most $(2md)^{2^r}$ and logarithmic height at most $(2md)^{6^r} h$.*

Proof By Theorem 6.1.2 (i), we have to study $V\mathbf{x} = \mathbf{0}$, restricted to vectors $\mathbf{x} \in R^n$ consisting of polynomials in R of degree at most $(2d)^{2^r}$. Let $\mathbf{y}$

be the tuple of coefficients of the polynomials in $\mathbf{x}$. Then $\mathbf{y} \in \mathbb{Q}^{n^*}$, where $n^* \leq n(2md)^{r \cdot 2^r}$. Further, $V\mathbf{x}$ consists of m polynomials in $\mathbb{Q}[X_1, \ldots, X_r]$ of degree at most $d + (2md)^{2^r}$ all whose coefficients have to be set to 0. The total number of coefficients of $V\mathbf{x}$ is $m^* \leq m(d + (2md)^{2^r})^r$. Thus, the system of equations $V\mathbf{x} = 0$ in polynomials in R of degree at most $(2md)^{2^r}$ reduces to a system of equations $U\mathbf{y} = \mathbf{0}$ in $\mathbf{y} \in \mathbb{Q}^{n^*}$, where $U \in \mathbb{Z}^{m^*, n^*}$. By Lemma 6.1.1 (i), the solution space of this system is generated by vectors $\mathbf{y}$ in $\mathbb{Z}^{n^*}$ of logarithmic height at most $\frac{1}{2}m^* \log m^* + m^* h(U) \leq (2md)^{6^r} h =: T$. Hence, the corresponding vectors $\mathbf{x}$ consist of polynomials in $\mathbb{Z}[X_1, \ldots, X_r]$ of logarithmic height at most T.   □

Theorem 6.1.5 *Let* $r \geq 1$, *and let* V *be an* $m \times n$-*matrix and* $\mathbf{b}$ *a nonzero* m-*dimensional column vector consisting of polynomials in* $\mathbb{Z}[X_1, \ldots, X_r]$ *of degree at most* d *and logarithmic height at most* h, *where* $d \geq 1$, $h \geq 1$.
(i) The solution set of $\mathbf{x} \in \mathbb{Z}[X_1, \ldots, X_r]^n$ *with* $V\mathbf{x} = \mathbf{0}$ *is generated by vectors* $\mathbf{x} = (x_1, \ldots, x_n) \in \mathbb{Z}[X_1, \ldots, X_r]^n$ *with*

$$\deg x_i \leq (2md)^{\exp((2r)^r)}, \quad h(x_i) \leq (2md)^{\exp((6r)^r)} h \ \text{for} \ i = 1, \ldots, n.$$

(ii) Assume that

$$V\mathbf{x} = \mathbf{b} \tag{6.1.3}$$

is solvable in $\mathbf{x} \in \mathbb{Z}[X_1, \ldots, X_r]^n$. *Then* (6.1.3) *has a solution* $\mathbf{x} = (x_1, \ldots, x_n) \in \mathbb{Z}[X_1, \ldots, X_r]^n$ *with*

$$\left. \begin{aligned} \deg x_i &\leq d_0 := (2md)^{\exp O(r \log^* r)} h, \\ h(x_i) &\leq h_0 := (2md)^{\exp O(r \log^* r)} h^{r+1} \end{aligned} \right\} \text{for} \ i = 1, \ldots, n. \tag{6.1.4}$$

Proof (i) This follows from Aschenbrenner (2004, Theorem 4.1) except for the height bound. The height bound can be derived from Lemma 6.1.1 (i), with similar computations as in the proof of Corollary 6.1.4.

(ii) This follows from Aschenbrenner (2004, Theorem 6.5) except for the height bound. To derive such a height bound, let us restrict to solutions $\mathbf{x} = (x_1, \ldots, x_n)$ of (6.1.3) with $\deg x_i \leq d_0$ for $i = 1, \ldots, n$, and denote by $\mathbf{y}$ the vector of coefficients of the polynomials $x_1, \ldots, x_n$. Then (6.1.3) translates into a system of linear equations $U\mathbf{y} = \mathbf{b}^*$, which is solvable over $\mathbb{Z}$. Here, the number m^* of equations, that is, the number of rows of U, is $\leq (d_0 + d)^r$. Further, $h(U, \mathbf{b}^*) \leq h$. By Lemma 6.1.1 (ii), $U\mathbf{y} = \mathbf{b}^*$ has a solution $\mathbf{y}$ with coordinates in $\mathbb{Z}$ of logarithmic height at most

$$m^* h + \frac{1}{2} m^* \log m^* \leq (2d)^{\exp O(r \log^* r)} h^{r+1} =: h_0.$$

It follows that (6.1.3) has a solution $\mathbf{x} \in \mathbb{Z}[X_1, \ldots, X_r]^n$ satisfying (6.1.4).   □

Aschenbrenner (2004) gives an example, which shows that the upper bound for the degrees of the x_i cannot depend on d and r only.

Part (ii) of Theorem 6.1.5 gives an effective criterion for ideal membership in $\mathbb{Z}[X_1,\ldots,X_r]$:

Corollary 6.1.6 *Given $b, f_1,\ldots,f_M \in \mathbb{Z}[X_1,\ldots,X_r]$, it can be decided effectively whether b belongs to the ideal $\mathcal{I} := (f_1,\ldots,f_M)$ of $\mathbb{Z}[X_1,\ldots,X_r]$.*

Proof By Theorem 6.1.5, if $b \in \mathcal{I}$, then there are $x_1,\ldots,x_M \in \mathbb{Z}[X_1,\ldots,X_r]$ with upper bounds for the degrees and heights as in (6.1.4) with $m = 1, n = M$, such that $b = \sum_{i=1}^{M} x_i f_i$. It requires only a finite computation to check whether such x_i exist. $\square$

Theorem 6.1.7 *Let $f_1,\ldots,f_M$ be polynomials in $\mathbb{Z}[X_1,\ldots,X_r]$ of total degrees at most d and logarithmic heights at most h. Let $\overline{\mathcal{I}}$ be the ideal of $\mathbb{Q}[X_1,\ldots,X_r]$ generated by $f_1,\ldots,f_M$. Then $\mathcal{I} := \overline{\mathcal{I}} \cap \mathbb{Z}[X_1,\ldots,X_r]$ is an ideal generated by polynomials of total degrees at most $d + (2d)^{(2r)^r}$ and logarithmic heights at most $(6r)^r \log(2d) + h$.*

Proof The upper bound for the degrees follows from Aschenbrenner (2004, Theorem 4.7). But in his proof, he uses Corollary 3.5 of his paper, some details of the proof of which he has left to the reader and which were not fully obvious to us. So we provide an argument avoiding Aschenbrenner's Corollary 3.5. Observe that $\mathcal{I}$ is the union of the ideals $\mathcal{I}_a$ ($a \in \mathbb{Z}$), where $\mathcal{I}_a$ consists of the polynomials $x \in \mathbb{Z}[X_1,\ldots,X_r]$ such that ax is in the ideal of $\mathbb{Z}[X_1,\ldots,X_r]$ generated by $f_1,\ldots,f_M$. We can find these x by solving the equation

$$x_1 f_1 + \cdots + x_M f_M - ax = 0 \quad \text{in} \ (x, x_1, \ldots, x_M) \in \mathbb{Z}[X_1,\ldots,X_M]^{M+1}.$$

By Theorem 6.1.5, the solutions $(x, x_1, \ldots, x_M)$ of this equation form a module over $\mathbb{Z}[X_1,\ldots,X_r]$, generated by tuples of polynomials of total degree at most $C := (2d)^{(2r)^r}$. Hence, for every positive integer a, $\mathcal{I}_a$ is generated by polynomials $g_1 f_1 + \cdots + g_M f_M$, where $g_1,\ldots,g_M \in a^{-1}\mathbb{Z}[X_1,\ldots,X_r]$ and $\deg g_i \le C$ for $i = 1,\ldots,M$. It follows that $\mathcal{I} = \cup_a \mathcal{I}_a$ is generated by polynomials $g_1 f_1 + \cdots + g_M f_M$, where

$$\begin{cases} g_1,\ldots,g_M \in \mathbb{Q}[X_1,\ldots,X_r], \\ g_1 f_1 + \cdots + g_M f_M \in \mathbb{Z}[X_1,\ldots,X_r], \\ \deg g_1,\ldots,\deg g_M \le C. \end{cases} \tag{6.1.5}$$

The $\mathbb{Q}$-vector space $\mathcal{V}$ of $g_1 f_1 + \cdots + g_M f_M$ with $g_1,\ldots,g_M$ satisfying (6.1.5) but without the condition $g_1 f_1 + \cdots + g_M f_M \in \mathbb{Z}[X_1,\ldots,X_r]$ is contained in the vector space of polynomials in $\mathbb{Q}[X_1,\ldots,X_r]$ of degree $\le C + d$, therefore, $\mathcal{V}$ has dimension $N \le \binom{C+d+r}{r}$. Further, $\mathcal{V}$ is generated by the polynomials

$X_1^{j_1} \cdots X_r^{j_r} f_i$ ($i = 1, \ldots, M$, $j_1 + \cdots + j_r \leq C$); hence, we can select a basis $b_1, \ldots, b_N$ of $\mathcal{V}$ from this set. Notice that $b_1, \ldots, b_N$ belong to $\mathbb{Z}[X_1, \ldots, X_r]$ and have logarithmic heights $\leq h$. By Cassels (1959, Chap. V, Lemma 8), the $\mathbb{Z}$-module of polynomials $g_1 f_1 + \cdots + g_M f_M$ with (6.1.5) has a basis $c_1, \ldots, c_N$ with $c_i = \sum_{j=1}^{i} \xi_{i,j} b_j$ for $j = 1, \ldots, N$, where $\xi_{i,j} \in \mathbb{Q}$ and $|\xi_{i,j}| \leq 1$ for all i and j. These polynomials $c_1, \ldots, c_N$ generate $\mathcal{I}$, and they have total degrees at most $C + d$ and logarithmic heights at most

$$h + \log N \leq h + \log \binom{C+d+r}{r} \leq h + (6r)^r \log 2d.$$

This proves our theorem. $\qquad\qquad\qquad\qquad\qquad\qquad\qquad\qquad\qquad\square$

6.2 Finitely Generated Fields over $\mathbb{Q}$

To a field $K = \mathbb{Q}(z_1, \ldots, z_r)$ that is finitely generated over $\mathbb{Q}$, we may associate the polynomial ideal

$$\mathcal{I} := \{f \in \mathbb{Q}[X_1, \ldots, X_r] : f(z_1, \ldots, z_r) = 0\}.$$

By Hilbert's Basis Theorem, the ideal $\mathcal{I}$ is finitely generated, that is, $\mathcal{I} = (f_1, \ldots, f_M)$ with $f_1, \ldots, f_M \in \mathbb{Q}[X_1, \ldots, X_r]$. Thus, K is isomorphic to the quotient field of

$$\mathbb{Q}[X_1, \ldots, X_r]/(f_1, \ldots, f_M), \qquad\qquad (6.2.1)$$

and $z_1, \ldots, z_r$ may be identified with the residue classes of $X_1, \ldots, X_r$ modulo $(f_1, \ldots, f_M)$. We call $(f_1, \ldots, f_M)$ an *ideal representation* for K. We say that K is *given effectively* if an ideal representation for it is given.

Notice that for polynomials $f_1, \ldots, f_M \in \mathbb{Q}[X_1, \ldots, X_r]$ to form an ideal representation of a field, it is necessary and sufficient that $(f_1, \ldots, f_M)$ be a prime ideal of $\mathbb{Q}[X_1, \ldots, X_r]$. This can be verified effectively; see, for instance, Seidenberg (1974, Section 46, p. 293) (there, in fact, Seidenberg gives a method to determine the prime ideals associated to a given ideal $\mathcal{I}$, which certainly enables one to decide whether $\mathcal{I}$ is a prime ideal).

Let $K = \mathbb{Q}(z_1, \ldots, z_r)$ be a field with given ideal representation $\mathcal{I} = (f_1, \ldots, f_M)$. We say that $y \in K$ is *given/can be computed* (in terms of $z_1, \ldots, z_r$), if polynomials g and $h \in \mathbb{Q}[X_1, \ldots, X_r]$ are given/can be computed such that $y = g(z_1, \ldots, z_r)/h(z_1, \ldots, z_r)$. Thanks to Theorem 6.1.2, we can verify whether an expression $g(z_1, \ldots, z_r)/h(z_1, \ldots, z_r)$ is well-defined (i.e., $h(z_1, \ldots, z_r) \neq 0$ or equivalently, $h \notin \mathcal{I}$) and whether two expressions $g_i(z_1, \ldots, z_r)/h_i(z_1, \ldots, z_r)$ ($i = 1, 2$) are equal (i.e., $g_1 h_2 - g_2 h_1 \in \mathcal{I}$).

We note that if $y_1, \ldots, y_m$ are given in terms of $z_1, \ldots, z_r$, then for any given polynomial $h \in \mathbb{Q}[Y_1, \ldots, Y_m]$, it can be decided whether $h(y_1, \ldots, y_m) \neq 0$. Moreover, for any two given g and $h \in \mathbb{Q}[Y_1, \ldots, Y_m]$ with $h(y_1, \ldots, y_m) \neq 0$, one can compute $g(y_1, \ldots, y_m)/h(y_1, \ldots, y_m)$ in terms of $z_1, \ldots, z_r$.

Finally, if $y_1, \ldots, y_m$ are elements of K given in terms of $z_1, \ldots, z_r$, then we say that y *is given/can be computed* in terms of $y_1, \ldots, y_m$, if g and $h \in \mathbb{Q}[Y_1, \ldots, Y_m]$ are given/can be computed, such that $h(y_1, \ldots, y_m) \neq 0$ and $y = g(y_1, \ldots, y_m)/h(y_1, \ldots, y_m)$.

Theorem 6.2.1 *(i) For any $r \geq 1$ and any effectively given field $K = \mathbb{Q}(z_1, \ldots, z_r)$, we can:*
(i) determine a permutation $x_1, \ldots, x_q$ and $y_1, \ldots, y_t$ of $z_1, \ldots, z_r$ such that $x_1, \ldots, x_q$ are algebraically independent of $\mathbb{Q}$, and $y_1, \ldots, y_t$ are algebraic over $\mathbb{Q}(x_1, \ldots, x_q)$;
(ii) determine the monic minimal polynomial of y_1 over $\mathbb{Q}(x_1, \ldots, x_q)$ with coefficients given in terms of $x_1, \ldots, x_q$, and for $i = 2, \ldots, t$, determine the monic minimal polynomial of y_i over $\mathbb{Q}(x_1, \ldots, x_q, y_1, \ldots, y_{i-1})$ with coefficients given in terms of $x_1, \ldots, x_q$ and $y_1, \ldots, y_{i-1}$.

Proof Repeated application of Seidenberg (1974, Section 23 on p. 284 and Section 25 on p. 285). □

Theorem 6.2.2 *For any effectively given field $K = \mathbb{Q}(z_1, \ldots, z_r)$ and any $y_1, \ldots, y_s$ and $y \in K$ given in terms of $z_1, \ldots, z_r$, we can:*
(i) determine a finite set of generators for the ideal

$$\{f \in \mathbb{Q}[X_1, \ldots, X_s] : f(y_1, \ldots, y_s) = 0\};$$

(ii) decide whether $y \in \mathbb{Q}(y_1, \ldots, y_s)$ and, if so, determine g and $h \in \mathbb{Z}[X_1, \ldots, X_s]$ such that $y = g(y_1, \ldots, y_s)/h(y_1, \ldots, y_s)$.

Proof (i) Seidenberg (1974, Section 27, p. 287).

(ii) By (i), one can compute a finite set of generators for the ideal of $f \in \mathbb{Q}[X_1, \ldots, X_{s+1}]$ such that $f(y_1, \ldots, y_s, y) = 0$. Using Theorem 6.2.1, one can decide whether y is algebraic over $\mathbb{Q}(y_1, \ldots, y_s)$, and if so, compute its monic minimal polynomial over $\mathbb{Q}(y_1, \ldots, y_s)$, and check if it has degree 1. □

Theorem 6.2.3 *For any effectively given field $K = \mathbb{Q}(z_1, \ldots, z_r)$ and any polynomial $\mathcal{F} \in K[X_1, \ldots, X_m]$ with coefficients given in terms of $z_1, \ldots, z_r$, we can determine a factorization of $\mathcal{F}$ into irreducible polynomials of $K[X_1, \ldots, X_m]$, whose coefficients are all given in terms of $z_1, \ldots, z_r$. In particular, we can decide whether $\mathcal{F}$ is irreducible.*

Proof This follows from Seidenberg (1974, Sections 33–35, p. 289). For $m = 1$, a more precise quantitative version can be deduced from Proposition 8.2.3. □

Theorem 6.2.4 *For any effectively given field* $K = \mathbb{Q}(z_1, \ldots, z_r)$ *and any monic irreducible polynomial* $\mathcal{F} \in K[X]$ *with coefficients given in terms of* $z_1, \ldots, z_r$, *we can:*
(i) determine a finite set of generators for the ideal

$$\{f \in \mathbb{Q}[X_1, \ldots, X_r, Y] : f(z_1, \ldots, z_r, y) = 0\}$$

where y *is a root of* $\mathcal{F}$;
(ii) for any $a \in K(y)$ *given in terms of* $z_1, \ldots, z_r, y$, *determine* $b_0, \ldots, b_{\mathcal{F}-1} \in K$, *given in terms of* $z_1, \ldots, z_r$, *such that* $a = \sum_{i=0}^{\deg \mathcal{F}-1} b_i y^i$.

Proof Put $L := K(y)$, $d := [L : K]$. Let $(f_1, \ldots, f_M)$ be an ideal representation for K. We may express $\mathcal{F}$ as $X^d + (a_1/b)X^{d-1} + \cdots + (a_d/b)$ where $a_1, \ldots, a_d, b$ are given as polynomials with integer coefficients in $z_1, \ldots, z_r$.

Let $y' := by$. Then $K(y') = L$, and y' has minimal polynomial $X^d + a_1 X^{d-1} + \cdots + b^{d-1}a_d$ over K. We can write $b^{i-1}a_i = h_i(z_1, \ldots, z_r)$ with $h_i \in \mathbb{Z}[X_1, \ldots, X_r]$ for $i = 1, \ldots, d$. Then the ideal of polynomials $Q \in \mathbb{Q}[X_1, \ldots, X_r, Y]$ with $Q(z_1, \ldots, z_r, y') = 0$ is generated by $f_1, \ldots, f_M$ and $Y^d + \sum_{i=1}^d h_i Y^{d-i}$, and so these polynomials provide an ideal representation for L. Using Theorem 6.2.2, we can compute a finite set of generators for the ideal of $f \in \mathbb{Q}[X_1, \ldots, X_r, Y]$ with $f(z_1, \ldots, z_r, y) = 0$.

Using division by $\mathcal{F}$ with remainder, from an expression of $a \in L$ in terms of $z_1, \ldots, z_r$ and y, we can compute an expression $\sum_{i=0}^{d-1} b_i y^i$, with $b_i \in K$ given in terms of $z_1, \ldots, z_r$. $\qquad\square$

Corollary 6.2.5 *For any effectively given field* $K = \mathbb{Q}(z_1, \ldots, z_r)$ *and any polynomial* $\mathcal{F} \in K[X]$ *with coefficients given in terms of* $z_1, \ldots, z_r$, *we can determine effectively an ideal representation for the splitting field of* $\mathcal{F}$ *over* K.

Proof We proceed by induction on $n := \deg \mathcal{F}$. For $n = 1$, our assertion is clear. Let $n \geq 2$. By Theorem 6.2.3, we can compute an irreducible factor $\mathcal{F}_1 \in K[X]$ of $\mathcal{F}$ in terms of $z_1, \ldots, z_r$ and then adjoin a zero y_1 of $\mathcal{F}_1$ to K. By Theorem 6.2.4, we can compute an ideal representation for $K_1 := K(y_1)$, and then by the induction hypothesis an ideal representation for the splitting field of $\mathcal{F}(X)/(X - y_1)$ over K_1. This is then the splitting field of $\mathcal{F}$ over K. $\qquad\square$

Corollary 6.2.6 *For any effectively given ideal representations for* $K = \mathbb{Q}(z_1, \ldots, z_r)$ *and a finite extension* $L = \mathbb{Q}(z_1, \ldots, z_r, y_1, \ldots, y_n)$ *of* K, *we can:*
(i) determine effectively an element y *of* L *in terms of* $z_1, \ldots, z_r$ *and* $y_1, \ldots, y_n$ *such that* $L = K(y)$, *together with the monic minimal polynomial of* y *over* K, *with coefficients given in terms of* $z_1, \ldots, z_r$;
(ii) for any $a \in L$ *given in terms of* $z_1, \ldots, z_r$ *and* $y_1, \ldots, y_n$, *determine effectively* $b_0, \ldots, b_{[L:K]-1} \in K$ *in terms of* $z_1, \ldots, z_r$ *such that* $a = \sum_{i=0}^{[L:K]-1} b_i y^i$.

Proof Let K be the effectively given field, put $K_0 := K$ and for $i = 1, \ldots, n$, define $K_i := K(y_1, \ldots, y_i)$, put $d_i := [K_i : K_{i-1}]$, and denote by G_i the monic minimal polynomial of y_i over K_{i-1}. The coefficients of G_i can be computed in terms of $z_1, \ldots, z_r$ and $y_1, \ldots, y_{i-1}$ by means of Theorem 6.2.4. Put $d := [L : K]$. Then

$$\{\omega_1, \ldots, \omega_d\} := \{y_1^{k_1} \cdots y_n^{k_n} : 0 \le k_j < d_j, \ j = 1, \ldots, n\}$$

is a K-basis of $L = K_n$. Given an element α of L in terms of $z_1, \ldots, z_r$ and $y_1, \ldots, y_n$, we can compute by means of Theorem 6.2.4 an expression of α as a K-linear combination of $\omega_1, \ldots, \omega_d$ with coefficients given in terms of $z_1, \ldots, z_r$. As is well-known, there are integers $c_1, \ldots, c_d$ of absolute values at most d^2, such that $y := c_1\omega_1 + \cdots + c_d\omega_d$ is a primitive element of L over K. For each choice of the c_i, we can check whether y is primitive by expressing 1 and $y, \ldots, y^{d-1}$ as K-linear combinations of $\omega_1, \ldots, \omega_d$ and check if they are linearly independent of K. Having found such an y, we can express $\omega_1, \ldots, \omega_d$, and thus every element of L, as K-linear combinations of 1 and $y, \ldots, y^{d-1}$ with coefficients given in terms of $z_1, \ldots, z_r$. In particular, we can express y^d as such a linear combination, and thus find the monic minimal polynomial of y. $\square$

Remark From Corollary 8.3.4, one can deduce quantitative versions of Corollaries 6.2.5 and 6.2.6.

6.3 Finitely Generated Integral Domains over $\mathbb{Z}$

We need some analogues of the results previously mentioned for finitely generated integral domains $\mathbb{Z}[z_1, \ldots, z_r]$ of characteristic 0. We start with recalling some basic concepts introduced in Section 2.1.

To an integral domain $A = \mathbb{Z}[z_1, \ldots, z_r]$ of characteristic 0, we may associate the polynomial ideal

$$\mathcal{I} := \{f \in \mathbb{Z}[X_1, \ldots, X_r] : f(z_1, \ldots, z_r) = 0\}.$$

By Hilbert's Basis Theorem, there are finitely many polynomials $f_1, \ldots, f_M \in \mathbb{Z}[X_1, \ldots, X_r]$ such that $\mathcal{I} = (f_1, \ldots, f_M)$. Thus, A is isomorphic to

$$\mathbb{Z}[X_1, \ldots, X_r]/(f_1, \ldots, f_M), \tag{6.3.1}$$

and $z_1, \ldots, z_r$ may be identified with the residue classes of $X_1, \ldots, X_r$ modulo $(f_1, \ldots, f_M)$. We call $(f_1, \ldots, f_M)$ an *ideal representation* for A. We say that A is *effectively given* if an ideal representation for it is given.

Notice that for polynomials $f_1, \ldots, f_M$ and $\in \mathbb{Z}[X_1, \ldots, X_r]$ to form an ideal representation of an integral domain, it is necessary and sufficient that

$\mathcal{I} := (f_1, \ldots, f_M)$ be a prime ideal of $\mathbb{Z}[X_1, \ldots, X_r]$ and $\mathcal{I} \cap \mathbb{Z} = (0)$. This is equivalent to $\overline{\mathcal{I}} := \mathcal{I} \cdot \mathbb{Q}[X_1, \ldots, X_r]$ being a prime ideal of $\mathbb{Q}[X_1, \ldots, X_r]$, $\overline{\mathcal{I}} \cap \mathbb{Z}[X_1, \ldots, X_r] = \mathcal{I}$, and $1 \notin \overline{\mathcal{I}}$. For instance, by Seidenberg (1974, Section 46, p. 293), one can check that $\overline{\mathcal{I}}$ is a prime ideal in $\mathbb{Q}[X_1, \ldots, X_r]$, and by Theorem 6.1.2 (ii), one can check that $1 \notin \overline{\mathcal{I}}$. To verify that $\overline{\mathcal{I}} \cap \mathbb{Z}[X_1, \ldots, X_r] = \mathcal{I}$, one can compute a set of generators for $\overline{\mathcal{I}} \cap \mathbb{Z}[X_1, \ldots, X_r]$ using Theorem 6.1.7, and then check, using Theorem 6.1.5, whether these generators belong to $\mathcal{I}$.

Let $A = \mathbb{Z}[z_1, \ldots, z_r]$ be an integral domain with given ideal representation $\mathcal{I} = (f_1, \ldots, f_M)$. We say that $y \in A$ *is given/can be computed* (as a polynomial in $z_1, \ldots, z_r$), if a polynomial $g \in \mathbb{Z}[X_1, \ldots, X_r]$ *is given/can be computed* such that $y = g(z_1, \ldots, z_r)$. Thanks to Corollary 6.1.6, we can verify whether two expressions $g_i(z_1, \ldots, z_r)$ $(i = 1, 2)$ are equal (i.e., $g_1 - g_2 \in \mathcal{I}$).

Finally, if $y_1, \ldots, y_m$ are the given elements of A, then we say that y is given/can be computed as a polynomial in terms of $y_1, \ldots, y_m$, if $g \in \mathbb{Z}[Y_1, \ldots, Y_m]$ are given/can be computed, such that $y = g(y_1, \ldots, y_m)$.

Theorem 6.3.1 *For any effectively given integral domain* $A = \mathbb{Z}[z_1, \ldots, z_r]$ *of characteristic* 0 *and any given monic irreducible polynomial* $\mathcal{F} \in A[X]$ *with coefficients given as polynomials in* $z_1, \ldots, z_r$, *we can:*
(i) determine effectively a finite set of generators for the ideal

$$\{f \in \mathbb{Z}[X_1, \ldots, X_r, Y] : f(z_1, \ldots, z_r, y) = 0\}$$

where y *is a root of* $\mathcal{F}$;
(ii) for any $a \in A[y]$ *given as polynomial in* $z_1, \ldots, z_r, y$, *determine effectively* $b_0, \ldots, b_{\deg \mathcal{F} - 1} \in A$, *given as polynomials in* $z_1, \ldots, z_r$, *such that* $a = \sum_{i=0}^{\deg \mathcal{F} - 1} b_i y^i$.

Proof Similar to Theorem 6.2.4.   □

Theorem 6.3.2 *For any effectively given integral domain* $A = \mathbb{Z}[z_1, \ldots, z_r]$ *of characteristic* 0 *finitely generated over* $\mathbb{Z}$, *any* $(m \times n)$ *matrix* V *with entries in the quotient field* K *of* A, *and any column vector* $\mathbf{b} \in K^n$, *all with entries given in terms of* $z_1, \ldots, z_r$ *we can:*
(i) determine effectively a finite set of generators, with coordinates given as polynomials in $z_1, \ldots, z_r$, *for the* A-*module* $\{\mathbf{x} \in A^n : V\mathbf{x} = \mathbf{0}\}$;
(ii) decide whether $V\mathbf{x} = \mathbf{b}$ *is solvable in* $\mathbf{x} \in A^n$ *and, if so, determine effectively a solution with coordinates given as polynomials in* $z_1, \ldots, z_r$.

Proof After multiplication with a suitable nonzero element of A, we may assume that V and $\mathbf{b}$ have their entries in A and are given as polynomials with integer coefficients in $z_1, \ldots, z_r$. Suppose that A is given by an ideal representation $\mathcal{I} = (f_1, \ldots, f_M)$. By choosing representatives for the entries of V and $\mathbf{b}$

in $R := \mathbb{Z}[X_1, \ldots, X_r]$, we can rewrite the systems of linear equations in (i) and (ii) as systems of linear congruence equations modulo $\mathcal{I}$ in unknowns from R. By writing the elements of $\mathcal{I}$ as R-linear combinations of $f_1, \ldots, f_M$, we can rewrite these congruence systems as systems of linear equations as considered in Theorem 6.1.5 and apply the latter. □

Theorem 6.3.3 *For any effectively given field* $K = \mathbb{Q}(z_1, \ldots, z_r)$ *and any* $y_1, \ldots, y_s$ *and* $y \in K$ *given in terms of* $z_1, \ldots, z_r$, *we can:*
(i) determine effectively a finite set of generators for the ideal

$$\mathcal{J} := \{f \in \mathbb{Z}[X_1, \ldots, X_s] : f(y_1, \ldots, y_s) = 0\};$$

(ii) decide whether $y \in \mathbb{Z}[y_1, \ldots, y_s]$ *and, if so, determine effectively* $g \in \mathbb{Z}[X_1, \ldots, X_s]$ *such that* $y = g(y_1, \ldots, y_s)$.

Proof (i) Theorem 6.2.2 (i) provides an algorithm to compute a finite set of generators for the ideal

$$\overline{\mathcal{J}} := \{f \in \mathbb{Q}[X_1, \ldots, X_s] : f(y_1, \ldots, y_s) = 0\}$$

and subsequently, by means of Theorem 6.1.7, one can determine a finite set of generators for $\overline{\mathcal{J}} \cap \mathbb{Z}[X_1, \ldots, X_s] = \mathcal{J}$.

(ii) By Theorem 6.2.2 (ii), it can be decided whether $y \in \mathbb{Q}(y_1, \ldots, y_s)$ and, if so, elements a, b of $\mathbb{Z}[y_1, \ldots, y_s]$ can be computed, both represented as polynomials with integer coefficients in $y_1, \ldots, y_s$, such that $y = a/b$. By Theorem 6.3.2, it can be decided whether $a/b \in \mathbb{Z}[y_1, \ldots, y_s]$ and, if so, a polynomial $g \in \mathbb{Z}[X_1, \ldots, X_s]$ can be computed such that $a/b = g(y_1, \ldots, y_s)$. □

Let $A = \mathbb{Z}[z_1, \ldots, z_r]$ be an effectively given integral domain finitely generated over $\mathbb{Z}$, and K its quotient field. We consider finitely generated A-modules contained in K (so in other words, fractional ideals of A). The A-module generated by elements $y_1, \ldots, y_m$ is denoted by $(y_1, \ldots, y_m)$. We say that such a module is given/can be determined in terms of $z_1, \ldots, z_r$, if a finite set of generators for it is given/can be determined in terms of $z_1, \ldots, z_r$.

We say that a finitely generated A-module $\mathcal{M} \subset K$ is given if a finite set of A-module generators for $\mathcal{M}$ is given.

Theorem 6.3.4 *For any two given* A-*submodules* $\mathcal{M}_1$ *and* $\mathcal{M}_2$ *of* K, *one can*
(i) decide whether $\mathcal{M}_1 \subseteq \mathcal{M}_2$;
(ii) compute a finite set of A-*module generators for* $\mathcal{M}_1 \cap \mathcal{M}_2$.

Proof Let $\mathcal{M}_1 = (a_1, \ldots, a_u)$ and $\mathcal{M}_2 = (b_1, \ldots, b_v)$ with the a_i and $b_j \in K$ given in terms of $z_1, \ldots, z_r$. Then (i) comes down to check whether $a_1, \ldots, a_u \in$

$\mathcal{M}_2$, which is a special case of part (ii) of Theorem 6.3.2. To determine a finite set of A-module generators for $\mathcal{M}_1 \cap \mathcal{M}_2$ using part (i) of Theorem 6.3.2, one first determines a finite set of A-module generators for the solution set $(x_1, \ldots, x_u, y_1, \ldots, y_v) \in A^{u+v}$ of $\sum_{i=1}^{u} x_i a_i = \sum_{j=1}^{v} y_j b_j$ and then for each generator one takes the coordinates $x_1, \ldots, x_u$, and subsequently $\sum_{i=1}^{u} x_i a_i$. $\square$

The quotient module of two A-modules $\mathcal{M}_1$ and $\mathcal{M}_2$ with $\mathcal{M}_1 \subseteq \mathcal{M}_2$ is given by $\mathcal{M}_2/\mathcal{M}_1 := \{a + \mathcal{M}_1 : a \in \mathcal{M}_2\}$, with the usual addition and scalar multiplication of cosets. By a full system of representatives for $\mathcal{M}_2/\mathcal{M}_1$, we mean a subset of $\mathcal{M}_2$ consisting of precisely one element from each of the cosets $a + \mathcal{M}_1$ ($a \in \mathcal{M}_2$).

Theorem 6.3.5 *For any effectively given integral domain* $A = \mathbb{Z}[z_1, \ldots, z_r]$ *finitely generated over* $\mathbb{Z}$ *and any two given finitely generated* A-*modules* $\mathcal{M}_1$ *and* $\mathcal{M}_2$ *with* $\mathcal{M}_1 \subseteq \mathcal{M}_2$ *contained in the quotient field of* A, *it can be decided whether* $\mathcal{M}_2/\mathcal{M}_1$ *is finite. If this is the case, a full system of representatives for* $\mathcal{M}_2/\mathcal{M}_1$ *can be determined in terms of* $z_1, \ldots, z_r$.

Proof The proof is too lengthy to be inserted here. See, for instance, Evertse and Győry (2017b, Proposition 3.6). $\square$

Let A be an integral domain, K its quotient field, and G a finite extension of K. Then we denote by A_G the integral closure of A in G. In particular, A_K is the integral closure of A in its quotient field K. Recall that G is effectively given if an irreducible polynomial $P \in K[X]$ is given such that $G \cong K[X]/(P)$. The irreducibility of P can be checked for instance by means of Theorem 6.2.3.

Theorem 6.3.6 *Assume that* A *and a finite extension* G *of its quotient field* K *are effectively given. Then one can compute a finite set of* A-*module generators for* A_G. *Moreover, one can compute an ideal representation for* A_G.

Proof A method to compute a finite set of A-module generators for A_G can be derived by combining results of Nagata (1956), de Jong (1998), Matsumura (1986), and Matsumoto (2000); see for more details Evertse and Győry (2017a, Corollary 10.7.18). Then an ideal representation for A_G can be computed using Theorem 6.3.3. $\square$

We finish with two consequences related to Theorems 1.6.1 and 1.6.3.

Corollary 6.3.7 *Assume that* A *is effectively given. Let* n *be an integer* ≥ 2. *Then one can effectively decide whether the quotient* A-*module* $(\frac{1}{n}A \cap A_K)/A$ *is finite and, if so, compute a full system of representatives for* $(\frac{1}{n}A \cap A_K)/A$.

Proof Immediate consequence of Theorems 6.3.4–6.3.6. $\square$

Let, again A be effectively given, and denote by K its quotient field. Recall that a finite étale K-algebra Ω is effectively given if a separable polynomial $P \in K[X]$ is given such that $\Omega \cong K[X]/(P)$. The separability of P can be checked for instance by means of Theorem 6.2.3.

Let $n := \deg P$, and denote by θ the residue class of X modulo P. Then $\{1, \theta, \ldots, \theta^{n-1}\}$ is a K-basis of Ω, and every element of Ω can be expressed uniquely as $\sum_{i=0}^{n-1} a_i \theta^i$ with all $a_i \in K$. We say that such an element is given if the a_i are given in terms of $z_1, \ldots, z_r$.

We say that a finitely generated A-module $\mathcal{M} \subset \Omega$ is effectively given, if $\omega_1, \ldots, \omega_u$ are given such that $\mathcal{M} = \{\sum_{i=1}^{u} x_i \omega_i : x_1, \ldots, x_m \in A\}$.

Corollary 6.3.8 *Assume that A, a finite étale K-algebra Ω, and a finitely generated A-module $\mathcal{M} \subset \Omega$ are effectively given.*
(i) For any given $\alpha \in \Omega$, it can be decided whether $\alpha \in \mathcal{M}$.
(ii) A set of A-module generators for $\mathcal{M} \cap K$ can be determined effectively in terms of $z_1, \ldots, z_r$.

Proof Let P, n, and θ be as above. Further, let $\{\omega_1, \ldots, \omega_u\}$ be an effectively given set of A-module generators for $\mathcal{M}$. Then $\omega_1, \ldots, \omega_u$ can be expressed as K-linear combinations of 1 and $\theta, \ldots, \theta^{n-1}$, with coefficients given in terms of $z_1, \ldots, z_r$. Then we may express elements of $\mathcal{M}$ as $\sum_{k=0}^{n-1} \ell_k(\mathbf{x}) \theta^k$ with $\mathbf{x} \in A^u$, where $\ell_0, \ldots, \ell_{n-1}$ are linear forms from $K[X_1, \ldots, X_u]$ with coefficients given in terms of $z_1, \ldots, z_r$.

(i) Let $\alpha = \sum_{k=0}^{n-1} a_k \theta^k$, where $a_0, \ldots, a_{n-1} \in K$ are effectively given in terms of $z_1, \ldots, z_r$. Clearly, $\alpha \in \mathcal{M}$ if and only if there is $\mathbf{x} \in A^u$ with $\ell_k(\mathbf{x}) = a_k$ for $k = 0, \ldots, n-1$, and this can be ckecked by means of Theorem 6.3.2 (ii).

(ii) By Theorem 6.3.2 (i), we can compute a set of A-module generators, say $\{\mathbf{x}_1, \ldots, \mathbf{x}_v\}$, for

$$\{\mathbf{x} \in A^u : \ell_1(\mathbf{x}) = \cdots = \ell_{n-1}(\mathbf{x}) = 0\}.$$

Then $\{\ell_0(\mathbf{x}_1), \ldots, \ell_0(\mathbf{x}_v)\}$ is a set of A-module generators for $\mathcal{M} \cap K$. □

Corollary 6.3.9 *Assume that A, a finite étale K-algebra Ω, and a finitely generated A-module $\mathcal{O} \subset \Omega$ are effectively given.*
(i) It can be decided whether $\mathcal{O}$ is an A-order of Ω.
(ii) If $\mathcal{O}$ is an A-order of Ω, one can decide whether the quotient A-module $(\mathcal{O} \cap K)/A$ is finite and, if so, compute a full system of representatives for $(\mathcal{O} \cap K)/A$.

Proof (i) Let $\{\omega_1, \ldots, \omega_u\}$ be a set of A-module generators for $\mathcal{O}$, and let $\ell_0, \ldots, \ell_{n-1}$ be the linear forms from the proof of Corollary 6.3.8.

We first have to verify that the linear forms $\ell_0, \ldots, \ell_{n-1}$ have rank n over K, to make sure that $\mathcal{O}$ contains a K-basis of Ω; this is simply a matter of computing a determinant. The next thing to verify is whether $1 \in \mathcal{O}$ and $\omega_i \omega_j \in \mathcal{O}$ for $i, j = 1, \ldots, u$; this can be done using Corollary 6.3.8 (i).

(ii) Using Corollary 6.3.8 (ii), we can compute a finite set of A-module generators for $\mathcal{O} \cap K$. With these generators for $\mathcal{O} \cap K$ and Theorem 6.3.5, we can check whether $(\mathcal{O} \cap K)/A$ is finite and, if so, compute a full system of representatives. □

7

The Effective Specialization Method

In this chapter, we present our general effective specialization method and make it ready for application to our Diophantine equations under consideration.

The general idea of our method is to reduce our given Diophantine equations over a finitely generated integral domain A of characteristic 0 to Diophantine equations of the same type over function fields and over number fields by means of an effective specialization method. In the first step, we extend our equations to equations of the same form over a finitely generated overring B of A of a special type which is more convenient to deal with.

As was mentioned in the Introduction and Chapter 3, such an effective specialization argument was elaborated by Győry (1983, 1984b) for decomposable form equations and discriminant equations over finitely generated integral domains of the same special type as B. To extend this to arbitrary finitely generated domains A, what was needed was an algorithm that selects those solutions from the overring B that belong to A, but that was missing at the time of Győry's work. A later effective result by Aschenbrenner (2004) on systems of linear equations over polynomial rings over $\mathbb{Z}$ enabled us in our paper Evertse and Győry (2013) to surmount this difficulty and extend the method to the case of arbitrary finitely generated domains A.

We follow closely our paper Evertse and Győry (2013). Save for some small modifications, Lemmas 7.2.3, 7.2.4, 7.2.6, 7.3.1, 7.3.3, and 7.4.2–7.4.7, as well as Propositions 7.2.5 and 7.2.7 are taken from that paper. For convenience of the reader, we reproduce here their proofs.

7.1 Notation

As in the previous chapters, for a polynomial f with coefficients in $\mathbb{Z}$, we denote by $\deg f$, $h(f)$ its total degree and its logarithmic height, i.e., the logarithm

of the maximum of the absolute values of its coefficients. Further, we define
$\log^* u := \max(1, \log u)$ for $u > 0$.

Let $A = \mathbb{Z}[z_1, \ldots, z_r]$ be an integral domain of characteristic 0 finitely generated over $\mathbb{Z}$ and denote by K the quotient field of A. We assume that $r > 0$. We have

$$A \cong \mathbb{Z}[X_1, \ldots, X_r]/\mathcal{I}, \tag{7.1.1}$$

where $\mathcal{I}$ is the ideal of polynomials $f \in \mathbb{Z}[X_1, \ldots, X_r]$ such that $f(z_1, \ldots, z_r) = 0$. The ideal $\mathcal{I}$ is finitely generated. We assume that

$$\mathcal{I} = (f_1, \ldots, f_M) \quad \text{with } \deg f_i \leq d, \ h(f_i) \leq h \text{ for } i = 1, \ldots, M, \tag{7.1.2}$$
$$\text{where } d \geq 1, h \geq 1.$$

A *representative* for $\alpha \in A$ is a polynomial $\widetilde{\alpha} \in \mathbb{Z}[X_1, \ldots, X_r]$ such that $\alpha = \widetilde{\alpha}(z_1, \ldots, z_r)$, or, with the representation (7.1.2) for A, $\alpha = \widetilde{\alpha} \pmod{\mathcal{I}}$.

We assume that K has transcendence degree $q \geq 0$ over $\mathbb{Q}$. For $q > 0$, we assume without loss of generality that $z_1, \ldots, z_q$ are algebraically independent of $\mathbb{Q}$, and that $z_1 = X_1, \ldots, z_q = X_q$. Write $t := r - q$ and rename $z_{q+1}, \ldots, z_r$ as $y_1, \ldots, y_t$. Put

$$A_0 := \mathbb{Z}[X_1, \ldots, X_q], \ K_0 := \mathbb{Q}(X_1, \ldots, X_q) \quad \text{if } q > 0,$$
$$A_0 := \mathbb{Z}, \ K_0 := \mathbb{Q} \quad \text{if } q = 0, \tag{7.1.3}$$

so that

$$A = A_0[y_1, \ldots, y_t], \quad K = K_0(y_1, \ldots, y_t), \quad [K : K_0] < \infty.$$

For $a \in A_0$, we denote by $\deg a$, $h(a)$ the total degree and logarithmic height of a if $q > 0$, while we put $\deg a := 0$ and $h(a) := \log |a|$ if $q = 0$.

Recall that A_0 is a unique factorization domain with unit group $A_0^* = \{\pm 1\}$. This implies that any finite set $a_1, \ldots, a_r$ of nonzero elements of A_0 has an up to sign unique greatest common divisor $b := \gcd(a_1, \ldots, a_r)$ such that $c \in A_0$ divides $a_1, \ldots, a_r$ if and only if c divides b.

7.2 Construction of a More Convenient Ground Domain *B*

In this section, we prove in a more general form that there are $w \in A$, $g \in A_0 \backslash \{0\}$ such that

$$A \subseteq B := A_0[w, g^{-1}]$$

and w has minimal polynomial $\mathcal{F}(X) = X^D + \mathcal{F}_1 X^{D-1} + \cdots + \mathcal{F}_D$ over K_0 with $\mathcal{F}_i \in A_0$ for $i = 1, \ldots, D$. Further, we give explicit upper bounds in terms of r, q, d and h for D and the degrees and logarithmic heights of $g, \mathcal{F}_1, \ldots, \mathcal{F}_D$. Moreover, we require that $\mathcal{A} \subset B^*$ for some prescribed finite set $\mathcal{A}$.

We shall need several lemmas.

Lemma 7.2.1 *Let* $b_1, \ldots, b_n \in A_0$ *and* $b = b_1 \cdots b_n$. *Then*

$$|h(b) - \sum_{i=1}^{n} h(b_i)| \leq q \deg b.$$

Proof Consequence of Corollary 4.1.6. □

Write $\mathbf{Y} := (X_{q+1}, \ldots, X_r)$ and $K_0(\mathbf{Y}) := K_0(X_{q+1}, \ldots, X_r)$. Given $f \in \mathbb{Z}[X_1, \ldots, X_r]$, we write f^* for f but viewed as a polynomial in the variables $\mathbf{Y} = (X_{q+1}, \ldots, X_r)$, with coefficients in A_0. Given $f \in K_0(\mathbf{Y})$, we denote by $\deg_{\mathbf{Y}} f$ its total degree with respect to $\mathbf{Y}$; recall that the total degree $\deg b$ of $b \in A_0$ is taken with respect to $X_1, \ldots, X_q$. With this notation, (7.1.1) and (7.1.2) can be rewritten as

$$\left. \begin{array}{l} A \cong A_0[\mathbf{Y}]/(f_1^*, \ldots, f_M^*), \\[2mm] \deg_{\mathbf{Y}} f_i^* \leq d \text{ for } i = 1, \ldots, M, \\[2mm] \text{the coefficients of } f_1^*, \ldots, f_M^* \text{ in } A_0 \text{ have total degrees} \\[2mm] \text{at most } d \text{ and logarithmic heights at most } h. \end{array} \right\} \tag{7.2.1}$$

Lemma 7.2.2 *Let* $\Bbbk$ *be an algebraically closed field of characteristic* 0, *let* s *be a positive integer, and let* $\mathcal{X}$ *be an algebraic subset of* $\Bbbk^s$ *given by polynomials of total degree at most* d. *Further, let* $\mathcal{Y}$ *be an algebraic subset of* $\mathcal{X}$ *such that* $\mathcal{X} \backslash \mathcal{Y}$ *is finite. Then* $\mathcal{X} \backslash \mathcal{Y}$ *has cardinality at most* d^s.

Proof See Corollary 7.5.3 of Evertse and Győry (2015). It is proved by repeated application of the version of Bezout's theorem from algebraic geometry as stated in Hartshorne (1977, Chapter 1, Theorem 7.7). □

Let $D := [K : K_0]$ and let $\sigma_1, \ldots, \sigma_D$ denote the K_0-isomorphic embeddings of K in an algebraic closure $\overline{K_0}$ of K_0.

Lemma 7.2.3 *(i) We have* $D \leq d^t$.

(ii) There exist rational integers $a_1, \ldots, a_t$ *with* $|a_i| \leq D^2$ *for* $i = 1, \ldots, t$, *such that for* $v := a_1 y_1 + \cdots + a_t y_t$ *we have* $K = K_0(v)$.

Proof (i) The images of $(y_1, \ldots, y_t)$ under $\sigma_1, \ldots, \sigma_D$ belong to

$$\mathcal{W} := \{\mathbf{y} \in \overline{K_0}^t : f_1^*(\mathbf{y}) = \cdots = f_M^*(\mathbf{y}) = 0\}.$$

Conversely, each assignment $\mathbf{Y} = (X_{q+1}, \ldots, X_r) \mapsto \mathbf{y}$ with $\mathbf{y} \in \mathcal{W}$ yields a K_0-isomorphic embedding of K in $\overline{K_0}$ since $K \cong K_0[\mathbf{Y}]/(f_1^*, \ldots, f_M^*)$. Thus

$|\mathcal{W}| = D < \infty$. Now, Lemma 7.2.2 with $\Bbbk = \overline{K}_0$, $\mathcal{X} = \mathcal{W}, \mathcal{Y} = \emptyset$ gives $|\mathcal{W}| \leq d^t$. Hence $D \leq d^t$.

(ii) For integers $a_1, \ldots, a_t$, the quantity $v := a_1 y_1 + \cdots + a_t y_t$ generates K over K_0 if and only if $a_1 \sigma_1(y_1) + \cdots + a_t \sigma_t(y_t)$ are distinct for $i = 1, \ldots, D$. There are integers a_j with $|a_j| \leq D^2$, $j = 1, \ldots, t$, for which this holds. $\square$

Lemma 7.2.4 *There are* $\mathcal{G}_0, \ldots, \mathcal{G}_D \in A_0$ *such that*

$$\sum_{i=0}^{D} \mathcal{G}_i v^{D-i} = 0, \quad \mathcal{G}_0 \cdot \mathcal{G}_D \neq 0 \tag{7.2.2}$$

and

$$\deg \mathcal{G}_i \leq (2d)^{\exp O(r)}, \quad h(\mathcal{G}_i) \leq (2d)^{\exp O(r)} h \tag{7.2.3}$$

for $i = 0, \ldots, D$.

Proof We write $\mathbf{Y} := (X_{q+1}, \ldots, X_r)$ and $\mathbf{Y}^{\mathbf{u}} := X_{q+1}^{u_1} \cdots X_{q+t}^{u_t}$, $|\mathbf{u}| := u_1 + \cdots + u_t$ for tuples of nonnegative integers $\mathbf{u} = (u_1, \ldots, u_t)$. Further, we define $W := \sum_{j=1}^{t} a_j X_{q+j}$.

Since v has degree D over K_0, elements $\mathcal{G}_0, \ldots, \mathcal{G}_D$ of A_0 as in (7.2.2) exist. By (7.2.1), there are $g_1^*, \ldots, g_M^* \in A_0[\mathbf{Y}]$ with the property

$$\sum_{i=0}^{D} \mathcal{G}_i W^{D-i} = \sum_{j=1}^{M} g_j^* f_j^*. \tag{7.2.4}$$

By Theorem 6.1.2 (ii), applied with the field $F = K_0$, there are polynomials $g_j^* \in K_0[\mathbf{Y}]$ satisfying (7.2.4) of degrees at most $(2 \max(d, D))^{2^t} \leq (2^{d^t})^{2^t} =: d'$ in $\mathbf{Y}$. Multiplying $\mathcal{G}_0, \ldots, \mathcal{G}_D$ with an appropriate nonzero factor from A_0, we may assume that g_j^* are polynomials in $A_0[\mathbf{Y}]$ of degree at most d' in $\mathbf{Y}$. Considering (7.2.4) with such polynomials g_j^*, we obtain

$$\sum_{i=0}^{D} \mathcal{G}_i W^{D-i} = \sum_{j=1}^{M} \left(\sum_{|\mathbf{u}| \leq d'} g_{j,\mathbf{u}} \mathbf{Y}^{\mathbf{u}} \right) \left(\sum_{|\mathbf{v}| \leq d} f_{j,\mathbf{v}} \mathbf{Y}^{\mathbf{v}} \right), \tag{7.2.5}$$

where $g_{j,\mathbf{u}} \in A_0$ and $f_j^* = \sum_{|\mathbf{v}| \leq d} f_{j,\mathbf{v}} \mathbf{Y}^{\mathbf{v}}$ with $f_{j,\mathbf{v}} \in A_0$. Here, $\mathcal{G}_0, \ldots, \mathcal{G}_D$ and the polynomials $g_{j,\mathbf{u}}$ are viewed as the unknowns of (7.2.5). Thus, (7.2.5) has solutions with $\mathcal{G}_0 \cdot \mathcal{G}_D \neq 0$.

Consider (7.2.5) as a system of linear equations $V\mathbf{x} = \mathbf{0}$ over K_0, where $\mathbf{x}$ consists of $\mathcal{G}_i$, $i = 0, \ldots, D$, and $g_{j,\mathbf{u}}$, $j = 1, \ldots, M$, $|\mathbf{u}| \leq d'$. Using Lemma 7.2.3, and (4.1.7) and (4.1.8), we get that the polynomial $W^{D-i} = \left(\sum_{k=1}^{t} a_k X_{q+k} \right)^{D-i}$ has logarithmic height at most $O(D \log(2D^2 t)) \leq (2d)^{O(t)}$. Together with (7.2.1) this gives that the entries of the matrix V are elements of A_0 of degrees at most d and logarithmic heights at most $h' := \max((2d)^{O(t)}, h)$.

Further, the number of rows of V is at most the number of monomials in $\mathbf{Y}$ of degree at most $d + d'$ which is bounded above by

$$m_0 := \binom{d + d' + t}{t} \leq (2d)^{\exp O(r)}.$$

In view of Corollary 6.1.4, the A_0-module of solutions of (7.2.5) is generated by vectors $\mathbf{x} = (\mathcal{G}_0, \ldots, \mathcal{G}_D, \{g_{i,\mathbf{u}}\})$, whose coordinates are elements from A_0 of degrees and logarithmic heights at most

$$(2m_0 d)^{2^q}, \quad (2m_0 d)^{6^q} h',$$

respectively. Among these vectors $\mathbf{x}$, there is one with $\mathcal{G}_0 \neq 0$ and also one with $\mathcal{G}_D \neq 0$ since otherwise (7.2.5) would have no solution with $\mathcal{G}_0 \cdot \mathcal{G}_D \neq 0$, contradicting what we already observed about (7.2.2) and (7.2.3). Either among these vectors $\mathbf{x}$, there is one with $\mathcal{G}_0 \mathcal{G}_D \neq 0$; or there is no such vector but then among these vectors there are $\mathbf{x}_1$ with $\mathcal{G}_0 = 0, \mathcal{G}_D \neq 0$ and $\mathbf{x}_2$ with $\mathcal{G}_0 \neq 0, \mathcal{G}_D = 0$, so that $\mathbf{x} := \mathbf{x}_1 + \mathbf{x}_2$ has $\mathcal{G}_0 \mathcal{G}_D \neq 0$. Using the above established upper bound for m_0, we infer that in both cases, the coordinates of $\mathbf{x}$ have degrees and logarithmic heights at most

$$(2d)^{\exp O(r)}, \quad (2d)^{\exp O(r)} h,$$

respectively. This completes the proof.   □

It will be more convenient to work with

$$w := \mathcal{G}_0 v = \mathcal{G}_0(a_1 y_1 + \cdots + a_t y_t) \text{ if } D \geq 2, \quad w := 1 \text{ if } D = 1.$$

Notice that by (7.2.3) and the estimates $|a_i| \leq D^2 \leq d^{2r}$ from Lemma 7.2.3, this w belongs to A and has a representative $\widetilde{w} \in \mathbb{Z}[X_1, \ldots, X_r]$ with

$$\deg \widetilde{w} \leq (2d)^{\exp O(r)}, \quad h(\widetilde{w}) \leq (2d)^{\exp O(r)} h. \tag{7.2.6}$$

The following proposition follows at once from Lemmas 7.2.1, 7.2.3, and 7.2.4.

Proposition 7.2.5 *We have $K = K_0(w)$, where $w \in A$, w is integral over A_0, and w has minimal polynomial $\mathcal{F}(X) = X^D + \mathcal{F}_1 X^{D-1} + \cdots + \mathcal{F}_D$ over K_0 such that*

$$\mathcal{F}_i \in A_0, \quad \deg \mathcal{F}_i \leq (2d)^{\exp O(r)}, \quad h(\mathcal{F}_i) \leq (2d)^{\exp O(r)} h$$

for $i = 1, \ldots, D$.

In what follows, we fix such a $w \in A$. Since $A_0 = \mathbb{Z}[X_1, \ldots, X_q]$ is a unique factorization domain, the greatest common divisor of a finite set of elements of

A_0 is well defined and uniquely determined up to sign. With every $\alpha \in K$, we associate an up to sign unique tuple $P_{\alpha,0}, \ldots, P_{\alpha,D-1}, Q_\alpha$ from A_0 such that

$$\alpha = Q_\alpha^{-1} \sum_{j=0}^{D-1} P_{\alpha,j} w^j \text{ with } Q_\alpha \neq 0, \ \gcd(P_{\alpha,0}, \ldots, P_{\alpha,D-1}, Q_\alpha) = 1. \quad (7.2.7)$$

We keep the notation from (7.1.2). Set

$$\overline{\deg}\,\alpha := \max(\deg P_{\alpha,0}, \ldots, \deg P_{\alpha,D-1}, \deg Q_\alpha),$$

$$\overline{h}\,(\alpha) := \max(h(P_{\alpha,0}), \ldots, h(P_{\alpha,D-1}), h(Q_\alpha)).$$

Lemma 7.2.6 *Let $\alpha \in K^*$ and let (a,b) be a pair of representatives for α with $a,b \in \mathbb{Z}[X_1, \ldots, X_r]$, $b \neq \mathcal{I}$. Put*

$$d_0 := \max(d, \deg a, \deg b), \quad h_0 := \max(h, h(a), h(b)).$$

Then

$$\overline{\deg}\,\alpha \leq (2d_0)^{\exp O(r)}, \quad \overline{h}(\alpha) \leq (2d_0)^{\exp O(r)} h_0.$$

Proof Consider the linear equation

$$Q = \sum_{j=0}^{D-1} P_j w^j \quad (7.2.8)$$

in unknowns $P_0, \ldots, P_{D-1}, Q \in A_0$. Since $\alpha \in K = K_0(w)$ and w has degree D over K_0, equation (7.2.8) has a solution with $Q \neq 0$. Put again $\mathbf{Y} := (X_{q+1}, \ldots, X_r)$ and set $Y := \mathcal{G}_0\left(\sum_{j=1}^t a_j X_{q+j}\right)$. According to our general convention, we write a^*, b^* for a, b, viewed as polynomials in $\mathbf{Y}$ with coefficients in A_0. By (7.2.1), there exist $g_j^* \in A_0[\mathbf{Y}]$ such that

$$Qa^* - b^* \sum_{j=0}^{D-1} P_j Y^j = \sum_{j=1}^{M} g_j^* f_j^*. \quad (7.2.9)$$

By Theorem 6.1.2 (ii), this identity holds with polynomials $g_j^* \in K_0[\mathbf{Y}]$ of degree at most $(2\max(d_0, D))^{2^t} \leq (2d_0)^{t \cdot 2^t}$ in $\mathbf{Y}$; by multiplying the tuple $(P_0, \ldots, P_{D-1}, Q)$ with a suitable nonzero element of A_0, we can make it so that the g_j^* belong to $A_0[\mathbf{Y}]$. Now, as in the proof of Lemma 7.2.4, we can rewrite (7.2.9) as a system of linear equations over K_0 and then Corollary 6.1.4 can be applied. It follows that (7.2.8) is satisfied by $P_0, \ldots, P_{D-1}, Q \in A_0$ with $Q \neq 0$ and

$$\deg P_0, \ldots, \deg P_{D-1}, \deg Q \ \leq (2d_0)^{\exp O(r)},$$

$$h(P_0), \ldots, h(P_{D-1}), h(Q) \ \leq (2d_0)^{\exp O(r)} h_0.$$

Dividing $P_0, \ldots, P_{D-1}, Q$ by their greatest common divisor and using Lemma 7.2.1, we get $P_{\alpha,0}, \ldots, P_{\alpha,D-1}, Q_\alpha \in A_0$ satisfying (7.2.7) and

$$\deg P_{\alpha,0}, \ldots, \deg P_{\alpha,D-1}, \deg Q_\alpha \leq (2d_0)^{\exp O(r)},$$
$$h(P_{\alpha,0}), \ldots, h(P_{\alpha,D-1}), h(Q_\alpha) \leq (2d_0)^{\exp O(r)} h_0.$$

This proves our lemma.   □

Proposition 7.2.7 *Let w be as in Proposition 7.2.5, and let A be a finite (possibly empty) subset of K^* of cardinality $k \geq 0$. For $\alpha \in A$, let (a_α, b_α) be a pair of representatives of α with $a_\alpha, b_\alpha \in \mathbb{Z}[X_1, \ldots, X_r]$, $b_\alpha \notin \mathcal{I}$. Put*

$$d_1 := \max(d, \max_{\alpha \in A}(\deg a_\alpha, \deg b_\alpha))$$

and

$$h_1 := \max(h, \max_{\alpha \in A}(h(a_\alpha), h(b_\alpha))).$$

Then, there is a nonzero $g \in A_0$ such that

$$A \subseteq B := A_0[w, g^{-1}], \quad A \subset B^* \tag{7.2.10}$$

and

$$\left.\begin{array}{ll} \deg g & \leq (k+1)(2d_1)^{\exp O(r)}, \\ h(g) & \leq (k+1)(2d_1)^{\exp O(r)} h_1. \end{array}\right\} \tag{7.2.11}$$

Proof Take

$$g := \prod_{i=1}^{t} Q_{y_i} \cdot \prod_{\alpha \in A}(Q_\alpha \cdot Q_{\alpha^{-1}}),$$

where, as above, $A = A_0[y_1, \ldots, y_t]$. In general, we have $Q_\beta \cdot \beta \in A_0[w]$ for $\beta \in K^*$. Hence, we have $g\beta \in A_0[w]$ for $\beta = y_1, \ldots, y_t$ and for each α, α^{-1} with $\alpha \in A$. This implies (7.2.10). The inequalities (7.2.11) follow at once from Lemmas 7.2.6 and 7.2.1.   □

We shall use Proposition 7.2.7 in various special cases. Before stating the first, we introduce some further notation and prove a lemma.

We recall that $a_0, a_1, \ldots, a_n \in A$ are the coefficients of the binary form $F(X,Y)$ in Section 2.3, resp. of the polynomial $F(X)$ in Section 2.4, while $\delta \in A \setminus \{0\}$ is the term occurring in the Thue equation (2.3.1) and the superelliptic equation (2.4.3). Further, $\widetilde{a}_0, \widetilde{a}_1, \ldots, \widetilde{a}_n, \widetilde{\delta}$ denote their representatives in $\mathbb{Z}[X_1, \ldots, X_r]$ with degrees at most d and logarithmic heights at most h, where $d \geq 1$, $h \geq 1$. Denote by $\widetilde{F}$ the binary form $F(X,Y)$, resp. the polynomial $F(X)$ with coefficients $a_0, a_1, \ldots, a_n$ replaced by $\widetilde{a}_0, \widetilde{a}_1, \ldots, \widetilde{a}_n$, and by $D_{\widetilde{F}}$ the discriminant of $\widetilde{F}$. Then, the assumption $D_F \neq 0$ implies $D_{\widetilde{F}} \notin \mathcal{I}$.

With the above notation and assumptions from Sections 2.3 and 2.4, the following lemma holds.

Lemma 7.2.8 *For the discriminant $D_{\widetilde{F}}$ we have the following inequalities:*

$$\deg D_{\widetilde{F}} \le (2n - 2)d, \tag{7.2.12}$$

$$h(D_{\widetilde{F}}) \le (2n - 2)[\log(2n^2 \tbinom{d+r}{r})) + h]. \tag{7.2.13}$$

This is Lemma 3.2 of Bérczes, Evertse, and Győry (2014).

Proof Recall that $D_{\widetilde{F}}$ can be expressed as

$$D_{\widetilde{F}} = \pm \begin{vmatrix} \widetilde{a}_0 & \widetilde{a}_1 & \cdots & \cdots & \widetilde{a}_n & & & \\ & \ddots & & & & & \ddots & \\ & & \widetilde{a}_0 & \widetilde{a}_1 & \cdots & \cdots & \widetilde{a}_n \\ \widetilde{a}_1 & 2\widetilde{a}_2 & \cdots & n\widetilde{a}_n & & & \\ n\widetilde{a}_0 & (n-1)\widetilde{a}_1 & \cdots & \widetilde{a}_{n-1} & & & \\ & \ddots & & & & \ddots & \\ & & & n\widetilde{a}_0 & (n-1)\widetilde{a}_1 & \cdots & \widetilde{a}_{n-1} \end{vmatrix} \tag{7.2.14}$$

with on the first $n - 2$ rows of the determinant $\widetilde{a}_0, \ldots, \widetilde{a}_n$, on the $(n-1)$st row $\widetilde{a}_1, 2\widetilde{a}_2, \ldots, n\widetilde{a}_n$ and on the last $n-1$ rows $n\widetilde{a}_0, \ldots, \widetilde{a}_{n-1}$; see e.g. Section 1.4 in Evertse and Győry (2017b). Now, Lemma 7.2.8 follows at once from Lemma 4.1.7, using that the determinant has $(2n-2)! \le (2n-2)^{2n}$ terms and that each of the $\widetilde{a}_i$ has at most $\binom{d+r}{r}$ nonzero coefficients. $\square$

We can now apply Proposition 7.2.7 to the numbers $\alpha_1 = \delta$, $\alpha_2 = \delta^{-1}$, $\alpha_3 = D_F$, and $\alpha_4 = D_F^{-1}$. Then, the pairs $(\widetilde{\delta}, 1)$, $(1, \widetilde{\delta})$, $(D_{\widetilde{F}}, 1)$, and $(1, D_{\widetilde{F}})$ represent these numbers. Using the upper bounds for $\deg D_{\widetilde{F}}$, $h(D_{\widetilde{F}})$ provided by Lemma 7.2.8 as well as $\deg \widetilde{\delta} \le d$, $h(\widetilde{\delta}) \le h$ that we assumed in Sections 2.3 and 2.4, we obtain immediately from Proposition 7.2.7 the following.

Proposition 7.2.9 *There is a nonzero $g \in A_0$ such that*

$$A \subseteq B := A_0[w, g^{-1}], \quad \delta, D_F \in B^* \tag{7.2.15}$$

and

$$\deg g \le (nd)^{\exp O(r)}, \quad h(g) \le (nd)^{\exp O(r)} h. \tag{7.2.16}$$

This is Proposition 3.3 of Bérczes, Evertse, and Győry (2014).

7.3 Comparison of Different Degrees and Heights

We keep the above notation. Namely, A is finitely generated over $\mathbb{Z}$, i.e. $A \cong \mathbb{Z}[X_1, \ldots, X_r]/(f_1, \ldots, f_M)$, where $f_1, \ldots, f_M \in \mathbb{Z}[X_1, \ldots, X_r]$. In this section, we compare, for $\alpha \in A \backslash \{0\}$, certain degrees and heights related to α and an appropriate representative $\widetilde{\alpha} \in \mathbb{Z}[X_1, \ldots, X_r]$ of α. Lemma 7.2.6 provided upper bounds for $\overline{\deg}\, \alpha$ and $\overline{h}(\alpha)$ in terms of the degrees and heights of a, b, where (a, b) is a pair of representatives for α. Conversely, we have the following.

Lemma 7.3.1 *Let* $\lambda \in K^*$ *and let* α *be a nonzero element of* A. *Let* (a, b) *with* $a, b \in \mathbb{Z}[X_1, \ldots, X_r]$ *be a pair of representatives for* λ. *Put*

$$d_2 := \max(1, \deg f_1, \ldots, \deg f_M, \deg a, \deg b, \overline{\deg}\, \lambda\alpha),$$

$$h_2 := \max(1, h(f_1), \ldots, h(f_M), h(a), h(b), \overline{h}(\lambda\alpha)).$$

Then, α *has a representative* $\widetilde{\alpha} \in \mathbb{Z}[X_1, \ldots, X_r]$ *such that*

$$\deg \widetilde{\alpha} \le (2d_2)^{\exp O(r \log^* r)} h_2,$$

$$h(\widetilde{\alpha}) \le (2d_2)^{\exp O(r \log^* r)} h_2^{r+1}.$$

If moreover $\alpha \in A^*$, *then* α^{-1} *has a representative* $\widetilde{\alpha}' \in \mathbb{Z}[X_1, \ldots, X_r]$ *with*

$$\deg \widetilde{\alpha}' \le (2d_2)^{\exp O(r \log^* r)} h_2,$$

$$h(\widetilde{\alpha}') \le (2d_2)^{\exp O(r \log^* r)} h_2^{r+1}.$$

In the special case with $\lambda = 1$ and $a = b = 1$, we get the following corollary which will be useful in some applications.

Corollary 7.3.2 *Let* $\alpha \in A \backslash \{0\}$, *and let*

$$d_2' := \max(1, \deg f_1, \ldots, \deg f_M, \overline{\deg}\, \alpha),$$

$$h_2' := \max(1, h(f_1), \ldots, h(f_M), \overline{h}(\alpha)).$$

Then, α *has a representative* $\widetilde{\alpha} \in \mathbb{Z}[X_1, \ldots, X_r]$ *such that*

$$\deg \widetilde{\alpha} \le (2d_2')^{\exp O(r \log^* r)} h_2',$$

$$h(\widetilde{\alpha}) \le (2d_2')^{\exp O(r \log^* r)} h_2'^{r+1}.$$

Proof of Lemma 7.3.1 With the identification of z_i with X_i for $i = 1, \ldots, q$ we may view A_0 as a subring of $\mathbb{Z}[X_1, \ldots, X_r]$. Let $Y := \mathcal{G}_0\left(\sum_{i=1}^t a_i X_{q+i}\right)$. We have

$$\lambda\alpha = Q^{-1} \sum_{i=0}^{D-1} P_i w^i \tag{7.3.1}$$

with $P_0,\ldots,P_{D-1},Q \in A_0$ and $\gcd(P_0,\ldots,P_{D-1},Q) = 1$. In view of (7.3.1), $\widetilde{\alpha} \in \mathbb{Z}[X_1,\ldots,X_r]$ is a representative for α if and only if there exist $g_1,\ldots,g_m \in \mathbb{Z}[X_1,\ldots,X_r]$ such that

$$\widetilde{\alpha} \cdot (Q \cdot a) + \sum_{i=1}^{m} g_i f_i = b \sum_{i=0}^{D-1} P_i Y^i. \tag{7.3.2}$$

We may consider (7.3.2) as an inhomogeneous linear equation over $\mathbb{Z}[X_1,\ldots,X_r]$ in the unknowns $\widetilde{\alpha}, g_1,\ldots,g_m$. By Lemmas 7.2.3, 7.2.4, 7.2.5, and 7.2.6, the degrees and logarithmic heights of Qa and $b\sum_{i=0}^{D-1} P_i Y^i$ are bounded above by

$$(2d_2)^{\exp O(r)}, \quad (2d_2)^{\exp O(r)} h_2,$$

respectively. Theorem 6.1.5 implies that (7.3.2) has a solution with upper bounds for $\deg \widetilde{\alpha}$, $h(\widetilde{\alpha})$, as stated in the lemma.

Now suppose that $\alpha \in A^*$. Then, (7.3.1) gives as above that $\widetilde{\alpha}' \in \mathbb{Z}[X_1,\ldots,X_r]$ is a representative for α^{-1} if and only if there are $g_1',\ldots,g_m' \in \mathbb{Z}[X_1,\ldots,X_r]$ such that

$$\widetilde{\alpha}' b \sum_{i=0}^{D-1} P_i Y^i + \sum_{i=1}^{m} g_i' f_i = Qa.$$

Similarly as above, this equation has a solution with upper bounds for $\deg \widetilde{\alpha}'$, $h(\widetilde{\alpha}')$ as stated in the lemma. □

We next deduce some estimates for the $\overline{\deg}$ of elements from K, by applying the results from Chapter 5. Let as above $K_0 = \mathbb{Q}(X_1,\ldots,X_q)$, $K = K_0(y)$, $A_0 = \mathbb{Z}[X_1,\ldots,X_q]$, $B = \mathbb{Z}[X_1,\ldots,X_q,w,g^{-1}]$. Choose an algebraic closure $\overline{K}_0$ of K_0. Then, there are precisely D K_0-isomorphic embeddings of K into $\overline{K}_0$ which we denote by $\alpha \mapsto \alpha^{(j)}$, $j = 1,\ldots,D$.

For $i = 1,\ldots,q$, let $\Bbbk_i$ be the algebraic closure of $\mathbb{Q}(X_1,\ldots,X_{i-1},X_{i+1},\ldots,X_q)$ in $\overline{K}_0$. Then, A_0 is contained in $\Bbbk_i(X_i)$. Consider the function field

$$L_i := \Bbbk_i(X_i, w^{(1)},\ldots,w^{(D)}).$$

This is the splitting field of the polynomial $\mathcal{F}(X) = X^D + \mathcal{F}_1 X^{D-1} + \cdots + \mathcal{F}_D$ over $\Bbbk_i(X_i)$. The subring

$$B_i := \Bbbk_i[X_i, w^{(1)},\ldots,w^{(D)}, g^{-1}]$$

of L_i contains $B = \mathbb{Z}[X_1,\ldots,X_q,w,g^{-1}]$ as a subring. Define

$$\Delta_i := [L_i : \Bbbk_i(X_i)].$$

We shall apply some estimates from Section 5.1 with $X_i,\Bbbk_i,L_i$ instead of $z,\Bbbk,K$. The height H_{L_i} is taken with respect to $L_i/\Bbbk_i$. For $P \in A_0$, we denote by $\deg_{X_i} P$ the degree of P in the variable X_i. We recall that g from Proposition

7.2.7, and the coefficients $\mathcal{F}_1, \ldots, \mathcal{F}_D$ of the polynomial $\mathcal{F}(X)$ from Proposition 7.2.7 are contained in A_0.

Lemma 7.3.3 *For $\alpha \in K$, we have*

$$\overline{\deg}\,\alpha \le \sum_{i=1}^{q} \Delta_i^{-1} \sum_{j=1}^{D} H_{L_i}(\alpha^{(j)}) + qD \max(\deg \mathcal{F}_1, \ldots, \deg \mathcal{F}_D).$$

Remark It will be convenient to have estimates in which only d and r occur. Inserting the bounds for $\deg \mathcal{F}_i$ from Proposition 7.2.5 and the estimate $D \le d^t$ from Lemma 7.2.3, we obtain

$$\overline{\deg}\,\alpha \le (2d)^{\exp O(r)} + rd^r \max_{i,j} \Delta_i^{-1} H_{L_i}(\alpha^{(j)}), \qquad (7.3.3)$$

where the maximum is taken over $i = 1, \ldots, u$, $j = 1, \ldots, D$.

Proof of Lemma 7.3.3 Put

$$d^* := \max(\deg \mathcal{F}_1, \ldots, \deg \mathcal{F}_D).$$

We have

$$\alpha = Q^{-1} \sum_{j=0}^{D-1} P_j w^j$$

for certain $P_0, \ldots, P_{D-1}, Q \in A_0$ with $\gcd(Q, P_0, \ldots, P_{D-1}) = 1$. It is clear that

$$\overline{\deg}\,\alpha \le \sum_{i=1}^{q} \mu_i, \qquad (7.3.4)$$

where $\mu_i := \max(\deg_{X_i} Q, \deg_{X_i} P_0, \ldots, \deg_{X_i} P_{D-1})$.

Using the height properties listed in Section 5.1, we now estimate $\mu_1, \ldots, \mu_q$ from above. Fix $i \in \{1, \ldots, q\}$. By taking conjugates over K_0, we infer

$$\alpha^{(k)} = Q^{-1} \sum_{j=0}^{D-1} P_j \cdot (w^{(k)})^j, \quad \text{for } k = 1, \ldots, D.$$

Let Ω be the $D \times D$ matrix with rows

$$(1, \ldots, 1), (w^{(1)}, \ldots, w^{(D)}), \ldots, ((w^{(1)})^{D-1}, \ldots, (w^{(D)})^{D-1}).$$

By Cramer's rule, we get $P_j/Q = \delta_j/\delta$, where $\delta = \det \Omega$ and δ_j is the determinant of the matrix obtained by replacing the jth row of Ω by $(\alpha^{(1)}, \ldots, \alpha^{(D)})$.

Gauss' Lemma implies that $P_0, \ldots, P_{D-1}, Q$ are relatively prime in the ring $\Bbbk_i[X_i]$. Hence by (5.1.10) (with X_i in place of z), we obtain

$$\mu_i = H_{\Bbbk_i(X_i)}^{\hom}(Q, P_0, \ldots, P_{D-1}).$$

But $(\delta, \delta_1, \ldots, \delta_D)$ is a scalar multiple of $(Q, P_0, \ldots, P_{D-1})$. Combining (5.1.9) and (5.1.11) and inserting $[L_i : \Bbbk_i(X_i)] = \Delta_i$, we deduce that

$$\mu_i = \Delta_i^{-1} H_{L_i}^{\text{hom}}(Q, P_0, \ldots, P_{D-1}) = \Delta_i^{-1} H_{L_i}^{\text{hom}}(\delta, \delta_1, \ldots, \delta_D). \tag{7.3.5}$$

We now estimate from above the right-hand side. It follows that for every valuation v of $L_i / \Bbbk_i$,

$$- \min(v(\delta), v(\delta_1), \ldots, v(\delta_D))$$

$$\leq -D \sum_{j=1}^{D} \min(0, v(w^{(j)})) - \sum_{j=1}^{D} \min(0, v(\alpha^{(j)})),$$

and then summation over v gives

$$H_{L_i}^{\text{hom}}(\delta, \delta_1, \ldots, \delta_D) \leq D \sum_{j=1}^{D} H_{L_i}(w^{(j)}) + \sum_{j=1}^{D} H_{L_i}(\alpha^{(j)}). \tag{7.3.6}$$

A combination of (5.1.14), (5.1.11), and (5.1.10) yields

$$\Delta_i^{-1} \sum_{j=1}^{D} H_{L_i}(w^{(j)}) = \Delta_i^{-1} H_{L_i}^{\text{hom}}(\mathcal{F}) = H_{\Bbbk_i(X_i)}^{\text{hom}}(\mathcal{F})$$

$$= \max(\deg_{X_i} \mathcal{F}_1, \ldots, \deg_{X_i} \mathcal{F}_D) \leq d^*. \tag{7.3.7}$$

Together with (7.3.5) and (7.3.6), this gives

$$\mu_i \leq Dd^* + \Delta_i^{-1} \sum_{j=1}^{D} H_{L_i}(\alpha^{(j)}).$$

Now, these bounds for $i = 1, \ldots, q$ together with (7.3.4) imply our lemma. $\quad\square$

We have the following converse of Lemma 7.3.3.

Lemma 7.3.4 *Let $\alpha \in K^*$ and $\alpha^{(1)}, \ldots, \alpha^{(D)}$ be as in Lemma 7.3.3. Then*

$$\max_{i,j} \Delta_i^{-1} H_{L_i}(\alpha^{(j)}) \leq 2D \,\overline{\deg}\, \alpha + D \max(\deg \mathcal{F}_1, \ldots, \deg \mathcal{F}_D). \tag{7.3.8}$$

This is a slight refinement of Lemma 4.4 in Bérczes, Evertse, and Győry (2014).

Remark Inserting the bounds for $\deg \mathcal{F}_i$ from Proposition 7.2.5 and the estimate $D \leq d^t$ from Lemma 7.2.3, we obtain

$$\max_{i,j} \Delta_i^{-1} H_{L_i}(\alpha^{(j)}) \leq (2d)^{\exp O(r)} + 2d^r \,\overline{\deg}\, \alpha. \tag{7.3.9}$$

Proof of Lemma 7.3.4 Define again

$$d^* := \max(\deg \mathcal{F}_1, \ldots, \deg \mathcal{F}_D).$$

Consider the representation of α of the form (7.2.7). Then, we have

$$\alpha^{(j)} = Q_\alpha^{-1} \sum_{k=0}^{D-1} P_{\alpha,k}(w^{(j)})^k, \text{ for } j = 1, \ldots, D,$$

since $P_{\alpha,k}$ and Q_α are in K_0. Using (5.1.7) and (5.1.8), we get

$$H_{L_i}(\alpha^{(j)}) \le \sum_{k=0}^{D-1} H_{L_i}(P_{\alpha,k}/Q_\alpha) + \sum_{k=0}^{D-1} k H_{L_i}(w^{(j)}). \qquad (7.3.10)$$

However, we have

$$\begin{aligned}
H_{L_i}(P_{\alpha,k}/Q_\alpha) &\le \Delta_i H_{k_i(X_i)}(P_{\alpha,k}/Q_\alpha) \le \Delta_i (\deg_{X_i} P_{\alpha,k} + \deg_{X_i} Q_\alpha) \\
&\le \Delta_i (\deg P_{\alpha,k} + \deg Q_\alpha) \\
&\le 2\Delta_i \overline{\deg} \, \alpha. \qquad (7.3.11)
\end{aligned}$$

Further, it follows from the proof of (7.3.7) and from Lemma 7.2.3(i) that

$$\sum_{k=0}^{D-1} k H_{L_i}(w^{(j)}) \le D\Delta_i \max_{1 \le k \le D} \deg_{X_i} \mathcal{F}_k \le D\Delta_i d^*. \qquad (7.3.12)$$

Now, (7.3.10), (7.3.11), and (7.3.12) give (7.3.8).   □

7.4 Specializations

In this section, we first prove some results about our specialization homomorphisms from B to $\overline{\mathbb{Q}}$, where B denotes the overring of A from Proposition 7.2.7.

If $q = 0$, no specialization argument is needed. Hence, in this section, we assume that $q > 0$. We start with some auxiliary results that are used in the construction of our specializations.

Lemma 7.4.1 *Let* $\alpha_1, \ldots, \alpha_m \in \overline{\mathbb{Q}}$ *with* $G(X) := (X - \alpha_1) \cdots (X - \alpha_m) \in \mathbb{Z}[X]$. *Then*

$$\left| h(G) - \sum_{i=1}^{m} h(\alpha_i) \right| \le m \log 2.$$

Proof This is a special case of Corollary 4.1.5.   □

Lemma 7.4.2 *Let* $\alpha_1, \ldots, \alpha_m \in \overline{\mathbb{Q}}$ *be distinct and suppose that* $G(X) := (X - \alpha_1) \cdots (X - \alpha_m) \in \mathbb{Z}[X]$. *Let* $q, p_0, \ldots, p_{m-1}$ *be integers with* $\gcd(q, p_0, \ldots, p_{m-1}) = 1$ *and put*

$$\beta_i := \sum_{j=0}^{m-1} (p_j/q)\alpha_i^j \text{ for } i = 1, \ldots, m.$$

Then

$$\log \max(|q|, |p_0|, \ldots, |p_{m-1}|) \le 2m^2 + (m-1)h(G) + \sum_{i=1}^{m} h(\beta_i).$$

Proof For $m = 1$, the assertion is obvious, hence we assume $m \ge 2$. Let $L = \mathbb{Q}(\alpha_1, \ldots, \alpha_m)$. Denote by Ω the $m \times m$ matrix with rows $(\alpha_1^i, \ldots, \alpha_m^i)$ for $i = 0, \ldots, m-1$. By Cramer's rule, we get $p_i/q = \delta_i/\delta, i = 0, \ldots, m-1$, where $\delta = \det \Omega$ and δ_i is the determinant of the matrix obtained by replacing the ith row of Ω by $(\beta_i, \ldots, \beta_m)$. Put

$$\mu := \log \max(|q|, |p_0|, \ldots, |p_{m-1}|).$$

Since $(\delta, \delta_0, \ldots, \delta_{m-1})$ is a scalar multiple of $(q, p_0, \ldots, p_{m-1})$, we have by (4.1.4) and (4.1.6)

$$\mu = h^{\mathrm{hom}}(q, p_0, \ldots, p_{m-1}) = h^{\mathrm{hom}}(\delta, \delta_0, \ldots, \delta_{m-1})$$

$$= \frac{1}{d} \sum_{v \in \mathcal{M}_L} \log \max(|\delta|_v, |\delta_0|_v, \ldots, |\delta_{m-1}|_v). \tag{7.4.1}$$

Estimating the determinants using Hadamard's inequality for the infinite places and the ultrametric inequality for the finite places, we get

$$\max(|\delta|_v, |\delta_0|_v, \ldots, |\delta_{m-1}|_v)$$

$$\le m^{ms(v)/2} \cdot \prod_{i=1}^{m} \max(1, |\alpha_i|_v)^{m-1} \cdot \max(1, |\beta_i|_v)$$

for $v \in \mathcal{M}_L$, where $s(v) = 1$ if v is real, $s(v) = 2$ if v is complex, and $s(v) = 0$ if v is finite. Together with (7.4.1), this gives

$$\mu \le \tfrac{1}{2} m \log m + \sum_{i=1}^{m} ((m-1)h(\alpha_i) + h(\beta_i)).$$

Combining this with Corollary 4.1.5, we get Lemma 7.4.2. $\qquad \square$

Let again $A_0 = \mathbb{Z}[X_1, \ldots, X_q]$. Given $b \in A_0$, $\mathbf{u} = (u_1, \ldots, u_q) \in \mathbb{Z}^q$ we denote by $b(\mathbf{u})$ the image of b under $X_i \mapsto u_i$ $(i = 1, \ldots, q)$.

Lemma 7.4.3 *Let $b \in A_0$ have degree $\mathcal{D}$. Let $\mathcal{N}$ be a finite subset of $\mathbb{Z}$ of cardinality $> \mathcal{D}$. Then*

$$|\{\mathbf{u} \in \mathcal{N}^q : b(\mathbf{u}) = 0\}| \le \mathcal{D}|\mathcal{N}|^{q-1}.$$

Proof We proceed by induction on q. For $q = 1$, the assertion is obvious. Let $q \ge 2$. Write

$$b = \sum_{i=0}^{\mathcal{D}_0} b_i X_q^i,$$

where $b_i \in \mathbb{Z}[X_1,\dots,X_{q-1}]$ and $b_{\mathcal{D}_0} \neq 0$. Then $\deg b_{\mathcal{D}_0} \leq \mathcal{D} - \mathcal{D}_0$. By the induction hypothesis, there are at most $(\mathcal{D} - \mathcal{D}_0)|\mathcal{N}|^{q-2} \cdot |\mathcal{N}|$ tuples $(u_1,\dots,u_{q-1},u_q) \in \mathcal{N}^q$ with $b_{\mathcal{D}_0}(u_1,\dots,u_{q-1}) = 0$ and u_q arbitrary. Further, there are at most $|\mathcal{N}|^{q-1} \cdot \mathcal{D}_0$ tuples $\mathbf{u} \in \mathcal{N}^q$ with $b_{\mathcal{D}_0}(u_1,\dots,u_{q-1}) \neq 0$ and $b(u_1,\dots,u_q) = 0$. Summing these two quantities implies that b has at most $\mathcal{D}|\mathcal{N}|^{q-1}$ zeros in $\mathcal{N}^q$. $\qquad\square$

For $f \in A_0 = \mathbb{Z}[X_1,\dots,X_q]$ and $p \in \mathcal{M}_{\mathbb{Q}} := \{\infty\} \cup \{\text{primes}\}$, we define $|f|_p$ to be the maximum of the $|\cdot|_p$-values of the coefficients of f.

Lemma 7.4.4 *Let $b_1, b_2 \in A_0$ have degrees $\mathcal{D}_1, \mathcal{D}_2$, respectively, and let N be an integer $\geq \max(1, \mathcal{D}_1, \mathcal{D}_2)$. Define*

$$S := \{\mathbf{u} \in \mathbb{Z}^q \ : \ |\mathbf{u}| \leq N, \ b_2(\mathbf{u}) \neq 0\}.$$

Then, S is nonempty, and

$$|b_1|_p \leq U_p^q \max\{|b_1(\mathbf{u})|_p \ : \ \mathbf{u} \in S\} \ for \ p \in \mathcal{M}_{\mathbb{Q}}, \qquad (7.4.2)$$

where $U_\infty = (4N)^{\mathcal{D}_1}$, $U_p = (2N)^{\mathcal{D}_1}$ if p is a prime with $p \leq 2N$, and $U_p = 1$ if p is a prime $>2N$.

Proof We proceed by induction on q, starting with $q = 0$. In the case $q = 0$, we interpret b_1, b_2 as nonzero constants. Then, the lemma is trivial. Let $q \geq 1$. The lemma is obviously true if $\mathcal{D}_1 = 0$ so we assume $\mathcal{D}_1 \geq 1$. Put

$$C_p := \max\{|b_1(\mathbf{u})|_p \ : \ \mathbf{u} \in S\} \ \text{for} \ p \in \mathcal{M}_{\mathbb{Q}}.$$

Write

$$b_1 = \sum_{j=0}^{\mathcal{D}_1'} b_{1,j} X_q^j, \quad b_2 = \sum_{j=0}^{\mathcal{D}_2'} b_{2,j} X_q^j,$$

where the $b_{1,j}$, $b_{2,j}$ belong to $\mathbb{Z}[X_1,\dots,X_{q-1}]$ and $b_{1,\mathcal{D}_1'}, b_{2,\mathcal{D}_2'} \neq 0$. By induction hypothesis, the set

$$S' := \{\mathbf{u}' \in \mathbb{Z}^{q-1} \ : \ |\mathbf{u}'| \leq N, \ b_{2,\mathcal{D}_2'}(\mathbf{u}') \neq 0\}$$

is nonempty and moreover,

$$\max_{0 \leq j \leq \mathcal{D}_1'} |b_{1,j}|_p \leq U_p^{q-1} C_p' \ \text{for} \ p \in \mathcal{M}_{\mathbb{Q}},$$
$$\text{where } C_p' := \max\{|b_{1,j}(\mathbf{u}')|_p \ : \ \mathbf{u}' \in S', \ j = 0,\dots,\mathcal{D}_1'\}. \qquad (7.4.3)$$

We fix $p \in \mathcal{M}_{\mathbb{Q}}$ and estimate C_p' from above in terms of C_p. Take $\mathbf{u}' \in S'$ such that $C_p' = \max_{0 \leq j \leq \mathcal{D}_1'} |b_{1,j}(\mathbf{u}')|_p$. There exist at least $2N+1-\mathcal{D}_2' \geq \mathcal{D}_1'+1$

integers u_q with $|u_q| \leq N$ such that $b_2(\mathbf{u}', u_q) \neq 0$. Let $a_0, \ldots, a_{\mathcal{D}'_1}$ be distinct integers from this set. Using Lagrange's interpolation formula, we obtain

$$b_1(\mathbf{u}', X_q) = \sum_{j=0}^{\mathcal{D}'_1} b_{1j}(\mathbf{u}') X_q^j = \sum_{j=0}^{\mathcal{D}'_1} b_1(\mathbf{u}', a_j) \Big(\prod_{\substack{i=0 \\ i \neq j}}^{\mathcal{D}'_1} \frac{X_q - a_i}{a_j - a_i} \Big).$$

First, consider $p = \infty$. The coefficients of a polynomial $\prod_{k=1}^m (X - c_k)$ with $c_1, \ldots, c_m \in \mathbb{C}$ have absolute values at most $\prod_{k=1}^m (1 + |c_k|)$. Hence

$$\mathcal{C}'_\infty = \max_{0 \leq j \leq \mathcal{D}'_1} |b_{1j}(\mathbf{u}')| \leq \mathcal{C}_\infty \sum_{j=0}^{\mathcal{D}'_1} \prod_{\substack{i=0 \\ i \neq j}}^{\mathcal{D}'_1} (1 + |a_i|)$$

$$\leq \mathcal{C}_\infty (\mathcal{D}'_1 + 1)(N + 1)^{\mathcal{D}'_1} \leq U_\infty \mathcal{C}_\infty.$$

Now, let p be a prime and let k be the largest integer such that $p^k \leq 2N$. Then, for all i, j with $0 \leq i < j \leq \mathcal{D}'_1$, we have

$$|a_i - a_j|_p \geq p^{-k} \geq \begin{cases} (2N)^{-1} & \text{if } p \leq 2N, \\ 1 & \text{if } p > 2N, \end{cases}$$

and thus,

$$\mathcal{C}'_p = \max_{0 \leq j \leq \mathcal{D}'_1} |b_{1j}(\mathbf{u}')|_p \leq \mathcal{C}_p \max_{0 \leq j \leq \mathcal{D}'_1} \prod_{\substack{i=0 \\ i \neq j}}^{\mathcal{D}'_1} |a_j - a_i|_p^{-1} \leq U_p \mathcal{C}_p.$$

So $\mathcal{C}'_p \leq U_p \mathcal{C}_p$ for all $p \in \mathcal{M}_{\mathbb{Q}}$. Combining this with (7.4.3), we obtain (7.4.2). $\qquad \square$

We now define our specializations $B \to \overline{\mathbb{Q}}$ and prove some properties. These specializations were introduced by Győry (1983, 1984a) and, in a refined form, by Evertse and Győry (2013); see Chapter 3.

We recall that in this section $q > 0$ is assumed. Apart from that, we keep the notation and assumptions from Section 7.2. In particular,

$$A_0 = \mathbb{Z}[X_1, \ldots, X_q], \quad K_0 = \mathbb{Q}(X_1, \ldots, X_q),$$
$$K = \mathbb{Q}(X_1, \ldots, X_q, w), \quad B = \mathbb{Z}[X_1, \ldots, X_q, w, g^{-1}],$$

where $g \in A_0$ is the polynomial from Proposition 7.2.7, w is integral over A_0 and w has minimal polynomial

$$\mathcal{F}(X) := X^D + \mathcal{F}_1 X^{D-1} + \cdots + \mathcal{F}_D \in A_0[X]$$

over K_0. By construction, $\mathcal{A} \subset B^*$, where $\mathcal{A}$ is the finite set from Proposition 7.2.7. In the case $D = 1$, we take $w = 1$, $\mathcal{F} = X - 1$.

Let d_1, h_1 be the quantities from Proposition 7.2.7 and k the cardinality of $\mathcal{A}$. Further, define

$$\begin{cases} d_3 := \max(d, \deg \mathcal{F}_1, \ldots, \deg \mathcal{F}_D), & d_4 := \max(d_3, \deg g), \\ h_3 := \max(h, h(\mathcal{F}_1), \ldots, h(\mathcal{F}_D)), & h_4 := \max(h_3, h(g)). \end{cases} \tag{7.4.4}$$

By Propositions 7.2.5 and 7.2.7, we have

$$\begin{cases} d_3 \le (2d)^{\exp O(r)}, & d_4 \le (k+1)(2d_1)^{\exp O(r)}, \\ h_3 \le (2d)^{\exp O(r)} h, & h_4 \le (k+1)(2d_1)^{\exp O(r)} h_1. \end{cases} \tag{7.4.5}$$

Further, we will frequently use the consequence of Lemma 7.2.3 (i),

$$D \le d^r. \tag{7.4.6}$$

Let $\mathbf{u} = (u_1, \ldots, u_q) \in \mathbb{Z}^q$. Then, the substitution $X_1 \mapsto u_1, \ldots, X_q \mapsto u_q$ defines a ring homomorphism (specialization) from a subring of K_0 to $\mathbb{Q}$

$$\varphi_{\mathbf{u}} : \alpha \mapsto \alpha(\mathbf{u}) : \{b_1/b_2 \, : \, b_1, b_2 \in A_0, \ b_2(\mathbf{u}) \ne 0\} \to \mathbb{Q}.$$

We want to define ring homomorphisms from B to $\overline{\mathbb{Q}}$, and for this, we have to impose some restrictions on $\mathbf{u}$. Let $\Delta_{\mathcal{F}}$ denote the discriminant of $\mathcal{F}$ (with $\Delta_{\mathcal{F}} := 1$ if $D = \deg \mathcal{F} = 1$), and let

$$\mathcal{T} := \Delta_{\mathcal{F}} \mathcal{F}_D \cdot g. \tag{7.4.7}$$

Then $\mathcal{T} \in A_0$. Since $\Delta_{\mathcal{F}}$ is a polynomial of degree $2D - 2$ with integer coefficients in $\mathcal{F}_1, \ldots, \mathcal{F}_D$, we deduce easily that

$$\deg \mathcal{T} \le (2D - 1)d_3 + d_4 \le 2Dd_4. \tag{7.4.8}$$

Assume that

$$\mathcal{T}(\mathbf{u}) \ne 0.$$

Then $g(\mathbf{u}) \ne 0$ and the polynomial

$$\mathcal{F}_{\mathbf{u}} := X^D + \mathcal{F}_1(\mathbf{u})X^{D-1} + \cdots + \mathcal{F}_D(\mathbf{u})$$

has D distinct zeros which are all different from 0, say $w_1(\mathbf{u}), \ldots, w_D(\mathbf{u})$. Thus, for $j = 1, \ldots, D$, the assignment

$$X_1 \mapsto u_1, \ldots, X_q \mapsto u_q, \ w \mapsto w_j(\mathbf{u})$$

defines a ring homomorphism $\varphi_{\mathbf{u}, j}$ from B to $\overline{\mathbb{Q}}$; in the case $D = 1$, it is just $\varphi_{\mathbf{u}}$. The image of $\alpha \in B$ under $\varphi_{\mathbf{u}, j}$ is denoted by $\alpha_j(\mathbf{u})$. We recall that the elements α of B can be expressed as

$$\alpha = \sum_{i=0}^{D-1} (P_i/Q)w^i \text{ with relatively prime } P_0, \ldots, P_{D-1}, Q \in A_0. \tag{7.4.9}$$

Since $\alpha \in B$, the denominator Q must divide a power of g. Hence $Q(\mathbf{u}) \neq 0$. Thus, we have

$$\alpha_j(\mathbf{u}) = \sum_{i=0}^{D-1} (P_i(\mathbf{u})/Q(\mathbf{u}))w_j(\mathbf{u})^i, \ j = 1,\dots,D. \tag{7.4.10}$$

Clearly, $\varphi_{\mathbf{u},j}$ is the identity on $B \cap \mathbb{Q}$. Hence, if $\alpha \in B \cap \overline{\mathbb{Q}}$, then $\varphi_{\mathbf{u},j}(\alpha)$ has the same minimal polynomial as α and so it is conjugate to α.

For $\mathbf{u} = (u_1,\dots,u_q) \in \mathbb{Z}^q$, we put $|\mathbf{u}| := \max(|u_1|,\dots,|u_q|)$. It is easy to show that for any $b \in A_0, \mathbf{u} \in \mathbb{Z}^q$,

$$\log|b(\mathbf{u})| \leq q \log \deg b + h(b) + \deg b \log \max(1,|\mathbf{u}|). \tag{7.4.11}$$

In particular,

$$h(\mathcal{F}_\mathbf{u}) \leq q \log d_3 + h_3 + d_3 \log \max(1,|\mathbf{u}|) \tag{7.4.12}$$

and so by Corollary 4.1.5,

$$\sum_{j=1}^{D} h(w_j(\mathbf{u})) \leq D + 1 + q \log d_3 + h_3 + d_3 \log \max(1,|\mathbf{u}|). \tag{7.4.13}$$

Define the algebraic number fields $K_{\mathbf{u},j} := \mathbb{Q}(w_j(\mathbf{u})), \ j = 1,\dots,D$. We derive an upper bound for the discriminant $D_{K_{\mathbf{u},j}}$ of $K_{\mathbf{u},j}$.

Lemma 7.4.5 *Let $\mathbf{u} \in \mathbb{Z}^q$ with $\mathcal{T}(\mathbf{u}) \neq 0$. Then, for $j = 1,\dots,D$, the field $K_{\mathbf{u},j}$ has degree $[K_{\mathbf{u},j} : \mathbb{Q}] \leq D$ and absolute discriminant*

$$|D_{K_{\mathbf{u},j}}| \leq D^{2D-1}(d_3^q e^{h_3} \max(1,|\mathbf{u}|)^{d_3})^{2D-2}.$$

Remark Inserting (7.4.5) and (7.4.6), we obtain

$$\log|D_{K_{\mathbf{u},j}}| \leq (2d)^{\exp O(r)} (h + \log \max(1,|\mathbf{u}|)). \tag{7.4.14}$$

Proof Let $j \in \{1,\dots,D\}$. As observed above, $w_j(\mathbf{u})$ is a zero of $\mathcal{F}_\mathbf{u}$, which is a monic polynomial in $\mathbb{Z}[X]$ of degree D. Hence $[K_{\mathbf{u},j} : \mathbb{Q}] \leq D$. To estimate the discriminant of $K_{\mathbf{u},j}$, let $\mathcal{P}_j$ denote the monic minimal polynomial of $w_j(\mathbf{u})$ over $\mathbb{Q}$, which necessarily has its coefficients from $\mathbb{Z}$. Then, $D_{K_{\mathbf{u},j}}$ divides the discriminant of $\mathcal{P}_j$, which is the discriminant of the order $\mathbb{Z}[w_j(\mathbf{u})]$. Using the expression of the discriminant of a monic polynomial as the product of the squares of the differences of its zeros, it is easy to see that the discriminant of $\mathcal{P}_j$ divides that of $\mathcal{F}_\mathbf{u}$ in the ring of algebraic integers and so also in $\mathbb{Z}$. Denoting the latter discriminant by Δ, we infer that $D_{K_{\mathbf{u},j}}$ divides Δ in $\mathbb{Z}$.

It remains to estimate from above $|\Delta|$. We can express this as a determinant similar to (7.2.14), replacing n by D and $\widetilde{a}_0,\dots,\widetilde{a}_n$ by the coefficients of $\mathcal{F}_\mathbf{u}$. Hadamard's inequality gives that the absolute value of this determinant can be estimated from above by the product of the Euclidean norms of its rows. Letting H denote the maximum of the absolute values of the coefficients of $\mathcal{F}_{\mathbf{u}}$, this leads to

$$|\Delta| \le (D+1)^{(D-2)/2}(1^2 + \cdots + D^2)^{D/2}H^{2D-2}$$
$$= (D+1)^{(D-2)/2}(\tfrac{1}{6}D(D+1)(2D+1))^{D/2}H^{2D-2}$$
$$\le D^{2D-1}H^{2D-2},$$

provided that $D \ge 3$. For $D = 1, 2$, the inequality $|\Delta| \le D^{2D-1}H^{2D-2}$ can be verified by direct computation. Inserting (7.4.12), i.e., $H \le d_3^q e^{h_3} \max(1, |\mathbf{u}|)^{d_3}$, we arrive at

$$|\Delta| \le D^{2D-1}(d_3^q e^{h_3} \max(1, |\mathbf{u}|)^{d_3})^{2D-2}.$$

This implies our lemma. $\square$

Finally, we state and prove two lemmas which relate $\overline{h}(\alpha)$ to the heights of $\alpha_j(\mathbf{u})$ for $\alpha \in B, \mathbf{u} \in \mathbb{Z}^q$.

Lemma 7.4.6 *Let* $\mathbf{u} \in \mathbb{Z}^q$ *with* $\mathcal{T}(\mathbf{u}) \ne 0$. *Further, let* $\alpha \in B$. *Then for* $j = 1, \dots, D$

$$h(\alpha_j(\mathbf{u})) \le D^2 + q(D \log d_3 + \log \overline{\deg} \, \alpha)$$
$$+ Dh_3 + \overline{h}(\alpha) + (Dd_3 + \overline{\deg} \, \alpha) \log \max(1, |\mathbf{u}|).$$

Remark Inserting (7.4.5) and (7.4.6), we derive the estimate

$$h(\alpha_j(\mathbf{u})) \le \overline{h}(\alpha) + (2d)^{\exp O(r)}(h + (\overline{\deg} \, \alpha + 1) \log \max(1, |\mathbf{u}|)). \quad (7.4.15)$$

Proof Let $P_0, \dots, P_{D-1}, Q$ be as in (7.4.9). Let $L = \mathbb{Q}(w_j(\mathbf{u}))$. We denote by $\mathcal{M}_L$ the set of places of L, and $| \cdot |_v$ ($v \in \mathcal{M}_L$) the corresponding absolute values normalized as in Section 4.1. Then, for $v \in \mathcal{M}_L$, we have

$$|\alpha_j(\mathbf{u})|_v \le D^{s(v)} T_v \max(1, |w_j(\mathbf{u})|_v)^{D-1},$$

where $s(v) = 1$ if v is real, $s(v) = 2$ if v is complex, $s(v) = 0$ if v is finite, and

$$T_v = \max(1, |P_0(\mathbf{u})/Q(\mathbf{u})|_v, \dots, |P_{D-1}(\mathbf{u})/Q(\mathbf{u})|_v).$$

Hence

$$h(\alpha_j(\mathbf{u})) \le \log D + \frac{1}{[L : \mathbb{Q}]} \sum_{v \in \mathcal{M}_L} \log T_v + (D-1)h(w_j(\mathbf{u})). \quad (7.4.16)$$

We infer that

$$\frac{1}{[L:\mathbb{Q}]} \sum_{v \in \mathcal{M}_L} \log T_v = h(P_0(\mathbf{u})/Q(\mathbf{u}), \ldots, P_{D-1}(\mathbf{u})/Q(\mathbf{u}))$$

$$= h^{\mathrm{hom}}(Q(\mathbf{u}), P_0(\mathbf{u}), \ldots, P_{D-1}(\mathbf{u}))$$

$$\leq \log \max(|Q(\mathbf{u})|, |P_0(\mathbf{u})|, \ldots, |P_{D-1}(\mathbf{u})|)$$

$$\leq q \log \overline{\deg} \alpha + \overline{h}(\alpha) + \overline{\deg} \alpha \cdot \log \max(1, |\mathbf{u}|).$$

Combining this with (7.4.13) and (7.4.16), the lemma follows. □

Lemma 7.4.7 *Let* $\alpha \in B$, $\alpha \neq 0$, *and let* N *be an integer such that*

$$N \geq \max (\overline{\deg} \alpha, 2Dd_3 + 2(q+1)(d_4+1)).$$

Then the set

$$\mathcal{S} := \{\mathbf{u} \in \mathbb{Z}^q : |\mathbf{u}| \leq N, \ \mathcal{T}(\mathbf{u}) \neq 0\}$$

is nonempty and

$$\overline{h}(\alpha) \leq (6N)^{q+4}(h_4 + H),$$

where $H := \max\{h(\alpha_j(\mathbf{u})) : \mathbf{u} \in \mathcal{S}, j = 1, \ldots, D\}$.

Remark In view of (7.4.5) and (7.4.6), we may take here

$$N = \max(\overline{\deg} \alpha, (k+1)(2d_1)^{\exp O(r)}), \tag{7.4.17}$$

and get an upper bound

$$\overline{h}(\alpha) \leq (2d_1)^{\exp O(r)}((k+1) + \overline{\deg} \alpha)^{q+5}(h_1 + H), \tag{7.4.18}$$

where $k = |\mathcal{A}|$, with $\mathcal{A}$ the set from Proposition 7.2.7.

Proof Lemmas 7.4.4 and 7.4.6 and our assumption on N imply that $\mathcal{S}$ is nonempty. We proceed with estimating $\overline{h}(\alpha)$. Let $P_0, \ldots, P_{D-1}, Q \in A_0$ be as in (7.4.9). We analyze Q more closely. Let

$$g = \pm p_1^{k_1} \cdot \ldots \cdot p_m^{k_m} g_1^{\ell_1} \cdot \ldots \cdot g_n^{\ell_n}$$

be the up to the sign of the irreducible factors unique factorization of g in A_0, where $p_1, \ldots, p_m$ are distinct prime numbers and $g_1, \ldots, g_n$ are irreducible elements of A_0 of positive degree with $g_i \neq \pm g_j$ for all i, j with $1 \leq i < j \leq n$. By Corollary 4.1.5, we have

$$\sum_{i=1}^{n} \ell_i h(g_i) \leq qd_4 + h_4. \tag{7.4.19}$$

Since $\alpha \in B$, the polynomial Q is also composed of $p_1, \ldots, p_m, g_1, \ldots, g_n$. Thus

$$Q = aQ' \text{ with } a = \pm p_1^{k_1'} \cdot \ldots \cdot p_m^{k_m'}, \quad Q' = g_1^{\ell_1'} \cdot \ldots \cdot g_n^{\ell_n'} \tag{7.4.20}$$

for certain nonnegative integers $k_1', \ldots, k_m', \ell_1', \ldots, \ell_n'$. Clearly

$$\ell_1' + \cdots + \ell_n' \leq \deg Q \leq \overline{\deg} \, \alpha \leq N,$$

and by Lemma 7.2.1 and (7.4.19)

$$h(Q') \leq q \deg Q + \sum_{i=1}^{n} \ell_i' h(g_i) \leq N(q + qd_4 + h_4)$$
$$\leq N^2(h_4 + 1). \tag{7.4.21}$$

By virtue of (7.4.11), we have for $\mathbf{u} \in \mathcal{S}$

$$\log |Q'(\mathbf{u})| \leq q \log d_4 + h(Q') + \deg Q \log N$$
$$\leq \tfrac{3}{2} N \log N + N^2(h_4 + 1) \leq N^2(h_4 + 2).$$

Hence

$$h(Q'(\mathbf{u})\alpha_j(\mathbf{u})) \leq N^2(h_4 + 2) + H$$

for $\mathbf{u} \in \mathcal{S}$, $j = 1, \ldots, D$. Further, by (7.4.10) and (7.4.20), we have

$$Q'(\mathbf{u})\alpha_j(\mathbf{u}) = \sum_{i=0}^{D-1} (P_i(\mathbf{u})/a)w_j(\mathbf{u})^i.$$

Set

$$\delta(\mathbf{u}) := \gcd(a, P_0(\mathbf{u}), \ldots, P_{D-1}(\mathbf{u})).$$

Then, by applying Lemma 7.4.2 together with (7.4.12), we infer that

$$\log \left(\frac{\max(|a|, |P_0(\mathbf{u})|, \ldots, |P_{D-1}(\mathbf{u})|)}{\delta(\mathbf{u})} \right)$$
$$\leq 2D^2 + (D-1)h(\mathcal{F}_{\mathbf{u}}) + D(N^2(h_4 + 2) + H)$$
$$\leq 2D^2 + (D-1)(q \log d_4 + h_4 + d_4 \log N) + D(N^2(h_4 + 2) + H)$$
$$\leq N^3(h_4 + 2) + DH. \tag{7.4.22}$$

Our assumption that $Q, P_0, \ldots, P_{D-1}$ are relatively prime in A_0 implies that the greatest common divisor of a and the coefficients of $P_0, \ldots, P_{D-1}$ is 1. Let $p \in \{p_1, \ldots, p_m\}$ be one of the prime factors of a. There is $j \in \{0, \ldots, D-1\}$ such that $|P_j|_p = 1$. Our assumption on N and (7.4.8) implies that $N \geq$

$\max(\deg \mathcal{T}, \deg P_j)$. This means that Lemma 7.4.4 can be applied with $g_1 = P_j$ and $g_2 = \mathcal{T}$. It follows that

$$\max\{|P_j(\mathbf{u})|_p \,:\, \mathbf{u} \in \mathcal{S}\} \geq \begin{cases} (2N)^{-qN} & \text{if } p \leq 2N, \\ 1 & \text{if } p > 2N, \end{cases}$$

that is, there is $\mathbf{u}_p \in \mathcal{S}$ with

$$|P_j(\mathbf{u}_p)|_p \geq \begin{cases} (2N)^{-qN} & \text{if } p \leq 2N, \\ 1 & \text{if } p > 2N. \end{cases}$$

Thus,

$$|\delta(\mathbf{u}_p)|_p \geq \begin{cases} (2N)^{-qN} & \text{if } p \leq 2N, \\ 1 & \text{if } p > 2N. \end{cases} \tag{7.4.23}$$

For $\mathbf{u} \in \mathcal{S}$, let $\mathcal{P}_{\mathbf{u}}$ be the set of primes p dividing a with $\mathbf{u}_p = \mathbf{u}$ and $\mathcal{P}'_{\mathbf{u}}$ the set of primes $p \in \mathcal{P}_{\mathbf{u}}$ with $p \leq 2N$. Then, by (7.4.22) and (7.4.23), we have for $\mathbf{u} \in \mathcal{S}$,

$$\sum_{p \in \mathcal{P}_{\mathbf{u}}} \log |a|_p^{-1} \leq \log |a/\delta(\mathbf{u})| + \sum_{p \in \mathcal{P}'_{\mathbf{u}}} \log |\delta(\mathbf{u})|_p^{-1}$$
$$\leq N^3(h_4 + 2) + DH + |\mathcal{P}'_{\mathbf{u}}| \cdot qN \log 2N.$$

Summing over $\mathbf{u} \in \mathcal{S}$, using that $|\mathcal{S}| \leq (3N)^q$, we get

$$\log |a| \leq (3N)^q (N^3(h_4 + 2) + DH) + 2qN^2 \log 2N$$
$$\leq (3N)^{q+4}(h_4 + H). \tag{7.4.24}$$

Together with (7.4.20) and (7.4.21), this gives

$$h(Q) \leq (4N)^{q+4}(h_4 + H). \tag{7.4.25}$$

Further, the right-hand side of (7.4.24) provides an upper bound for $\log \delta(\mathbf{u})$ for $\mathbf{u} \in \mathcal{S}$. A combination of this with (7.4.22) gives

$$\log \max\{|P_j(\mathbf{u})| \,:\, \mathbf{u} \in \mathcal{S}, j = 0, \ldots, D-1\} \leq (5N)^{q+4}(h_4 + H).$$

Another application of Lemma 7.4.4 yields

$$h(P_j) \leq qN \log 4N + (5N)^{q+4}(h_4 + H) \leq (6N)^{q+4}(h_4 + H)$$

for $j = 0, \ldots, D-1$. Together with (7.4.25), this gives the upper bound for $\overline{h}(\alpha)$ as claimed in our lemma. $\qquad \square$

7.5 Multiplicative Independence

We prove a general effective multiplicative independence result for elements of a finitely generated field.

Recall that nonzero elements $\gamma_1, \ldots, \gamma_s$ of a field are called multiplicatively independent if there is no tuple $(b_1, \ldots, b_s) \in \mathbb{Z}^s$ with at least one of the b_i not equal to 0, such that $\gamma_1^{b_1} \cdots \gamma_s^{b_s} = 1$.

We start with a result over number fields and with the help of the specialization theory worked out above, we extend this to arbitrary finitely generated fields.

We state and prove a result on multiplicative dependence over number fields due to Loxton and van der Poorten (1983), which is not the strongest one available at present, but which amply suffices for our purposes. By d_L we denote the degree of a number field L and by w_L the number of its roots of unity. Further, $m(d_L)$ denotes the height lower bound from Lemma 4.1.2, with d replaced by d_L.

Lemma 7.5.1 *Let L be an algebraic number field, and let $\gamma_0, \ldots, \gamma_s$ be nonzero elements of L such that $\gamma_0, \ldots, \gamma_s$ are multiplicatively dependent, but any s elements among $\gamma_0, \ldots, \gamma_s$ are multiplicatively independent. Then, there are nonzero integers $k_0, \ldots, k_s$ such that*

$$\gamma_0^{k_0} \cdots \gamma_s^{k_s} = 1,$$
$$|k_i| \leq s! \cdot w_L m(d_L)^{-s} h(\gamma_0) \cdots h(\gamma_s)/h(\gamma_i) \text{ for } i = 0, \ldots, s.$$

Remark Loher and Masser (2004, Cor. 2.3) obtained the asymptotically sharper upper bound, based on an idea of Kunrui Yu,

$$|k_i| \leq 58(s!e^s/s^s)d_L^{s+1}(\log d_L)h(\gamma_0) \cdots h(\gamma_s)/h(\gamma_i) \text{ for } i = 0, \ldots, s.$$

Proof We follow Loxton and van der Poorten. The result is trivially true if $s = 0$ so we assume that $s \geq 1$. By assumption, there are nonzero integers $b_0, \ldots, b_s$ such that

$$\gamma_0^{b_0} \cdots \gamma_s^{b_s} = 1. \tag{7.5.1}$$

Without loss of generality,

$$|b_0| \cdot h(\gamma_0) \geq |b_i| \cdot h(\gamma_i) \quad \text{for } i = 1, \ldots, s. \tag{7.5.2}$$

The tuple $(b_0, \ldots, b_s)$ is uniquely determined up to a scalar factor because if $(b_0', \ldots, b_s')$ is any other tuple of nonzero integers with $\gamma_0^{b_0'} \cdots \gamma_s^{b_s'} = 1$, then

$$\gamma_1^{b_0'b_1 - b_0b_1'} \cdots \gamma_s^{b_0'b_s - b_0b_s'} = 1,$$

and thus, $b_0' b_i - b_0 b_i' = 0$ for $i = 1, \ldots, s$ by the multiplicative independence of $\gamma_1, \ldots, \gamma_s$.

Let $(\theta_k)_{k \geq 0}$ be a sequence of positive reals increasing to $m(d_L)$. For every k, consider the $(s + 1)$-dimensional symmetric convex body, consisting of the points $(x_0, \ldots, x_s) \in \mathbb{R}^{s+1}$ with

$$\sum_{i=1}^{s} h(\gamma_i) \left| x_i - \frac{b_i}{b_0} x_0 \right| \leq \theta_k, \qquad |x_0| \leq s! \theta_k^{-s} h(\gamma_1) \cdots h(\gamma_s).$$

This body has volume 2^{s+1}, so by Minkowski's convex body theorem, it contains a nonzero point $\mathbf{l}_k \in \mathbb{Z}^{s+1}$. But among the points $\mathbf{l}_k$ ($k \geq 0$), there are only finitely many distinct ones, since they all lie in a bounded set independent of k. Hence, there is a nonzero $\mathbf{l} = (l_0, \ldots, l_s) \in \mathbb{Z}^{s+1}$ belonging to the above defined convex bodies for infinitely many k. But then this point satisfies

$$\sum_{i=1}^{s} h(\gamma_i) \left| l_i - \frac{b_i}{b_0} l_0 \right| < m(d_L), \quad |l_0| \leq s! \cdot m(d_L)^{-s} h(\gamma_1) \cdots h(\gamma_s). \quad (7.5.3)$$

For $i = 0, \ldots, s$, choose β_i such that $\beta_i^{b_0} = \gamma_i$. By (7.5.1), $\zeta := \beta_0^{b_0} \cdots \beta_s^{b_s}$ is a root of unity. From the height properties (4.1.3), we infer

$$h(\gamma_0^{l_0} \cdots \gamma_s^{l_s}) = h(\gamma_0^{l_0} \cdots \gamma_s^{l_s} \zeta^{-l_0}) = h(\beta_1^{b_0 l_1 - b_1 l_0} \cdots \beta_s^{b_0 l_s - b_s l_0})$$

$$\leq \sum_{i=1}^{s} h(\beta_i) |b_0 l_i - b_i l_0| = \sum_{i=1}^{s} h(\gamma_i) |l_i - \frac{b_i}{b_0} l_0| < m(d_L).$$

So by Lemma 4.1.2, $\gamma_0^{l_0} \cdots \gamma_s^{l_s}$ is a root of unity. It follows that $\gamma_0^{k_0} \cdots \gamma_s^{k_s} = 1$, where $k_i := w_L l_i$ for $i = 0, \ldots, s$. Since we assumed that any s elements among $\gamma_0, \ldots, \gamma_s$ are multiplicatively independent, the integers $k_0, \ldots, k_s$ are all nonzero.

It remains to estimate $k_0, \ldots, k_s$. By (7.5.3), we have

$$|k_0| \leq s! w_L m(d_L)^{-s} h(\gamma_1) \cdots h(\gamma_s).$$

Further, $(k_0, \ldots, k_s)$ is up to a scalar multiple equal to $(b_0, \ldots, b_s)$, and so, in view of (7.5.2), we have for $i = 1, \ldots, s$,

$$|k_i| = \left| \frac{b_i}{b_0} k_0 \right| \leq \frac{h(\gamma_0)}{h(\gamma_i)} \cdot |k_0| \leq s! w_L m(d_L)^{-s} h(\gamma_0) \cdots h(\gamma_s)/h(\gamma_i).$$

This proves our lemma. $\qquad \square$

We prove a generalization for arbitrary finitely generated integral domains. As before, let $A = \mathbb{Z}[z_1, \ldots, z_r] \supseteq \mathbb{Z}$ be an integral domain finitely generated over $\mathbb{Z}$ with quotient field K and suppose that the ideal $\mathcal{I}$ of polynomials $f \in \mathbb{Z}[X_1, \ldots, X_r]$ with $f(z_1, \ldots, z_r) = 0$ is generated by $f_1, \ldots, f_M$. Let $\gamma_0, \ldots, \gamma_s$

be nonzero elements of K, and for $i = 0,\ldots,s$, let $(g_{i,1},g_{i,2})$ be a pair of representative for γ_i, i.e. elements of $\mathbb{Z}[X_1,\ldots,X_r]$ such that

$$\gamma_i = \frac{g_{i,1}(z_1,\ldots,z_r)}{g_{i,2}(z_1,\ldots,z_r)}.$$

Proposition 7.5.2 *Assume that $\gamma_0,\ldots,\gamma_s$ are multiplicatively dependent. Further, assume that $f_1,\ldots,f_M$ and $g_{i,1},g_{i,2}$ $(i = 0,\ldots,s)$ have degrees at most d and logarithmic heights at most h, where $d \geq 1, h \geq 1$. Then, there are integers $k_0,\ldots,k_s$, not all zero, such that*

$$\gamma_0^{k_0} \cdots \gamma_s^{k_s} = 1, \tag{7.5.4}$$

$$|k_i| \leq (2d)^{\exp O(r+s)} h^s \text{ for } i = 0,\ldots,s. \tag{7.5.5}$$

This is Lemma 7.2 of Evertse and Győry (2013).

Proof We may assume without loss of generality that any s elements among $\gamma_0,\ldots,\gamma_s$ are multiplicatively independent (if this is not the case, take a minimal multiplicatively independent subset of $\{\gamma_0,\ldots,\gamma_s\}$ and proceed further with this subset). We first assume that $q > 0$. We use an argument of van der Poorten and Schlickewei (1991). Keeping the above notation and assumptions from Chapter 7, we assume that $z_1 = X_1,\ldots,z_q = X_q$ is a transcendence basis of K and rename $z_{q+1},\ldots,z_r$ as $y_1,\ldots,y_t$, respectively. For brevity, we include the case $t = 0$ as well in our proof. But it should be possible to prove in this case a sharper result by means of a more elementary method. We keep the notation and assumptions from Section 7.2, in particular,

$$A_0 = \mathbb{Z}[X_1,\ldots,X_q], \quad K_0 = \mathbb{Q}(X_1,\ldots,X_q), \quad K = \mathbb{Q}(X_1,\ldots,X_q,w),$$

where w is integral over A_0 and w has minimal polynomial

$$\mathcal{F}(X) := X^D + \mathcal{F}_1 X^{D-1} + \cdots + \mathcal{F}_D \in A_0[X]$$

over K_0. In the case $D = 1$, we take $w = 1$, $\mathcal{F} = X - 1$. We construct a specialization such that among the images of $\gamma_0,\ldots,\gamma_s$ no s elements are multiplicatively dependent, and then apply Lemma 7.5.1.

Let $V \geq 2d$ be a positive integer. Later, we shall make our choice of V more precise. Define the set

$$\mathcal{V} := \{\mathbf{v} = (v_0,\ldots,v_s) \in \mathbb{Z}^{s+1}\backslash\{0\} :$$

$$|v_i| \leq V \text{ for } i = 0,\ldots,s, \text{ and with } v_i = 0 \text{ for some } i\}. \tag{7.5.6}$$

Then

$$\gamma_{\mathbf{v}} := \left(\prod_{i=0}^{s} \gamma_i^{v_i}\right) - 1 \ (\mathbf{v} \in \mathcal{V})$$

are nonzero elements of K. It is easy to show that for $\mathbf{v} \in \mathcal{V}$, $\gamma_{\mathbf{v}}$ has a pair of representatives $(g_{1,\mathbf{v}}, g_{2,\mathbf{v}})$ such that

$$\deg g_{1,\mathbf{v}}, \deg g_{2,\mathbf{v}} \leq sdV.$$

In the case $t > 0$, there exists by Proposition 7.2.7 a nonzero $g \in A_0$ such that

$$A \subseteq B := A_0[w, g^{-1}], \quad \gamma_{\mathbf{v}} \in B^* \text{ for } \mathbf{v} \in \mathcal{V}$$

and

$$\deg g \leq V^{s+1}(2sdV)^{\exp O(r)} \leq V^{\exp O(r+s)}.$$

In the case $t = 0$, this holds true as well, with $w = 1$ and $g = \prod_{\mathbf{v} \in \mathcal{V}}(g_{1,\mathbf{v}} \cdot g_{2,\mathbf{v}})$. We apply the theory of specializations explained in Section 7.4 above with this g. We put $\mathcal{T} := \Delta_{\mathcal{F}}\mathcal{F}_D \cdot g$, where $\Delta_{\mathcal{F}}$ denotes the discriminant of $\mathcal{F}$. Using Proposition 7.2.5 and inserting the bound $D \leq d^t$ from Lemma 7.2.3, we get for $t > 0$:

$$\begin{cases} d_3 := \max(d, \deg \mathcal{F}_1, \ldots, \deg \mathcal{F}_D) \leq (2d)^{\exp O(r)} \\ h_3 := \max(h, h(\mathcal{F}_1), \ldots, h(\mathcal{F}_D)) \leq (2d)^{\exp O(r)} h, \end{cases} \tag{7.5.7}$$

with the provision $\deg 0 = h(0) = -\infty$; this is true also if $t = 0$. Combining this with Lemma 7.2.8, we obtain

$$\deg \mathcal{T} \leq (2D - 1)d_3 + \deg g \leq V^{\exp O(r+s)}.$$

By Lemma 7.4.3, there exists $\mathbf{u} \in \mathbb{Z}^q$ with

$$\mathcal{T}(\mathbf{u}) \neq 0, \quad |\mathbf{u}| \leq V^{\exp O(r+s)}. \tag{7.5.8}$$

We proceed further with this $\mathbf{u}$.

As was seen above, $\gamma_{\mathbf{v}} \in B^*$ for $\mathbf{v} \in \mathcal{V}$. By our choice of $\mathbf{u}$, there are D distinct specialization maps $\varphi_{\mathbf{u},j}$ $(j = 1, \ldots, D)$ from B to $\overline{\mathbb{Q}}$. We fix one of these specializations, which we denote by $\varphi_{\mathbf{u}}$. Given $\alpha \in B$, we write $\alpha(\mathbf{u})$ for $\varphi_{\mathbf{u}}(\alpha)$. As the elements $\gamma_{\mathbf{v}}$ are all units in B, their images under $\varphi_{\mathbf{u}}$ are nonzero. Thus, we have

$$\prod_{i=0}^{s} \gamma_i(\mathbf{u})^{v_i} \neq 1 \text{ for } \mathbf{v} \in \mathcal{V} \tag{7.5.9}$$

where $\mathcal{V}$ is defined by (7.5.6).

We use Lemma 7.4.6 to estimate the heights $h(\gamma_i(\mathbf{u}))$ for $i = 0, \ldots, s$. Recall that by Lemma 7.2.6, we have

$$\overline{\deg} \gamma_i \leq (2d)^{\exp O(r)}, \quad \overline{h}(\gamma_i) \leq (2d)^{\exp O(r)} h$$

for $i = 0, \ldots, s$. By inserting these bounds and that for $|\mathbf{u}|$ from (7.5.8) into (7.4.15), we obtain for $i = 0, \ldots, s$,

$$h(\gamma_i(\mathbf{u})) \leq (2d)^{\exp O(r)}(1 + h + \log \max(1, |\mathbf{u}|))$$
$$\leq (2d)^{\exp O(r+s)}(1 + h + \log V). \tag{7.5.10}$$

We show that any s numbers among $\gamma_0(\mathbf{u}), \ldots, \gamma_s(\mathbf{u})$ are multiplicatively independent, provided V is chosen appropriately. Assume the contrary. By Lemma 7.5.1, there are integers $k_0, \ldots, k_s$, at least one of which is nonzero and at least one of which is 0, such that

$$\prod_{i=0}^{s} \gamma_i(\mathbf{u})^{k_i} = 1,$$
$$|k_i| \leq (2d)^{\exp O(r+s)}(1 + h + \log V)^{s-1} \text{ for } i = 0, \ldots, s. \tag{7.5.11}$$

We now choose V large enough such that this upper bound for the numbers $|k_i|$ is smaller than V. This is satisfied with

$$V = (2d)^{\exp O(r+s)} h^{s-1} \tag{7.5.12}$$

where the constant in the O-symbol in (7.5.12) is sufficiently large compared with that of (7.5.11). But then we have $\prod_{i=0}^{s} \gamma_i(\mathbf{u})^{v_i} = 1$ for some $\mathbf{v} \in \mathcal{V}$, contrary to (7.5.9). Hence, we conclude that with the choice (7.5.12) for V, there exists $\mathbf{u} \in \mathbb{Z}^q$ with (7.5.8), such that any s numbers among $\gamma_0(\mathbf{u}), \ldots, \gamma_s(\mathbf{u})$ are multiplicatively independent. Of course, the numbers $\gamma_0(\mathbf{u}), \ldots, \gamma_s(\mathbf{u})$ are multiplicatively dependent, since they are the images under $\varphi_{\mathbf{u}}$ of $\gamma_0, \ldots, \gamma_s$ which are multiplicatively dependent. Substituting (7.5.12) into (7.5.10), we obtain

$$h(\gamma_i(\mathbf{u})) \leq (2d)^{\exp O(r+s)} h \text{ for } i = 0, \ldots, s. \tag{7.5.13}$$

Now, Lemma 7.5.1 implies that there are nonzero integers $k_0, \ldots, k_s$ such that

$$\prod_{i=0}^{s} \gamma_i(\mathbf{u})^{k_i} = 1, \tag{7.5.14}$$
$$|k_i| \leq (2d)^{\exp O(r+s)} h^s \text{ for } i = 0, \ldots, s. \tag{7.5.15}$$

Our assumption concerning $\gamma_0, \ldots, \gamma_s$ implies that there are nonzero integers $\ell_0, \ldots, \ell_s$ such that $\prod_{i=0}^{s} \gamma_i^{\ell_i} = 1$. Hence $\prod_{i=0}^{s} \gamma_i(\mathbf{u})^{\ell_i} = 1$. Together with (7.5.14), this yields

$$\prod_{i=1}^{s} \gamma_i(\mathbf{u})^{\ell_0 k_i - \ell_i k_0} = 1.$$

However, we have $\ell_0 k_i - \ell_i k_0 = 0$ for $i = 1, \ldots, s$ since $\gamma_1(\mathbf{u}), \ldots, \gamma_s(\mathbf{u})$ are multiplicatively independent, that is,

$$\ell_0 \cdot (k_0, \ldots, k_s) = k_0 \cdot (\ell_0, \ldots, \ell_s).$$

It follows that

$$\prod_{i=0}^{s} \gamma_i^{k_i} = \zeta$$

for some root of unity ζ. But $\varphi_{\mathbf{u}}(\zeta) = 1$ and it is conjugate to ζ. Hence $\zeta = 1$. So, in fact, we have $\prod_{i=0}^{s} \gamma_i^{k_i} = 1$ with nonzero integers k_i satisfying (7.5.15). This proves our Proposition, but under the assumption $q > 0$. If $q = 0$, then a much simpler argument, without specializations, gives $h(\gamma_i) \le (2d)^{\exp O(r+s)} h$ for $i = 0, \dots, s$ in place of (7.5.13). Then, the proof is finished in the same way as in the case $q > 0$. □

Corollary 7.5.3 *Let* $\gamma_0, \gamma_1, \dots, \gamma_s \in K^*$ *and suppose that* $\gamma_1, \dots, \gamma_s$ *are multiplicatively independent and*

$$\gamma_0 = \gamma_1^{k_1} \cdots \gamma_s^{k_s}$$

for certain integers $k_1, \dots, k_s$*. Let* d, h *be as in Proposition 7.5.2. Then*

$$|k_i| \le (2d)^{\exp O(r+s)} h^s \text{ for } i = 1, \dots, s.$$

Proof By Proposition 7.5.2 and by the multiplicative independence of $\gamma_1, \dots, \gamma_s$, there are integers $\ell_0, \dots, \ell_s$ such that

$$\prod_{i=0}^{s} \gamma_i^{\ell_i} = 1,$$

$$\ell_0 \ne 0, |\ell_i| \le (2d)^{\exp O(r+s)} h^s \text{ for } i = 0, \dots, s.$$

But then we have also

$$\prod_{i=1}^{s} \gamma_i^{\ell_0 k_i - \ell_i} = 1,$$

whence $\ell_0 k_i - \ell_i = 0$ for $i = 1, \dots, s$. It follows that

$$|k_i| = |\ell_i/\ell_0| \le (2d)^{\exp O(r+s)} h^s \text{ for } i = 1, \dots, s,$$

which is what we wanted to prove. □

8

Degree-Height Estimates

Let, as before, A be an integral domain of characteristic 0 that is finitely generated over $\mathbb{Z}$, K its quotient field, and $\overline{K}$ an algebraic closure of K. *We introduce so-called degree-height estimates for elements of $\overline{K}$*, which may be seen as an analogue for the naive height (height of the minimal polynomial over $\mathbb{Z}$) of an algebraic number. *Our goal is to give a degree-height estimate for $\beta \in \overline{K}$ in terms of degree-height estimates for $\alpha_1, \ldots, \alpha_m \in \overline{K}$, if β is related to the α_i by $P(\beta, \alpha_1, \ldots, \alpha_m) = 0$ for some given $P \in \mathbb{Z}[X, X_1, \ldots, X_m]$ that is monic in X*. Estimates of this type will be crucial in Chapter 10.

8.1 Definitions

We keep using the following notation: $A = \mathbb{Z}[z_1, \ldots, z_r]$ with $r > 0$ is an integral domain of characteristic 0, K is its quotient field, and $\overline{K}$ is an algebraic closure of K. Further, $\mathcal{I}$ is the ideal of $f \in \mathbb{Z}[X_1, \ldots, X_r]$ with $f(z_1, \ldots, z_r) = 0$, so that

$$A \cong \mathbb{Z}[X_1, \ldots, X_r]/\mathcal{I}. \tag{8.1.1}$$

We assume again that

$$\mathcal{I} = (f_1, \ldots, f_M) \text{ with } \deg f_i \leq d, \ h(f_i) \leq h \text{ for } i = 1, \ldots, M,$$
$$\text{where } d \geq 1, h \geq 1. \tag{8.1.2}$$

We now introduce the notion of degree-height estimate. Given a monic polynomial $G \in K[X]$, we call $(g_0, \ldots, g_n)$ a *tuple of representatives for G* if $g_0, \ldots, g_n \in \mathbb{Z}[X_1, \ldots, X_r]$, $g_0 \notin \mathcal{I}$ and

$$G = X^n + \frac{g_1(z_1, \ldots, z_r)}{g_0(z_1, \ldots, z_r)} X^{n-1} + \cdots + \frac{g_n(z_1, \ldots, z_r)}{g_0(z_1, \ldots, z_r)}.$$

We write

$$G \prec (d^*, h^*)$$

if G has a tuple of representatives $(g_0, \ldots, g_n)$ with total degree $\deg g_i \leq d^*$ and logarithmic height $h(g_i) \leq h^*$ for $i = 0, \ldots, n$, and call (d^*, h^*) a *degree-height estimate for G*.

In case that G is a monic polynomial in $A[X]$, we call $(g_1, \ldots, g_n)$ an *integral tuple of representatives* for G if $g_1, \ldots, g_n \in \mathbb{Z}[X_1, \ldots, X_r]$ and

$$G = X^n + g_1(z_1, \ldots, z_r)X^{n-1} + \cdots + g_n(z_1, \ldots, z_r).$$

We write

$$G \overset{\text{int}}{\prec} (d^*, h^*)$$

if G has an integral tuple of representatives $(g_1, \ldots, g_n)$ with $\deg g_i \leq d^*$, $h(g_i) \leq h^*$ for $i = 1, \ldots, n$.

Let $\alpha \in \overline{K}$. We denote the monic minimal polynomial of α over K by F_α. We denote by $d_K(\alpha)$ the degree of α over K, i.e., the degree of F_α. We define a tuple of representatives for α to be a tuple of representatives for F_α. We write

$$\alpha \prec (d^*, h^*) \quad \text{if } F_\alpha \prec (d^*, h^*)$$

and call (d^*, h^*) a *degree-height estimate for α*. In case that $F_\alpha \in A[X]$, an integral tuple of representatives for F_α is also called an integral tuple of representatives for α, and we write

$$\alpha \overset{\text{int}}{\prec} (d^*, h^*) \quad \text{if } F_\alpha \overset{\text{int}}{\prec} (d^*, h^*).$$

In particular, if $\alpha \in K$ then $\alpha \prec (d^*, h^*)$ if α has a pair of representatives each of which has total degree at most d^* and logarithmic height at most h^*, while if $\alpha \in A$, then $\alpha \overset{\text{int}}{\prec} (d^*, h^*)$ if α has a representative of total degree at most d^* and logarithmic height at most h^*.

We should mention here that Moriwaki (2000) developed a sophisticated height theory for points in projective space $\mathbb{P}^n(\overline{K})$, based on Arakelov intersection theory, which may be seen as an analogue of the theory of absolute Weil heights over $\mathbb{P}^n(\overline{\mathbb{Q}})$. We preferred to keep our presentation down to earth and to use the naive degree-height estimates introduced above. It would be of interest to figure out how our degree-height estimates relate to Moriwaki's height.

As mentioned above, our aim is to give a degree-height estimate for $\beta \in \overline{K}$ in terms of degree-height estimates for $\alpha_1, \ldots, \alpha_m \in \overline{K}$, if β is related to the α_i by $P(\beta, \alpha_1, \ldots, \alpha_m) = 0$ for some given $P \in \mathbb{Z}[X, X_1, \ldots, X_m]$ that is monic in X. We outline our procedure. Consider the polynomial

$$G(X) := \prod_{i_1=1}^{n_1} \cdots \prod_{i_m=1}^{n_m} P\left(X, \alpha_1^{(i_1)}, \ldots, \alpha_m^{(i_m)}\right),$$

where $\alpha_i^{(i_j)}$ $(j = 1, \ldots, n_i)$ are the conjugates of α_i over K, for $i = 1, \ldots, m$. The polynomial G is monic and by the theory of symmetric functions, its co-efficients belong to K and can be expressed in terms of the coefficients of the monic minimal polynomials of $\alpha_1, \ldots, \alpha_m$ over K. This enables us to derive a degree-height estimate for G. The polynomial G has β as a zero and thus, is a multiple of the monic minimal polynomial of β, but in general it is not equal to this minimal polynomial. To get a degree-height estimate for the minimal poly-nomial of β, hence of β itself, we use estimates for degree-height estimates of the factors in $K[X]$ of a given polynomial in $K[X]$. We derive such estimates in Section 8.2. In Section 8.3, we derive the degree-height estimate for β in the way explained above and give some further applications.

8.2 Estimates for Factors of Polynomials

We obtain explicit degree-height estimates for the monic divisors in $K[X]$ of a given monic polynomial in $K[X]$. Probably this would have been possible by making explicit arguments from Seidenberg (1974). We have chosen to use instead the specialization theory developed in Sections 7.2–7.4. We keep our assumptions that $A = \mathbb{Z}[z_1, \ldots, z_r]$ and that the ideal $\mathcal{I}$ of $f \in \mathbb{Z}[X_1, \ldots, X_r]$ with $f(z_1, \ldots, z_r) = 0$ is generated by polynomials $f_1, \ldots, f_M$ with (8.1.2). We assume again that $z_1 = X_1, \ldots, z_q = X_q$ is a transcendence basis of K, and write

$$A_0 = \mathbb{Z}[X_1, \ldots, X_q], \quad K_0 = \mathbb{Q}(X_1, \ldots, X_q).$$

We will work with a domain $B = A_0[w, g^{-1}]$ as in Proposition 7.2.7, where we take $\mathcal{A} = \{\Delta_{\mathcal{F}}\}$, with $\Delta_{\mathcal{F}}$ the discriminant of the polynomial $\mathcal{F}$ from Proposi-tion 7.2.5, so that $\Delta_{\mathcal{F}} \in B^*$. Since $g \in A_0$ and since w is integral over A_0, we have in fact

$$\Delta_{\mathcal{F}} \in K_0 \cap B^* = A_0[g^{-1}]^*. \tag{8.2.1}$$

We take the quantities d_1, h_1 defined in Proposition 7.2.7. In our situation, we have

$$d_1 = \max(d, \deg \Delta_{\mathcal{F}}), \quad h_1 = \max(h, h(\Delta_{\mathcal{F}})).$$

By estimates completely similar to those in Lemma 7.2.8, we have

$$\deg \Delta_{\mathcal{F}} \le (2D - 2)d^*,$$
$$h(\Delta_{\mathcal{F}}) \le (2D - 2)\left(\log \left(2D^2 \left(\binom{d_* + r}{r} \right) \right) + h^* \right),$$

where

$$D = [K : K_0], \quad d^* = \max(\deg \mathcal{F}_1, \ldots, \deg \mathcal{F}_D), \quad h^* = \max(h(\mathcal{F}_1), \ldots, h(\mathcal{F}_D)).$$

Invoking the estimates $D \leq d^r$ implied by Lemma 7.2.3 and those for $\deg \mathcal{F}_i$, $h(\mathcal{F}_i)$ from Lemma 7.2.5, we obtain

$$d_1 \leq (2d)^{\exp O(r)}, \quad h_1 \leq (2d)^{\exp O(r)} h. \qquad (8.2.2)$$

We start with some preparatory lemmas.

Lemma 8.2.1 *The above domain B is integrally closed.*

Proof Denote by $x \mapsto x^{(i)}$ ($i = 1, \ldots, D = [K : K_0]$) the K_0-isomorphic embeddings of K in an algebraic closure $\overline{K_0}$ of K_0. Let $\beta \in K$ be integral over B. Then β is integral over $A_0[g^{-1}]$. We have

$$\beta = \sum_{j=0}^{D-1} b_j w^j \quad \text{with } b_0, \ldots, b_{D-1} \in K_0$$

and thus,

$$\beta^{(i)} = \sum_{j=0}^{D-1} b_j (w^{(i)})^j \quad \text{for } i = 1, \ldots, D.$$

Viewing this as a system of linear equations in $b_0, \ldots, b_{D-1}$, we get by Cramer's rule,

$$b_j = \Delta_j / \Delta \quad \text{for } j = 0, \ldots, D - 1,$$

where $\Delta = \det \left((w^{(i)})^{j-1} \right)_{i=1,\ldots,D,\, j=0,\ldots,D-1}$ and where Δ_j is the determinant obtained by replacing $(w^{(i)})^{j-1}$ by $\beta^{(i)}$ for $i = 1, \ldots, D$. Using Vandermonde's identity $\Delta^2 = \prod_{1 \leq j < k \leq D} (w^{(j)} - w^{(k)})^2 = \Delta_{\mathcal{F}}$, we obtain

$$b_j = \Delta_j \cdot \Delta / \Delta_{\mathcal{F}} \quad \text{for } j = 0, \ldots, D - 1.$$

Clearly, $\Delta_j \Delta \in K$. Recall that the polynomial $\mathcal{F}$ is monic in $A_0[X]$, so w is integral over A_0. Hence, the $w^{(i)}$ ($i = 1, \ldots, D$) are integral over A_0. Further, β is integral over B, hence over $A_0[g^{-1}]$, and so the $\beta^{(i)}$ ($i = 1, \ldots, D$) are integral over $A_0[g^{-1}]$. It follows that $\Delta_j \Delta$ is integral over $A_0[g^{-1}]$, and so it belongs to $A_0[g^{-1}]$ since the latter is a localization of a unique factorization domain, hence integrally closed. Together with (8.2.1), this implies $b_j \in A_0[g^{-1}]$ for $j = 0, \ldots, D - 1$. We conclude that $\beta \in B$, as required. $\square$

Lemma 8.2.2 *Let $F \in B[X]$ be a monic polynomial, and $G \in K[X]$ a monic polynomial that divides F in $K[X]$. Then $G \in B[X]$.*

Proof For certain $\alpha_1, \ldots, \alpha_n \in \overline{K}$, we have $F = (X - \alpha_1) \cdots (X - \alpha_n)$, $G = (X - \alpha_1) \cdots (X - \alpha_m)$. Since $\alpha_1, \ldots, \alpha_m$ are integral over B, the coefficients of G are also integral over B. These must belong to B, since B is integrally closed. $\square$

We are now ready to prove our result concerning the degree-height estimates of the factors of a given polynomial.

Proposition 8.2.3 *Let $d_5 \geq d$, $h_5 \geq h$ and let $F \in K[X]$ be a monic polynomial of degree $n \geq 2$ with $F \prec (d_5, h_5)$. Then for each monic polynomial $G \in K[X]$ dividing F, we have*

$$G \prec ((nd_5)^{\exp O(r)}, (nd_5)^{\exp O(r)} h_5).$$

Proof We can write $F = X^n + (a_1/a_0)X^{n-1} + \cdots + (a_n/a_0)$ where $a_0, \ldots, a_n \in A$, $a_0 \neq 0$, and a_i has a representative $\widetilde{a}_i$ with $\deg \widetilde{a}_i \leq d_5$, $h(\widetilde{a}_i) \leq h_5$, for $i = 0, \ldots, n$. Define $F^*(X) := a_0^n F(X/a_0)$. Then

$$F^*(X) = X^n + a_1^* X^{n-1} + \cdots + a_n^* \quad \text{where } a_i^* := a_i a_0^{i-1} \text{ for } i = 1, \ldots, n.$$

Clearly, $F^* \in A[X] \subseteq B[X]$ and by Corollary 4.1.6, a_i^* $(i = 1, \ldots, n)$ has a representative $\widetilde{a_i^*}$ with

$$\deg \widetilde{a_i^*} \leq nd_5, \quad h(\widetilde{a_i^*}) \leq n(rd_5 + h_5) \quad \text{for } i = 1, \ldots, n. \tag{8.2.3}$$

Together with Lemma 7.2.6, this implies

$$\overline{\deg} \, a_i^* \leq (nd_5)^{\exp O(r)} \quad \text{for } i = 1, \ldots, n, \tag{8.2.4}$$

$$\overline{h}(a_i^*) \leq (nd_5)^{\exp O(r)} h_5 \quad \text{for } i = 1, \ldots, n. \tag{8.2.5}$$

Let $G \in K[X]$ be a monic divisor of F of degree m, say, and set $G^*(X) := a_0^m G(X/a_0)$. Then G^* is a monic divisor of F^* in $K[X]$, so by Lemma 8.2.2,

$$G^* \in B[X], \quad G^{**} := F^*/G^* \in B[X]. \tag{8.2.6}$$

Write

$$G^*(X) = X^m + b_1^* X^{m-1} + \cdots + b_m^*.$$

Then

$$G(X) = a_0^{-m} G^*(a_0 X)$$
$$= X^m + (b_1^* a_0^{-1})X^{m-1} + b_2^* a_0^{-2} X^{m-2} + \cdots + b_m^* a_0^{-m}. \tag{8.2.7}$$

We first estimate $\overline{\deg} \, b_i^*$ for $i = 1, \ldots, m$ by making a reduction to function-field heights. For $k = 1, \ldots, q$, let $\Bbbk_k$ be the algebraic closure in $\overline{K_0}$ of $\mathbb{Q}(X_1, \ldots, X_{k-1}, X_{k+1}, \ldots, X_q)$, and

$$L_k := \Bbbk_k(X_k, w^{(1)}, \ldots, w^{(D)}), \quad \Delta_k := [L_k : \Bbbk_k(X_k)],$$

where $w^{(1)}, \ldots, w^{(D)}$ are the conjugates of w over K_0. From (8.2.4) and (7.3.9), with the notation as in Lemma 7.3.4, we deduce

$$\Delta_k^{-1} H_{L_k}((a_i^*)^{(j)}) \leq (nd_5)^{\exp O(r)}$$

for $k = 1, \ldots, q$, $j = 1, \ldots, D$, $i = 1, \ldots, m$. The polynomial G^* divides F^* in $L_k[X]$, so by (5.1.15) and (5.1.14),

$$\Delta_k^{-1} H_{L_k}((b_i^*)^{(j)}) \le (nd_5)^{\exp O(r)},$$

which together with (7.3.3) yields

$$\overline{\deg}\, b_i^* \le (nd_5)^{\exp O(r)} \text{ for } i = 1, \ldots, m. \tag{8.2.8}$$

The next step is to estimate $\overline{h}(b_i^*)$ for $i = 1, \ldots, m$. Inequalities (8.2.5) and (7.4.15) imply that for $i = 1, \ldots, n$, $j = 1, \ldots, D$ and for each $\mathbf{u} \in \mathbb{Z}^q$ with $|\mathbf{u}| \le (nd_5)^{\exp O(r)}$, $\mathcal{T}(\mathbf{u}) \ne 0$,

$$h((a_i^*)_j(\mathbf{u})) \le (nd_5)^{\exp O(r)} h_5,$$

where $(a_i^*)_j(\mathbf{u})$ is the image of a_i^* under the specialization homomorphism $\varphi_{\mathbf{u},j}$. By (8.2.6), this homomorphism is also defined on the coefficients of G^*, G^{**}. By applying $\varphi_{\mathbf{u},j}$ to the coefficients of G^* and G^{**}, we see that the image of G^* under $\varphi_{\mathbf{u},j}$ divides the image of F^* in $K_{\mathbf{u},j}[X]$. Now we infer from Corollary 4.1.4 and inequality (4.1.5),

$$h((b_i^*)_j(\mathbf{u})) \le (nd_5)^{\exp O(r)} h_5$$

for $i = 1, \ldots, m$, $j = 1, \ldots, D$ and $\mathbf{u} \in \mathbb{Z}^q$ with $|\mathbf{u}| \le (nd_5)^{\exp O(r)}$, $\mathcal{T}(\mathbf{u}) \ne 0$, where $(b_i^*)_j(\mathbf{u})$ is the image of b_i^* under $\varphi_{\mathbf{u},j}$. An application of inequality (7.4.18) with $k = 1$, using (8.2.2) and (8.2.8), then gives

$$\overline{h}(b_i^*) \le (nd_5)^{\exp O(r)} h_5 \text{ for } i = 1, \ldots, m. \tag{8.2.9}$$

Inequalities (8.2.8) and (8.2.9) mean that there are $P_{i,0}, \ldots, P_{i,D-1}, Q_i \in A_0$ such that

$$b_i^* = Q_i^{-1} \sum_{j=0}^{D-1} P_{i,j} w^j \text{ with}$$

$$\deg P_{i,j}, \deg Q_i \le (nd_5)^{\exp O(r)}, \ h(P_{i,j}), h(Q_i) \le (nd_5)^{\exp O(r)} h_5 \tag{8.2.10}$$

for $i = 1, \ldots, n$, $j = 0, \ldots, D-1$. The $P_{i,j}$ and Q_i belong to $A_0 = \mathbb{Z}[X_1, \ldots, X_q] \subset \mathbb{Z}[X_1, \ldots, X_r]$. Let $\widetilde{w}$ be the representative for w from (7.2.6). From (8.2.10), (8.2.7) and some computations, using Lemma 4.1.7, it follows that $(\widetilde{b}_0, \ldots, \widetilde{b}_m)$, given by

$$\widetilde{b}_0 = \widetilde{a}_0{}^m Q_1 \cdots Q_m,$$

$$\widetilde{b}_i = \widetilde{a}_0{}^{m-i} \prod_{\substack{k=1 \\ k \ne i}}^{m} Q_k \sum_{j=0}^{D-1} P_{i,j} \widetilde{w}^j \text{ for } i = 1, \ldots, m$$

is a tuple of representatives for G with

$$\deg \widetilde{b_i} \le (nd_5)^{\exp O(r)}, \quad h(\widetilde{b_i}) \le (nd_5)^{\exp O(r)} h_5 \text{ for } i = 0, \ldots, m.$$

This proves Proposition 8.2.3.   □

8.3 Consequences

Given degree-height estimates for certain elements $\alpha_1, \ldots, \alpha_m$ of $\overline{K}$, and given $P \in \mathbb{Z}[X, X_1, \ldots, X_m]$, we derive a degree-height estimate for β satisfying $P(\beta, \alpha_1, \ldots, \alpha_m) = 0$. Further, we give a degree-height estimate for a primitive element of a given finite extension $K(\alpha_1, \ldots, \alpha_m)$ of K. Lastly, we give degree-height estimates for solutions of systems of linear equations with coefficients from $\overline{K}$. These results are all consequences of the work from the previous section, together with a simple estimate for symmetric polynomials that we deduce below. The quantities d and h satisfy (8.1.2).

Let $\mathbf{X}_i = (X_{i,1}, \ldots, X_{i,n_i})$ $(i = 1, \ldots, m)$ be blocks of variables. The block $\mathbf{Y}_i = (Y_{i,1}, \ldots, Y_{i,n_i})$ of elementary symmetric polynomials in $\mathbf{X}_i$ is given by

$$X^{n_i} - Y_{i,1} X^{n_i-1} + \cdots + (-1)^{n_i} Y_{i,n_i} = (X - X_{i,1}) \cdots (X - X_{i,n_i}).$$

Let R be any commutative ring with 1. A polynomial $F \in R[\mathbf{X}_1, \ldots, \mathbf{X}_m]$ is called *symmetric in* $\mathbf{X}_1, \ldots, \mathbf{X}_m$ if

$$F(\sigma_1(\mathbf{X}_1), \ldots, \sigma_m(\mathbf{X}_m)) = F(\mathbf{X}_1, \ldots, \mathbf{X}_m)$$

for each tuple $(\sigma_1, \ldots, \sigma_m)$, with σ_i a permutation of the variables in $\mathbf{X}_i$, for $i = 1, \ldots, m$. By the theory of symmetric polynomials, such F can be expressed as

$$F(\mathbf{X}_1, \ldots, \mathbf{X}_m) = F^{\text{sym}}(\mathbf{Y}_1, \ldots, \mathbf{Y}_m)$$

where F^{sym} is a polynomial with coefficients in R. Further, if F has total degree $\mathcal{D}$, then F^{sym} is an R-linear combination of monomials $\prod_{i=1}^{m} \prod_{h=1}^{n_i} Y_{i,h}^{k_{i,h}}$, with

$$\sum_{i=1}^{m} \sum_{h=1}^{n_h} h \cdot k_{i,h} \le \mathcal{D}. \tag{8.3.1}$$

We define the scalar product of any two tuples $\mathbf{a} = (a_i : i \in I)$, $\mathbf{b} = (b_i : i \in I)$ with entries in some commutative ring by $\langle \mathbf{a}, \mathbf{b} \rangle = \sum_{i \in I} a_i b_i$.

Lemma 8.3.1 *Let* $F \in R[\mathbf{X}_1, \ldots, \mathbf{X}_m]$ *be of total degree* $\mathcal{D}$ *and symmetric in the blocks* $\mathbf{X}_1, \ldots, \mathbf{X}_m$. *Denote by* $\mathbf{f}$ *the tuple of nonzero coefficients of* F. *Then*

$$F^{\text{sym}}(\mathbf{Y}_1, \ldots, \mathbf{Y}_m) = \sum_{\mathbf{k}} \langle \mathbf{s_k}, \mathbf{f} \rangle \prod_{i=1}^{m} \prod_{h=1}^{n_i} Y_{i,h}^{k_{i,h}},$$

where the sum is taken over all tuples $\mathbf{k} = (k_{1,1}, \ldots, k_{m,n_m})$ *of nonnegative integers with* (8.3.1), *and where* $\mathbf{s_k}$ *is a tuple with entries in* $\mathbb{Z}$ *of absolute value at most* $3^{\mathcal{D}+n_1+\cdots+n_m}$.

Proof For a tuple of nonnegative integers $\mathbf{j} = (j_{1,1}, \ldots, j_{m,n_m})$, we write $X^{\mathbf{j}} := \prod_{i=1}^{m} \prod_{h=1}^{n_m} X_{i,h}^{j_{i,h}}$, $Y^{\mathbf{j}} := \prod_{i=1}^{m} \prod_{h=1}^{n_m} Y_{i,h}^{j_{i,h}}$. For a tuple of nonnegative integers $\mathbf{k} = (k_{1,1}, \ldots, k_{m,n_m})$ with $k_{i,1} \geq \cdots \geq k_{i,n_i} \geq 0$ for $i = 1, \ldots, m$, let $\mathcal{J}_{\mathbf{k}}$ be the minimal set of tuples of nonnegative integers $\mathbf{j} = (j_{1,1}, \ldots, j_{m,n_m})$ such that $\mathbf{k} \in \mathcal{J}_{\mathbf{k}}$ and

$$F_{\mathbf{k}} := \sum_{\mathbf{j} \in \mathcal{J}_{\mathbf{k}}} X^{\mathbf{j}}$$

is symmetric in $\mathbf{X}_1, \ldots, \mathbf{X}_m$. Then

$$F = \sum_{\mathbf{k}} f_{\mathbf{k}} F_{\mathbf{k}},$$

where $f_{\mathbf{k}} \in R$ and the sum is taken over those tuples $\mathbf{k}$ with

$$k_{i,1} \geq \cdots \geq k_{i,n_i} \geq 0 \text{ for } i = 1, \ldots, m, \quad \sum_{i=1}^{m} \sum_{h=1}^{n_i} k_{i,h} \leq \mathcal{D}.$$

By the theory of symmetric polynomials, $F_{\mathbf{k}}^{\mathrm{sym}}$ has its coefficients in $\mathbb{Z}$. It suffices to show that these coefficients have absolute values at most $3^{\mathcal{E}+n}$, where $n := n_1 + \cdots + n_m$ and $\mathcal{E} := \sum_{i=1}^{m} \sum_{h=1}^{n_i} k_{i,h}$. We have

$$F_{\mathbf{k}}^{\mathrm{sym}}(\mathbf{Y}_1, \ldots, \mathbf{Y}_m) = \sum_{\mathbf{j}} f_{\mathbf{k},\mathbf{j}}^{\mathrm{sym}} Y^{\mathbf{j}},$$

where the sum is over the tuples $\mathbf{j} = (j_{1,1}, \ldots, j_{m,n_m})$ of nonnegative integers with $\sum_{i=1}^{m} \sum_{h=1}^{n_i} h \cdot j_{i,h} \leq \mathcal{E}$, and the $f_{\mathbf{k},\mathbf{j}}^{\mathrm{sym}}$ are integers. Let $\mathbf{D}_r$ denote the set of vectors $\mathbf{z} = (z_{1,1}, \ldots, z_{m,n_m}) \in \mathbb{C}^n$ with $|z_{i,j}| \leq r$ for $i = 1, \ldots, m$, $h = 1, \ldots, n_i$. Let γ be the circle with center 0 and radius 1 in the complex plane, traversed counterclockwise. Recall that if $x_{i,1}, \ldots, x_{i,n_i}, y_{i,1}, \ldots, y_{i,n_i} \in \mathbb{C}$ are such that $X^n - y_{i,1}X^{n-1} + \cdots + (-1)^n y_{i,n_i} = (X - x_{i,1}) \cdots (X - x_{i,n_i})$, then $\max_h |x_{i,h}| \leq 1 + \max_h |y_{i,h}|$. This leads to

$$|f_{\mathbf{k},\mathbf{j}}^{\mathrm{sym}}| = (2\pi)^{-n} \left| \oint_{\gamma} \cdots \oint_{\gamma} F_{\mathbf{k}}^{\mathrm{sym}}(\mathbf{z}) \prod_{i=1}^{m} \prod_{h=1}^{n_h} (z_{i,h}^{-j_{i,h}-1} dz_{i,h}) \right|$$

$$\leq \sup_{\mathbf{y} \in \mathbf{D}_1} |F_{\mathbf{k}}^{\mathrm{sym}}(\mathbf{y})| \leq \sup_{\mathbf{x} \in \mathbf{D}_2} |F_{\mathbf{k}}(\mathbf{x})|$$

$$\leq |\mathcal{J}_{\mathbf{k}}| \cdot 2^{\mathcal{E}} \leq \binom{n+\mathcal{E}-1}{\mathcal{E}} 2^{\mathcal{E}} \leq \sum_{k=0}^{n+\mathcal{E}-1} \binom{n+\mathcal{E}-1}{k} 2^k \leq 3^{\mathcal{E}+n},$$

as required. $\qquad\square$

Before proving the result mentioned in the beginning of this section, we make a simple observation. Let $\alpha \in \overline{K}^*$. If $(\widetilde{g_0}, \ldots, \widetilde{g_n})$ is a tuple of representatives for α, i.e., for its minimal polynomial F_α over K, then clearly its reverse $(\widetilde{g_n}, \ldots, \widetilde{g_0})$ is a tuple of representatives for α^{-1}. This shows that

$$\alpha \prec (d^*, h^*) \Leftrightarrow \alpha^{-1} \prec (d^*, h^*). \tag{8.3.2}$$

We now state and prove the main result of this section. In the somewhat elaborate computations, we use the properties of heights and lengths of polynomials, stated in (4.1.7), (4.1.8) and Lemma 4.1.7.

Proposition 8.3.2 *Let $P \in \mathbb{Z}[X, X_1, \ldots, X_m]$ be such that*

$$\deg_X P \geq 1, \quad P \text{ is monic in } X$$

and let $\alpha_1, \ldots, \alpha_m \in \overline{K}$ be such that

$$\deg_K \alpha_i = n_i, \quad \alpha_i \prec (d_6, h_6) \quad \text{for } i = 1, \ldots, m,$$

where $d_6 \geq d$, $h_6 \geq h$. Lastly, let $\beta \in \overline{K}$ satisfy

$$P(\beta, \alpha_1, \ldots, \alpha_m) = 0.$$

Then

$$\deg_K \beta \leq (\deg_X P) \cdot n_1 \cdots n_m,$$
$$\beta \prec \left(\mathcal{R}_1^{\exp O(r)}, \mathcal{R}_1^{\exp O(r)}(h(P) + h_6) \right),$$

where $\mathcal{R}_1 := 2m \cdot n_1 \cdots n_m \cdot \deg P \cdot d_6$.

Proof The estimate for $\deg_K \beta$ being clear, we proceed with computing a degree-height estimate for β.

For $i = 1, \ldots, m$, let $\mathbf{X}_i = (X_{i,1}, \ldots, X_{i,n_i})$ and let $\mathbf{Y}_i = (Y_{i,1}, \ldots, Y_{i,n_i})$ be the elementary symmetric polynomials in $\mathbf{X}_i$. Consider the polynomial

$$Q(X, \mathbf{X}_1, \ldots, \mathbf{X}_m) = \prod_{h_1=1}^{n_1} \cdots \prod_{h_m=1}^{n_m} P(X, X_{1,h_1}, \ldots, X_{m,h_m}).$$

This is symmetric in $\mathbf{X}_1, \ldots, \mathbf{X}_m$, hence

$$Q(X, \mathbf{X}_1, \ldots, \mathbf{X}_m) = Q^{\mathrm{sym}}(X, \mathbf{Y}_1, \ldots, \mathbf{Y}_m),$$

for some polynomial Q^{sym} with integer coefficients. For $i = 1, \ldots, m$, let $(g_{i,0}, \ldots, g_{i,n_i})$ be a tuple of representatives for F_{α_i} in $\mathbb{Z}[X_1, \ldots, X_r]$, with

$$\deg g_{i,h} \leq d_6, \quad h(g_{i,h}) \leq h_6 \quad \text{for } h = 1, \ldots, n_i. \tag{8.3.3}$$

Then the monic minimal polynomial of α_i over K is given by

$$F_{\alpha_i} := X^{n_i} + \sum_{h=1}^{n_i} \frac{g_{i,h}(z_1, \ldots, z_r)}{g_{i,0}(z_1, \ldots, z_r)} X^{n_i - h}.$$

Let $\alpha_i^{(h)}$ ($h = 1, \ldots, n_i$) be the conjugates of α_i over K. Further, let $\mathbf{b}_i :=$ $(b_{i,1}, \ldots, b_{i,n_i})$ be the tuple of elementary symmetric functions of $\alpha_i^{(1)}, \ldots, \alpha_i^{(n_i)}$. Then $b_{i,h} = (-1)^h g_{i,j}(z_1, \ldots, z_r)/g_{i,0}(z_1, \ldots, z_r)$ for $h = 1, \ldots, n_i$. Clearly,

$$\prod_{h_1=1}^{n_1} \cdots \prod_{h_m=1}^{n_m} P\left(X, \alpha_1^{(h_1)}, \ldots, \alpha_m^{(h_m)}\right) = Q^{\text{sym}}(X, \mathbf{b}_1, \ldots, \mathbf{b}_m) =: G(X)$$

is a monic polynomial in $K[X]$ with $G(\beta) = 0$. By replacing $Y_{i,h}$ by $(-1)^h g_{i,h}/g_{i,0}$ for $i = 1, \ldots, m$, $h = 1, \ldots, n_i$ in Q^{sym} we obtain a polynomial $\widetilde{G}$ in X with coefficients in $\mathbb{Q}(X_1, \ldots, X_r)$. By substituting z_i for X_i in $\widetilde{G}$ for $i = 1, \ldots, r$, we obtain again G. Clearing the denominators of $\widetilde{G}$, we get a polynomial

$$F := (g_{1,0} \cdots g_{m,0})^{\deg Q^{\text{sym}}} \widetilde{G}$$

$$= \sum_{k=0}^{\deg Q^{\text{sym}}} \left(\sum_{\mathbf{u}} a_k(\mathbf{u}) \prod_{i=1}^{m} \prod_{h=0}^{n_i} g_{i,h}^{u_{i,h}} \right) X^k \in \mathbb{Z}[X, X_1, \ldots, X_r], \qquad (8.3.4)$$

where the inner sum is taken over tuples $\mathbf{u} = (u_{1,0}, \ldots, u_{m,n_m})$ of nonnegative integers with $\sum_{i,h} u_{i,h} = m \deg Q^{\text{sym}}$ and the $a_k(\mathbf{u})$ are up to sign coefficients of Q^{sym}. According to the definition, we have

$$G \prec (\deg F, h(F)).$$

Since $G \in K[X]$ is monic and $G(\beta) = 0$, the monic minimal polynomial F_β of β over K divides G. So by Proposition 8.2.3,

$$\beta \prec ((2d^*)^{\exp O(r)}, (2d^*)^{\exp O(r)} h^*),$$

$$\text{where } d^* := \deg G \cdot \max(d, \deg F), \quad h^* := \max(h, h(F)). \qquad (8.3.5)$$

We estimate the right-hand side of (8.3.5) from above. Let $n := n_1 + \cdots + n_m + 1$ and $\mathcal{D} := \deg P$, $H := H(P)$. We have

$$\deg Q = \mathcal{D} \cdot n_1 \cdots n_m,$$

$$L(Q) \le L(P)^{n_1 \cdots n_m} \le \left(\binom{\mathcal{D}+n}{\mathcal{D}} H \right)^{n_1 \cdots n_m},$$

where in the second estimate we used (4.1.7) and (4.1.8). From Lemma 8.3.1, we infer

$$\deg Q^{\text{sym}} \le \mathcal{D} \cdot n_1 \cdots n_m, \qquad (8.3.6)$$

$$H(Q^{\text{sym}}) \le 3^{\mathcal{D} n_1 \cdots n_m + n} \left(\binom{\mathcal{D}+n}{\mathcal{D}} H \right)^{n_1 \cdots n_m}. \qquad (8.3.7)$$

Using (8.3.6), we get

$$\deg G \le \mathcal{D} \cdot n_1 \cdots n_m$$

and, using (8.3.3), (8.3.4), and (8.3.6),

$$\deg F \le m \deg Q^{\mathrm{sym}}(d_6 + 1) \le (m+1)\mathcal{D}n_1 \cdots n_m \cdot d_6,$$

leading to

$$d^* \le (m+1)(\mathcal{D}n_1 \cdots n_m)^2 d_6. \tag{8.3.8}$$

For the height of F we get, using (4.1.7), (4.1.8), (8.3.7), (8.3.3), and (8.3.4),

$$H(F) \le L(F) \le L(Q^{\mathrm{sym}})\left(\max_{i,h} L(g_{i,h})\right)^{m\mathcal{D}n_1\cdots n_m}$$

$$\le 3^{\mathcal{D}n_1\cdots n_m + n}\binom{\mathcal{D}n_1\cdots n_m + n}{n}\binom{\mathcal{D}+n}{\mathcal{D}}^{n_1\cdots n_m} H^{n_1\cdots n_m}\left(\binom{d_6+r}{r}\exp h_6\right)^{m\mathcal{D}n_1\cdots n_m}.$$

Using $n \le 2m \cdot n_1 \cdots n_m$, this leads to

$$h(F) \le 5m\mathcal{D} \cdot n_1 \cdots n_m(\log(\mathcal{D} + n) + r + d_6 + h(P) + h_6),$$

which is clearly an upper bound for h^*. By substituting this bound and the upper bound for d^* from (8.3.8) into (8.3.5), we arrive at

$$F_\beta < \left(\mathcal{R}_1^{\exp O(r)}, \mathcal{R}_1^{\exp O(r)}(h(P) + h_6)\right).$$

This proves Proposition 8.3.2.   □

Corollary 8.3.3 *Let $Q \in \mathbb{Z}[X_1, \ldots, X_m]$ be a nonconstant polynomial, and let $\alpha_1, \ldots, \alpha_m \in \overline{K}$ be such that*

$$\deg_K \alpha_i = n_i, \quad \alpha_i \prec (d_6, h_6) \quad \text{for } i = 1, \ldots, m,$$

where $d_6 \ge d$, $h_6 \ge h$. Then $\deg_K Q(\alpha_1, \ldots, \alpha_m) \le n_1 \cdots n_m$ and

$$Q(\alpha_1, \ldots, \alpha_m) \prec \left(\mathcal{R}_2^{\exp O(r)}, \mathcal{R}_2^{\exp O(r)}(h(Q) + h_6)\right),$$

where $\mathcal{R}_2 := 2m \cdot n_1 \cdots n_m \cdot \deg Q \cdot d_6$.

Proof Apply Proposition 8.3.2 with $P = X - Q(X_1, \ldots, X_m)$.   □

Corollary 8.3.4 *Let $\alpha_0, \alpha_1, \ldots, \alpha_m \in \overline{K}$ be such that*

$$\alpha_0 \in L := K(\alpha_1, \ldots, \alpha_m),$$

$$\deg_K \alpha_i = n_i, \quad \alpha_i \prec (d_6, h_6) \quad \text{for } i = 0, \ldots, m,$$

where $d_6 \ge d$, $h_6 \ge h$. Put $\mathcal{E} := [L : K]$ and

$$\mathcal{R}_3 := 2m \cdot n_1 \cdots n_m \cdot d_6, \quad \mathcal{R}_4 := 2m \cdot n_0^{n_0} \cdots n_m^{n_m} \cdot d_6.$$

(i) There is $\theta \in L$ such that $L = K(\theta)$, θ has monic minimal polynomial $F_\theta \in A[X]$ over K, and

$$\theta \overset{\text{int}}{\prec} \left(\mathcal{R}_3^{\exp O(r)}, \mathcal{R}_3^{\exp O(r)} h_6 \right).$$

(ii) We have $\alpha_0 = \sum_{j=0}^{\mathcal{E}-1} p_j \theta^j$, where

$$p_j \in K, \quad p_j \prec \left(\mathcal{R}_4^{\exp O(r)}, \mathcal{R}_4^{\exp O(r)} h_6 \right) \quad \text{for } j = 0, \ldots, \mathcal{E} - 1.$$

Proof We start with some preliminaries, before proving (i) and (ii).

Let $\sigma_1, \ldots, \sigma_\mathcal{E} : L \hookrightarrow \overline{K}$ denote the K-isomorphic embeddings of L. Denote by $\alpha_k^{(1)}, \ldots, \alpha_k^{(n_k)}$ the conjugates of α_k over K, for $k = 0, \ldots, m$.

There are rational integers $a_1, \ldots, a_m$ with $|a_i| \le \mathcal{E}^2$ for $i = 1, \ldots, m$, such that the quantities $\sum_{j=1}^m a_j \sigma_i(\alpha_j)$ $(i = 1, \ldots, \mathcal{E})$ are pairwise distinct. Let $\gamma := \sum_{j=1}^m a_j \alpha_j$. Then $\sigma_i(\gamma)$ $(i = 1, \ldots, \mathcal{E})$ are pairwise distinct, and thus, $L = K(\gamma)$. By applying Corollary 8.3.3 with $Q := a_1 X_1 + \cdots + a_m X_m$, we get

$$\gamma \prec \left(\mathcal{R}_3^{\exp O(r)}, \mathcal{R}_3^{\exp O(r)} h_6 \right).$$

That is, there are $g_i \in \mathbb{Z}[X_1, \ldots, X_r]$ $(i = 0, \ldots, \mathcal{E})$ such that for the monic minimal polynomial F_γ of γ over K, we have

$$F_\gamma = X^\mathcal{E} + (b_1/b_0) X^{\mathcal{E}-1} + \cdots + (b_\mathcal{E}/b_0), \quad \text{with}$$

$$b_i := g_i(z_1, \ldots, z_r),$$

$$\deg g_i \le \mathcal{R}_3^{\exp O(r)}, \quad h(g_i) \le \mathcal{R}_3^{\exp O(r)} h_6 \quad \text{for } i = 0, \ldots, \mathcal{E}. \qquad (8.3.9)$$

(i) Let $\theta := b_0 \gamma$. Then $L = K(\theta)$, θ has monic minimal polynomial

$$F_\theta = b_0^\mathcal{E} F_\theta(X/b_0) = X^\mathcal{E} + b_1 X^{\mathcal{E}-1} + b_0 b_2 X^{\mathcal{E}-2} + \cdots + b_0^{\mathcal{E}-1} b_\mathcal{E} \in A[X]$$

over K, and by Lemma 4.1.7, applied with the polynomials g_i,

$$\theta \overset{\text{int}}{\prec} \left(\mathcal{R}_3^{\exp O(r)}, \mathcal{R}_3^{\exp O(r)} h_6 \right).$$

This proves (i).

(ii) There are $q_j \in K$ $(j = 0, \ldots, \mathcal{E} - 1)$ such that

$$\alpha_0 = \sum_{j=0}^{\mathcal{E}-1} q_j \gamma^j.$$

Hence,

$$\sigma_i(\alpha_0) = \sum_{j=0}^{\mathcal{E}-1} q_j \cdot \sigma_i(\gamma)^j \quad \text{for } k = 1, \ldots, m, \, i = 1, \ldots, \mathcal{E}.$$

By Cramer's rule, we have

$$q_j = \Delta_j/\Delta \ \text{ for } j = 0,\dots,\mathcal{E}-1,$$

where $\Delta := \det(\sigma_i(\gamma)^j)_{i=1,\dots,\mathcal{E},\, j=0,\dots,\mathcal{E}-1}$, and Δ_j is the determinant obtained by replacing $\sigma_i(\gamma)^j$ by $\sigma_i(\alpha_0)$, for $i = 1,\dots,\mathcal{E}$. We clearly have

$$\alpha_0 = \sum_{j=0}^{\mathcal{E}-1} p_j \beta^j \text{ with } p_j = \Delta_j/\left(\Delta g_0^j\right) \text{ for } j = 0,\dots,\mathcal{E}-1. \tag{8.3.10}$$

Using $\sigma_i(\gamma) = \sum_{\ell=1}^{m} a_\ell \sigma_i(\alpha_\ell)$, we see that both Δ_j and Δ are polynomials with integer coefficients in $\alpha_0^{(1)},\dots,\alpha_0^{(n_0)},\dots,\alpha_m^{(1)},\dots,\alpha_m^{(n_m)}$. These polynomials have degrees at most $\mathcal{E}^2$. Further, a similar computation as carried out in the proof of Proposition 8.3.2, using (4.1.7) and (4.1.8), shows that these polynomials have logarithmic heights at most $O(\mathcal{E}^2 \log(2m\mathcal{E}^2))$ with the implied constant being absolute. Using Corollary 8.3.3 and $\mathcal{E} \le n_1 \cdots n_m$, this shows that

$$\Delta_0,\dots,\Delta_{\mathcal{E}-1},\, \Delta \prec \left(\mathcal{R}_4^{\exp O(r)},\, \mathcal{R}_4^{\exp O(r)} h_6\right),$$

while (8.3.9) gives $b_0 \overset{\text{int}}{\prec} (\mathcal{R}_3^{\exp O(r)}, \mathcal{R}_3^{\exp O(r)} h_6)$. Now applying (8.3.2) and Corollary 8.3.3 to (8.3.10), using the estimates for Δ_j, Δ, g_0 just established, we arrive at

$$p_j \prec \left(\mathcal{R}_4^{\exp O(r)},\, \mathcal{R}_4^{\exp O(r)} h_6\right) \ \text{ for } j = 0,\dots,\mathcal{E}-1.$$

This proves (ii). □

Corollary 8.3.5 *Let k,l be positive integers, $\mathcal{A} = (\alpha_{i,j})_{i=1,\dots,k\, j=1,\dots,l}$ a $k\times l$-matrix and $\mathbf{b} = (\beta_1,\dots,\beta_k)^T$ a k-dimensional column vector, both with entries in $\overline{K}$, satisfying*

$$\deg_K \alpha_{i,j} \le n_{i,j},\ \alpha_{i,j} \prec (d_7, h_7),\ \deg_K \beta_i \le n_i,\ \beta_i \prec (d_7, h_7)$$
$$\text{for } i = 1,\dots,k,\ j = 1,\dots,l,$$

where $d_7 \ge d$, $h_7 \ge h$. Suppose that

$$\mathcal{A}\mathbf{x} = \mathbf{b} \tag{8.3.11}$$

has a solution $\mathbf{x} = (x_1,\dots,x_l)^T \in \mathcal{A}^l$. Then (8.3.11) has such a solution with

$$x_j \overset{\text{int}}{\prec} \left(\mathcal{R}_5^{\exp O(r \log^* r)} h_7,\, \mathcal{R}_5^{\exp O(r \log^* r)} h_7^{r+1}\right) \text{ for } j = 1,\dots,l, \tag{8.3.12}$$

where $\mathcal{R}_5 := \left(\prod_{i=1}^{k} \prod_{j=1}^{l} n_{i,j}^{n_{i,j}}\right) \cdot \left(\prod_{i=1}^{k} n_i^{n_i}\right) 2kld_7$.

Proof Let L be the extension of K generated by the $\alpha_{i,j}$ and the β_i, for $i = 1,\ldots,k$, $j = 1,\ldots,l$, and put $\mathcal{E} := [L : K]$. By Corollary 8.3.4, there exists $\theta \in L$ such that $L = K(\theta)$, θ has monic minimal polynomial $F_\theta \in A[X]$, and moreover, there exist $a_{i,j,h}, b_{i,h} \in K$, for $i = 1,\ldots,k$, $j = 1,\ldots,l$, $h = 0,\ldots,\mathcal{E} - 1$ such that

$$\alpha_{i,j} = \sum_{h=0}^{\mathcal{E}-1} a_{i,j,h}\theta^h, \quad \beta_i = \sum_{h=0}^{\mathcal{E}-1} b_{i,h}\theta^h,$$

$$\text{for } i = 1,\ldots,k, \ j = 1,\ldots,l,$$

and

$$a_{i,j,h}, \ b_{i,h} \prec \left(\mathcal{R}_5^{\exp O(r)}, \ \mathcal{R}_5^{\exp O(r)}h_7\right)$$

$$\text{for } i = 1,\ldots,k, \ j = 1\ldots l, \ h = 0,\ldots,\mathcal{E} - 1.$$

This means that

$$a_{i,j,h} = \frac{g'_{i,j,h}(z_1,\ldots,z_r)}{g''_{i,j,h}(z_1,\ldots,z_r)}, \quad b_{i,h} = \frac{g'_{i,j}(z_1,\ldots,z_r)}{g''_{i,j}(z_1,\ldots,z_r)},$$

where $g'_{i,j,h}, g''_{i,j,h}, g'_{i,j}, g''_{i,j}$ are polynomials from $\mathbb{Z}[X_1,\ldots,X_r]$ of total degree at most $\mathcal{R}_5^{\exp O(r)}$ and logarithmic height at most $\mathcal{R}_5^{\exp O(r)}h_7$, for $i = 1,\ldots,k$, $j = 1,\ldots,l$, $h = 0,\ldots,\mathcal{E} - 1$. Take the product of the denominators of the $a_{i,j,h}, b_{i,j}$,

$$c_0 := \left(\prod_{i=1}^{k}\prod_{j=1}^{l}\prod_{h=0}^{\mathcal{E}-1} g''_{i,j,h}(z_1,\ldots,z_r)\right) \cdot \left(\prod_{i=1}^{k}\prod_{j=1}^{l} g''_{i,j}(z_1,\ldots,z_r)\right)$$

and put

$$a'_{i,j,h} := c_0 a_{i,j,h}, \ b'_{i,h} := c_0 b_{i,h} \ \text{for all } i,j,h.$$

Then the $a'_{i,j,h}, b'_{i,h}$ all belong to A and by Lemma 4.1.7, we have

$$a'_{i,j,h}, \ b'_{i,h} \overset{\text{int}}{\prec} \left(\mathcal{R}_5^{\exp O(r)}, \ \mathcal{R}_5^{\exp O(r)}h_7\right)$$

$$\text{for } i = 1,\ldots,k, \ j = 1,\ldots,l, \ h = 0,\ldots,\mathcal{E} - 1. \tag{8.3.13}$$

Writing

$$A_h = \left(a'_{i,j,h}\right)_{i=1,\ldots,k, \ j=1,\ldots,l}, \quad \mathbf{b}_h = \left(b'_{1,h},\ldots,b'_{k,h}\right)^T,$$

we get

$$c_0 A = \sum_{h=0}^{\mathcal{E}-1} \theta^h A_h, \quad c_0 \mathbf{b} = \sum_{h=0}^{\mathcal{E}-1} \theta^h \mathbf{b}_h.$$

Therefore, the solution set in $\mathbf{x} \in A^l$ of (8.3.11) is equal to that of

$$\sum_{h=0}^{\mathcal{E}-1} \theta^h \mathcal{A}_h \mathbf{x} = \sum_{h=0}^{\mathcal{E}-1} \theta^h \mathbf{b}_h,$$

whence to that of

$$\mathcal{A}_h \mathbf{x} = \mathbf{b}_h \quad \text{for } h = 0, \dots, \mathcal{E} - 1, \tag{8.3.14}$$

since $1, \theta, \dots, \theta^{\mathcal{E}-1}$ are linearly independent over K. Let $\widetilde{\mathcal{A}_h}$ and $\widetilde{\mathbf{b}_h}$ consist of representatives in $\mathbb{Z}[X_1, \dots, X_r]$ of the entries of $\mathcal{A}_h$ and $\mathbf{b}_h$, which by (8.3.13) we may choose of total degrees at most $\mathcal{R}_5^{\exp O(r)}$ and logarithmic heights at most $\mathcal{R}_5^{\exp O(r)} h_7$. Then system (8.3.14) is equivalent to

$$\widetilde{\mathcal{A}_h}\widetilde{\mathbf{x}} \equiv \widetilde{\mathbf{b}_h} \pmod{\mathcal{I}} \quad \text{for } h = 0, \dots, \mathcal{E} - 1,$$

to be solved in $\widetilde{\mathbf{x}} \in \mathbb{Z}[X_1, \dots, X_r]^l$, where the components of $\widetilde{\mathbf{x}}$ are representatives for the components of $\mathbf{x}$, and we can rewrite the latter system as

$$\widetilde{\mathcal{A}_h}\widetilde{\mathbf{x}} = \widetilde{\mathbf{b}_h} + \sum_{i=1}^{M} f_i \widetilde{\mathbf{y}_{i,h}} \quad (h = 0, \dots, \mathcal{E} - 1)$$

with solutions

$$(\widetilde{\mathbf{x}}, \widetilde{\mathbf{y}_{1,0}}, \dots, \widetilde{\mathbf{y}_{M,\mathcal{E}-1}}) \in \mathbb{Z}[X_1, \dots, X_r]^{l(M\mathcal{E}+1)},$$

where $\{f_1, \dots, f_M\}$ is the set of generators for $\mathcal{I}$ that we have chosen in the very beginning. Note that this system contains altogether

$$k\mathcal{E} \le k\left(\prod_{i=1}^{k}\prod_{j=1}^{l} n_{i,j}\right) \cdot \left(\prod_{i=1}^{k} n_i\right)$$

linear equations, and that each coefficient of this system is a polynomial in $\mathbb{Z}[X_1, \dots, X_r]$ of total degree at most $\mathcal{R}_5^{\exp O(r)}$ and logarithmic height at most $\mathcal{R}_5^{\exp O(r)} h_7$. So by Theorem 6.1.5, if this system has a solution with coordinates in $\mathbb{Z}[X_1, \dots, X_r]$, then it has such a solution, of which each coordinate has total degree and logarithmic height at most

$$\mathcal{R}_5^{\exp O(r \log^* r)} h_7 \quad \text{and} \quad \mathcal{R}_5^{\exp O(r \log^* r)} h_7^{r+1},$$

respectively. This implies that system (8.3.14), hence the original system (8.3.11) we started with, has a solution $\mathbf{x} = (x_1, \dots, x_l)^T \in A^l$ with (8.3.12). This completes the proof of Corollary 8.3.5. □

9

Proofs of the Results from Sections 2.2 to 2.5
Use of Specializations

In this chapter, we prove the general effective theorems from Chapter 2 on unit equations, Thue equations, hyper- and superelliptic equations, and the Catalan equation over finitely generated domains of the form $A = \mathbb{Z}[z_1, \ldots, z_r]$.

We consider our equations over a more convenient finitely generated over-ring B of A, constructed in Chapter 7. Then, to prove our theorems, we reduce our equations first to the function field case and then to the number field case by means of the effective specializations described in Chapter 7.

Sections 9.1, 9.2, and 9.3 are devoted to unit equations, Thue equations, and hyper- and superelliptic equations, while Section 9.4 deals with the Catalan equation. Using some results from Chapter 7, in Section 9.1, we first reduce the estimates of the sizes of appropriate representatives of the solutions x, y in B resp. in B^* to bounding the degrees $\overline{\deg} x, \overline{\deg} y$, and heights $\overline{h}(x), \overline{h}(y)$, introduced in Chapter 7. Then, by means of the effective results from Chapter 5 concerning the corresponding equations over function fields we derive in Section 9.2 bounds for $\overline{\deg} x, \overline{\deg} y$. Finally, combining the effective specializations presented in Chapter 7 with the corresponding effective results from Chapter 4 over number fields we give in Section 9.3 effective upper bounds for $\overline{h}(x), \overline{h}(y)$, which completes the proof of our general effective results over finitely generated domains.

As was pointed out in Section 2.5, in case of the Catalan equation, it is enough to derive an effective upper bound for the unknown exponents. In Section 9.4, we combine the corresponding effective results from Chapters 4 and 5 over number fields resp. function fields with a simplified version, used by Brindza (1993) and Koymans (2017), of our general method to bound the exponents under consideration.

9.1 A Reduction

For convenience, we repeat some notation and definitions. As before, let $A = \mathbb{Z}[z_1,\ldots,z_r]$ with $r > 0$ be an integral domain of characteristic 0, which is finitely generated over $\mathbb{Z}$, and let K denote its quotient field. Then we have

$$A \cong \mathbb{Z}[X_1,\ldots,X_r]/\mathcal{I}, \qquad (9.1.1)$$

where $\mathcal{I}$ is the ideal of polynomials $f \in \mathbb{Z}[X_1,\ldots,X_r]$ such that $f(z_1,\ldots,z_r) = 0$. The ideal $\mathcal{I}$ is finitely generated. Assume as before that

$$\mathcal{I} = (f_1,\ldots,f_M) \text{ with } \deg f_i \le d, h(f_i) \le h \text{ for } i = 1,\ldots,M,$$
$$\text{where } d \ge 1, h \ge 1. \qquad (9.1.2)$$

Suppose that K has transcendence degree $q \ge 0$ over $\mathbb{Q}$. If $q > 0$, we assume without loss of generality that $z_1 = X_1,\ldots,z_q = X_q$ is a transcendence basis for $K/\mathbb{Q}$. We define as before,

$$A_0 := \mathbb{Z}[X_1,\ldots,X_q], \quad K_0 := \mathbb{Q}(X_1,\ldots,X_q) \quad \text{if } q > 0,$$
$$A_0 := \mathbb{Z}, \qquad\qquad K_0 := \mathbb{Q} \qquad\qquad\quad \text{if } q = 0.$$

For $q \ge 0$, A_0 is a unique factorization domain. For $f \in A_0\backslash\{0\}$, we denote by $\deg f$ and $h(f)$ the (total) degree and logarithmic height of f, where in the case of $q = 0$, we put $\deg f := 0$ and $h(f) := \log|f|$ if $f \in \mathbb{Z}\backslash\{0\}$.

By Proposition 7.2.5, there is a $w \in A$, integral over A_0, such that $K = K_0(w)$ and w has minimal polynomial $\mathcal{F}(X) = X^D + \mathcal{F}_1 X^{D-1} + \cdots + \mathcal{F}_D$ over K_0 such that

$$\mathcal{F}_j \in A_0, \ \deg \mathcal{F}_j \le (2d)^{\exp O(r)}, \ h(\mathcal{F}_j) \le (2d)^{\exp O(r)} h \qquad (9.1.3)$$

for $j = 1,\ldots,D$ and, by Lemma 7.2.3,

$$D \le d^t \text{ where } t = r - q. \qquad (9.1.4)$$

In what follows, we fix such an element w. With every $\alpha \in K$, we associate an up to sign unique tuple $P_{\alpha,0},\ldots,P_{\alpha,D-1},Q_\alpha$ from A_0 such that (7.2.7) holds, i.e.,

$$\alpha = Q_\alpha^{-1} \sum_{j=0}^{D-1} P_{\alpha,j} w^j \text{ with } Q_\alpha \ne 0, \ \gcd(P_{\alpha,0},\ldots,P_{\alpha,D-1},Q_\alpha) = 1.$$

Then, as in Section 7.2, we define

$$\overline{\deg}\,\alpha := \max(\deg P_{\alpha,0},\ldots,\deg P_{\alpha,D-1},\deg Q_\alpha),$$
$$\overline{h}\,(\alpha) := \max(h(P_{\alpha,0}),\ldots,h(P_{\alpha,D-1}),h(Q_\alpha)).$$

We shall deal separately with unit equations, Thue equations, and hyper- and superelliptic equations.

9.1.1 Unit Equations

We shall deduce our general Theorem 2.2.1 on unit equations from the following.

Proposition 9.1.1 *Let* $g \in A_0 \setminus \{0\}$ *and put*

$$d_4 := \max(d, \deg g, \deg \mathcal{F}_1, \ldots, \deg \mathcal{F}_D),$$
$$h_4 := \max(h, h(g), h(\mathcal{F}_1), \ldots, h(\mathcal{F}_D)).$$
(9.1.5)

Define the domain $B := A_0[w, g^{-1}]$. *Then for every pair* (ε, η) *with*

$$\varepsilon + \eta = 1, \ \varepsilon, \eta \in B^*,$$
(9.1.6)

we have

$$\overline{\deg}\, \varepsilon, \overline{\deg}\, \eta \leq 4qD^2 d_4,$$
(9.1.7)

$$\overline{h}(\varepsilon), \overline{h}(\eta) \leq \exp O(2D(q + d_4)(\log^*(2D(q + d_4)))^2 \cdot h_4).$$
(9.1.8)

The proof of Proposition 9.1.1 is given in Subsections 9.2.1 and 9.3.1. In Subsection 9.2.1, we deduce the degree bound (9.1.7). Here, our main tool is Mason's effective result on S-unit equations in function fields; see Mason (1983) or Theorem 5.2.1 in Section 5.2. In Subsection 9.3.1, we prove (9.1.8) by combining (9.1.7) with our general specialization method from Evertse and Győry (2013), as presented in Chapter 7, and with an effective result of Győry and Yu (2006) on S-unit equations over number fields; see also Theorem 4.3.1. We now deduce Theorem 2.2.1 from Proposition 9.1.1.

Proof of Theorem 2.2.1 Let a, b, c be the coefficients in the unit equation (2.2.1), and $\tilde{a}, \tilde{b}, \tilde{c}$ the representatives for a, b, c from the statement of Theorem 2.2.1. Then by assumption

$$\max(\deg f_1, \ldots, \deg f_M, \deg \tilde{a}, \deg \tilde{b}, \deg \tilde{c}) \leq d$$

and

$$\max(h(f_1), \ldots, h(f_M), h(\tilde{a}), h(\tilde{b}), h(\tilde{c})) \leq h,$$

where $d \geq 1, h \geq 1$. Further, as was previously mentioned, by Proposition 7.2.5 and Lemma 7.2.3, we have $K = K_0(w)$ with $w \in A$, integral over A_0, and w has minimal polynomial $\mathcal{F}(X) = X^D + \mathcal{F}_1 X^{D-1} + \cdots + \mathcal{F}_D$ over K_0 such that (9.1.3) and (9.1.4) hold.

In view of Proposition 7.2.7, there is a nonzero $g \in A_0$ such that

$$A \subseteq B := A_0[w, g^{-1}], \quad a, b, c \in B^*$$

and

$$\deg g \leq (2d)^{\exp O(r)}, \quad h(g) \leq (2d)^{\exp O(r)}h. \tag{9.1.9}$$

Let (x, y) be a solution of Eq. (2.2.1), and put $x_1 := ax/c, y_1 := by/c$. Then $x_1 + y_1 = 1$ and $x_1, y_1 \in B^*$. By (7.4.5), we have $d_4 \leq (2d)^{\exp O(r)}, h_4 \leq (2d)^{\exp O(r)}h$. We apply now Proposition 9.1.1 with $\varepsilon = x_1, \eta = y_1$. It follows from Proposition 9.1.1 that

$$\overline{\deg} x_1 \leq 4qd^{2t}(2d)^{\exp O(r)} \leq (2d)^{\exp O(r)}, \tag{9.1.10}$$

$$\overline{h}(x_1) \leq \exp((2d)^{\exp O(r)}h). \tag{9.1.11}$$

We use Lemma 7.3.1 with $\lambda = a/c$, which is represented by $(\tilde{a}, \tilde{c})$. Choosing $a = \tilde{a}, b = \tilde{c}$, and $\alpha = x$ in Lemma 7.3.1, we have $\lambda\alpha = x_1$, and in view of (9.1.10) and (9.1.11), we have

$$d_2 \leq (2d)^{\exp O(r)}, \quad h_2 \leq \exp((2d)^{\exp O(r)}h).$$

We infer that x, x^{-1} have representatives $\tilde{x}, \tilde{x}'$ in $\mathbb{Z}[X_1, \ldots, X_r]$ such that

$$\deg \tilde{x}, \deg \tilde{x}', h(\tilde{x}), h(\tilde{x}') \leq \exp((2d)^{\exp O(r)}h).$$

In the same way, one can derive similar upper bounds for the degrees and logarithmic heights of representatives for y and y^{-1}. This completes the proof of Theorem 2.2.1. $\qquad\square$

In our proof of Theorem 2.2.3, we need the following lemma.

Lemma 9.1.2 *Let* $A \cong \mathbb{Z}[X_1, \ldots, X_r]/(f_1, \ldots, f_M)$ *be an integral domain. Let* $\beta \in A\backslash\{0\}$, *and let* $\widetilde{\beta} \in \mathbb{Z}[X_1, \ldots, X_r]$ *be a representative for* β. *Then*

$$A[\beta^{-1}] \cong \mathbb{Z}[X_1, \ldots, X_r, Y]/(f_1, \ldots, f_M, 1 - \widetilde{\beta}Y).$$

Proof For $f \in A[Y]$, define $f^* := Y^{\deg f}f(Y^{-1})$. Then for $f \in A[Y]$, we have

$$f(\beta^{-1}) = 0 \Leftrightarrow f^*(\beta) = 0 \Leftrightarrow f^* = (Y - \beta)h^* \text{ for some } h^* \in A[Y]$$
$$\Leftrightarrow f = (1 - \beta Y)h \text{ for some } h \in A[Y].$$

Now, via the ring homomorphism $A[Y] \mapsto A[\beta^{-1}] : f \mapsto f(\beta^{-1})$, we obtain $A[Y]/(1 - \beta Y) \cong A[\beta^{-1}]$. This implies

$$\mathbb{Z}[X_1, \ldots, X_r, Y]/(f_1, \ldots, f_M, 1 - \widetilde{\beta}Y) \cong A[Y]/(1 - \beta Y) \cong A[\beta^{-1}].$$

$\qquad\square$

Proof of Theorem 2.2.3 We keep the notation and assumptions from the statement of Theorem 2.2.3. For $i = 1, \ldots, s$, let

$$\alpha_i := g_{i,1}(z_1, \ldots, z_r), \ \beta_i := g_{i,2}(z_1, \ldots, z_r),$$

with $\alpha_i, \beta_i \in A$ so that $\gamma_i = \alpha_i / \beta_i$ and define the ring

$$\widetilde{A} := A[\alpha_1^{-1}, \beta_1^{-1}, \ldots, \alpha_s^{-1}, \beta_s^{-1}].$$

Then by repeatedly applying Lemma 9.1.2, we obtain

$$\widetilde{A} \cong \mathbb{Z}[X_1, \ldots, X_r, X_{r+1}, \ldots, X_{r+2s}]/\widetilde{\mathcal{I}},$$

with

$$\widetilde{\mathcal{I}} = (f_1, \ldots, f_M, g_{1,1}X_{r+1} - 1, g_{1,2}X_{r+2} - 1, g_{2,1}X_{r+3} - 1, g_{2,2}X_{r+4} - 1,$$
$$\ldots, g_{s,1}X_{r+2s-1} - 1, g_{s,2}X_{r+2s} - 1).$$

Let $(u_1, \ldots, v_s)$ be a solution of (2.2.2) and put

$$\varepsilon := \prod_{i=1}^{s} \gamma_i^{u_i}, \ \eta := \prod_{i=1}^{s} \gamma_i^{v_i}.$$

Then

$$a\varepsilon + b\eta = c, \ \varepsilon, \eta \in \widetilde{A}^*.$$

By Theorem 2.2.1, ε has a representative $\widetilde{\varepsilon} \in \mathbb{Z}[X_1, \ldots, X_{r+2s}]$ of degree and logarithmic height both bounded above by

$$\exp((2d)^{\exp O(r+s)} h).$$

Now Corollary 7.5.3 implies

$$|u_i| \leq \exp((2d)^{\exp O(r+s)} h) \text{ for } i = 1, \ldots, s.$$

For $|v_i|$ $(i = 1, \ldots, s)$, we derive a similar upper bound. This completes the proof of Theorem 2.2.3. □

9.1.2 Thue Equations

As in Section 2.3, let

$$F(X, Y) = a_0 X^n + a_1 X^{n-1}Y + \cdots + a_n Y^n \in A[X, Y]$$

be a binary form of degree $n \geq 3$ with discriminant $D_F \neq 0$ and let $\delta \in A \backslash \{0\}$. Recall that for $a_0, \ldots, a_n, \delta$, we have chosen representatives $\tilde{a}_0, \ldots, \tilde{a}_n, \tilde{\delta}$ such that $\tilde{\delta}$ and the discriminant $D_{\tilde{F}}$ of $\tilde{F} := \sum_{j=0}^{n} \tilde{a}_j X^{n-j}$ are not contained in $\mathcal{I}$, and that $f_1, \ldots, f_M, \tilde{a}_0, \ldots, \tilde{a}_n, \tilde{\delta}$ have degrees at most d and logarithmic heights at most h where $d \geq 1, h \geq 1$.

Theorem 2.3.1 will be deduced from the following.

Proposition 9.1.3 *Let $g \in A_0 \setminus \{0\}$ with the properties specified in Proposition 7.2.9, and consider the integral domain*

$$A \subseteq B := A_0[w, g^{-1}], \text{ where } \delta, D_F \in B^*. \qquad (9.1.12)$$

Then for the solutions of x, y of the equation

$$F(x, y) = \delta \quad \text{in } x, y \in B \qquad (9.1.13)$$

we have

$$\overline{\deg} x, \overline{\deg} y \leq (nd)^{\exp O(r)}, \qquad (9.1.14)$$

$$\overline{h}(x), \overline{h}(y) \leq \exp(n!(nd)^{\exp O(r)} h). \qquad (9.1.15)$$

Proposition 9.1.3 will be proved in Subsections 9.2.2 and 9.3.2.

We now deduce Theorem 2.3.1 from Proposition 9.1.3.

Proof of Theorem 2.3.1 Let x, y be a solution of Eq. (2.3.1) in A. In view of (9.1.12), x, y are also contained in $B := A_0[w, g^{-1}]$, where w, g have the properties specified in Proposition 7.2.5 resp. in Proposition 7.2.9. Then, by Proposition 9.1.3, the inequalities (9.1.14) and (9.1.15) hold. Applying now Corollary 7.3.2 to x and y, we infer that x, y have representatives $\tilde{x}, \tilde{y}$ in $\mathbb{Z}[X_1, \ldots, X_r]$, which satisfy (2.3.2). ☐

9.1.3 Hyper- and Superelliptic Equations

Recall that as in Section 2.4, $F(X) = a_0 X^n + a_1 X^{n-1} + \cdots + a_n \in A[X]$ is a polynomial with $a_0 \neq 0$ and with discriminant $D_F \neq 0$, that $\delta \in A \setminus \{0\}$ and that for $a_0, \ldots, a_n, \delta$, we have chosen representatives $\tilde{a}_0, \ldots, \tilde{a}_n, \tilde{\delta}$ from $\mathbb{Z}[X_1, \ldots, X_r]$ with degrees at most d and logarithmic heights at most h such that $\tilde{\delta}$ and the discriminant of $\tilde{F} := \sum_{j=0}^{n} \tilde{a}_j X^{n-j}$ are not in $\mathcal{I}$.

We shall deduce Theorem 2.4.1 from the following

Proposition 9.1.4 *Let $g \in A_0 \setminus \{0\}$ with the properties specified in Proposition 7.2.9 such that for the overring $B := A_0[w, g^{-1}]$ of A, we have $\delta, D_F \in B^*$. Further, let m be an integer ≥ 2, and assume that $n \geq 3$ if $m = 2$ and $n \geq 2$ if $m \geq 3$. Then for all solutions of x, y of the equation*

$$F(x) = \delta y^m \quad \text{in } x, y \in B, \qquad (9.1.16)$$

we have

$$\overline{\deg} x, \overline{\deg} y \le (nd)^{\exp O(r)}, \tag{9.1.17}$$

$$m \le (nd)^{\exp O(r)} \text{ if } y \notin \overline{\mathbb{Q}}, \tag{9.1.18}$$

$$\overline{h}(x), \overline{h}(y) \le \exp(m^3 (nd)^{\exp O(r)} h). \tag{9.1.19}$$

We prove (9.1.17) and (9.1.18) in Subsection 9.2.3 and Eq. (9.1.19) in Subsection 9.3.3.

We now deduce Theorem 2.4.1 from Proposition 9.1.4.

Proof of Theorem 2.4.1 Let x, y be a solution of Eq. (2.4.1). In view of $A \subseteq B$, the pair x, y is a solution also in B. Then, by Proposition 9.1.4, the inequalities (9.1.17) and (9.1.19) hold. Applying Corollary 7.3.2 to x and y, we infer that x, y have representatives $\tilde{x}, \tilde{y}$ in $\mathbb{Z}[X_1, \ldots, X_r]$ satisfying (2.4.2). □

Together with Proposition 9.1.4, the Proposition 9.1.5 implies Theorem 2.4.2.

Proposition 9.1.5 *Suppose that Eq. (9.1.16) has a solution of x, y with $y \in \overline{\mathbb{Q}}$ such that $y \ne 0$ and y is not a root of unity. Then*

$$m \le \exp\left((nd)^{\exp O(r)} h\right).$$

Proposition 9.1.5 will be proved at the end of Section 9.3.

Proof of Theorem 2.4.2 Immediate from (9.1.18) and (9.1.19). □

9.2 Bounding the Degrees

In this section, we prove separately the inequalities (9.1.7) from Proposition 9.1.1, (9.1.14) from Proposition 9.1.3, and (9.1.17) from Proposition 9.1.4. The main tools will be Theorem 5.2.1 on unit equations, Theorem 5.4.1 on Thue equations, and Theorems 5.5.1, 5.5.2 on hyper- and superelliptic equations over function fields.

We recall some notation and introduce further notation. The case $q = 0$ being trivial, in this section, we assume that $q > 0$. Let, as above, $K_0 = \mathbb{Q}(X_1, \ldots, X_q)$, $K = K_0(w), A_0 = \mathbb{Z}[X_1, \ldots, X_q], B = A_0[w, g^{-1}]$. Choose an algebraic closure $\overline{K}_0$ of K_0. Then there are D K_0-isomorphic embeddings of K into $\overline{K}_0$, which we denote by $\alpha \mapsto \alpha^{(j)}, j = 1, \ldots, D$.

As in Section 7.3, let $\Bbbk_i$ be an algebraic closure of $\mathbb{Q}(X_1, \ldots, X_{i-1}, X_{i+1}, \ldots, X_q)$ for $i = 1, \ldots, q$. Then, A_0 is contained in $\Bbbk_i[X_i]$. Consider the function field

$$L_i := \Bbbk_i(X_i, w^{(1)}, \ldots, w^{(D)}), \quad i = 1, \ldots, q,$$

where $w^{(1)} = w, \ldots, w^{(D)}$ denote the conjugates of w over K_0 in $\overline{K}_0$. Then L_i is the splitting field of the polynomial $\mathcal{F}(X) = X^D + \mathcal{F}_1 X^{D-1} + \cdots + \mathcal{F}_D$ over $\Bbbk_i(X_i)$ with $\mathcal{F}_j \in \Bbbk_i[X_i]$, $j = 1, \ldots, D$. The subring

$$B_i := \Bbbk_i[X_i, w^{(1)}, \ldots, w^{(D)}, g^{-1}]$$

of L_i contains $B = \mathbb{Z}[X_1, \ldots, X_q, w, g^{-1}]$ as a subring. Put $\Delta_i := [L_i : \Bbbk_i(X_i)]$.

Let $g_{L_i/\Bbbk_i}$ denote the genus of $L_i/\Bbbk_i$ and H_{L_i} the height taken with respect to $L_i/\Bbbk_i$. By Lemma 5.1.1, applied with $\Bbbk_i, X_i, \Bbbk_i(X_i), L_i$ instead of $\Bbbk, z, K, L$ and with $F = \mathcal{F} = X^D + \mathcal{F}_1 X^{D-1} + \cdots + \mathcal{F}_D$, and using (9.1.3), we obtain

$$g_{L_i/\Bbbk_i} \leq \Delta_i D \max_j \deg_{X_i} \mathcal{F}_j \leq \Delta_i D \max_j \deg \mathcal{F}_j. \qquad (9.2.1)$$

Let S_i denote the subset of valuations v of $L_i/\Bbbk_i$ such that $v(X_i) < 0$ or $v(g) > 0$. Every valuation of $\Bbbk_i(X_i)$ can be extended to at most Δ_i valuations of L_i. Thus, L_i has at most Δ_i valuations v with $v(X_i) < 0$ and at most $\Delta_i \deg g$ valuations with $v(g) > 0$. Hence, using also (9.1.9), we infer

$$|S_i| \leq \Delta_i + \Delta_i \deg_{X_i} g \leq \Delta_i(1 + \deg g) \leq \Delta_i(2d)^{\exp O(r)}. \qquad (9.2.2)$$

Since $w^{(1)}, \ldots, w^{(D)}$ lie in L_i and are all integral over $\Bbbk_i[X_i]$, they belong to $\mathcal{O}_{S_i}$, i.e., the ring of S_i-integers in L_i. Further, $g^{-1} \in \mathcal{O}_{S_i}$. Consequently, if $\alpha \in B = A_0[w, g^{-1}]$, then $\alpha^{(j)} \in \mathcal{O}_{S_i}$ for $j = 1, \ldots, D$, $i = 1, \ldots, q$.

9.2.1 Unit Equations

We now prove the upper bound of (9.1.7) in Proposition 9.1.1.

Proof of (9.1.7) Keeping the preceding notation, let (ε, η) be a solution of equation (9.1.6) $\varepsilon + \eta = 1$ in $\varepsilon, \eta \in B^*$. Then we have

$$\varepsilon^{(j)} + \eta^{(j)} = 1, \ \varepsilon^{(j)}, \eta^{(j)} \in \mathcal{O}_{S_i}^* \ \text{for } j = 1, \ldots, D$$

and $i = 1, \ldots, q$. We apply Theorem 5.2.1, insert the upper bounds (9.2.1), (9.2.2), and use $\deg g, \deg \mathcal{F}_j \leq d_4$ from (9.1.5) for $j = 1, \ldots, D$. It follows that for $j = 1, \ldots, D$, we have either $\varepsilon^{(j)} \in \Bbbk_i$ or

$$H_{L_i}(\varepsilon^{(j)}) \leq |S_i| + 2g_{L_i/\Bbbk_i} - 2 \leq 3\Delta_i D d_4.$$

Of course, the last upper bound is valid also if $\varepsilon^{(j)} \in \Bbbk_i$. Together with Lemma 7.3.3, this implies

$$\overline{\deg}\,\varepsilon \leq qDd_4 + q \cdot 3Dd_4 \leq 4qD^2 d_4.$$

For $\overline{\deg}\,\eta$, we obtain the same upper bound. This proves (9.1.7). $\square$

9.2.2 Thue Equations

Keeping the notation introduced at the beginning of Section 9.2, we prove the upper bound (9.1.14) from Proposition 9.1.3.

Proof of (9.1.14) Let x, y be a solution of Eq. (9.1.13). Put $F' := \delta^{-1} F$, and let $F'^{(j)}$ denote the binary form obtained by taking the jth conjugates of the coefficients of F'. Let $i \in \{1, \ldots, q\}$, $j \in \{1, \ldots, D\}$. Then $F'^{(j)} \in L_i[X, Y]$, and we get

$$F'^{(j)}(x^{(j)}, y^{(j)}) = 1 \text{ with } x^{(j)}, y^{(j)} \in \mathcal{O}_{S_i}.$$

By Theorem 5.4.1, we obtain

$$\max(H_{L_i}(x^{(j)}), H_{L_i}(y^{(j)})) \leq (8n + 62) H_{L_i}(F'^{(j)}) + 8 g_{L_i/k_i} + 4|S_i|. \quad (9.2.3)$$

We estimate from preceding the parameters occurring in this bound. We start with $H_{L_i}(F'^{(j)})$. Recall that $F'(X, Y) = \delta^{-1}(a_0 X^n + a_1 X^{n-1} Y + \cdots + a_n Y^n)$. Using the properties of heights from Section 5.1 and inequality (7.3.9), we infer that

$$H_{L_i}(F'^{(j)}) = H_{L_i}(a_0^{(j)}, \ldots, a_n^{(j)}) \leq H_{L_i}(a_0^{(j)}) + \cdots + H_{L_i}(a_n^{(j)})$$
$$\leq \Delta_i (2d^r \, (\overline{\deg} \, a_0 + \cdots + \overline{\deg} \, a_n) + n(2d)^{\exp O(r)}).$$

By Lemma 7.2.6,

$$\overline{\deg} \, a_i \leq (2d)^{\exp O(r)} \text{ for } i = 1, \ldots, n.$$

Thus,

$$H_{L_i}(F'^{(j)}) \leq \Delta_i((n+1)(2d)^{\exp O(r)} + n(2d)^{\exp O(r)})$$
$$\leq \Delta_i(nd)^{\exp O(r)}. \quad (9.2.4)$$

Next, we estimate the genus g_{L_i/k_i}. Using (9.2.1), Proposition 7.2.5, and Lemma 7.2.3, we deduce that

$$g_{L_i/k_i} \leq \Delta_i D \max_j \deg \mathcal{F}_j \leq \Delta_i (2d)^{\exp O(r)}. \quad (9.2.5)$$

Lastly, we estimate $|S_i|$. Combining Proposition 7.2.9 with (9.2.2), we obtain

$$|S_i| \leq \Delta_i(nd)^{\exp O(r)}. \quad (9.2.6)$$

Inserting the bound (9.2.4), (9.2.5), and (9.2.6) into (9.2.3), we infer

$$\max(H_{L_i}(x^{(j)}), H_{L_i}(y^{(j)})) \leq \Delta_i(nd)^{\exp O(r)}. \quad (9.2.7)$$

In view of Lemma 7.3.3, the inequalities of (9.2.7), $D \leq d^r$, $q \leq r$, and Propositions 7.2.5 and 7.2.9, we obtain

$$\overline{\deg} \, x \leq qD(nd)^{\exp O(r)} + \sum_{i=1}^{q} \Delta_i^{-1} \sum_{j=1}^{D} H_{L_i}(x^{(j)}) \leq (nd)^{\exp O(r)}$$

and similarly for $\overline{\deg} \, y$. This proves (9.1.14). $\qquad \square$

9.2.3 Hyper- and Superelliptic Equations

Using the notation of Subsection 9.1.3, we prove the bounds in (9.1.17) and (9.1.18) from Proposition 9.1.4.

Proof of (9.1.17) We follow the proof of (9.1.14) in Proposition 9.1.3, and use the same notation. In particular, $\Bbbk_i, L_i, S_i, g_{L_i/\Bbbk_i}, \Delta_i$ have the same meaning and for $\alpha \in B$ and $j = 1, \ldots, D$, the jth conjugate $\alpha^{(j)}$ is the one corresponding to $w^{(j)}$. Put $F' := \delta^{-1}F$, and let $F'^{(j)}$ denote the polynomial obtained by taking the jth conjugates of the coefficients of F'.

We keep the argument together for both hyper- and superelliptic equations by using the worse bounds everywhere. Let $x, y \in B$ be a solution of (9.1.16), where $m, n \geq 2$ and $n \geq 3$ if $m = 2$. Then we get

$$F'^{(j)}(x^{(j)}) = (y^{(j)})^m, \quad x^{(j)}, y^{(j)} \in \mathcal{O}_{S_i}.$$

Combining Theorems 5.5.1 and 5.5.2, we obtain the generous bound

$$H_{L_i}(x^{(j)}), mH_{L_i}(y^{(j)}) \leq 16n^2(H_{L_i}(F'^{(j)}) + g_{L_i/\Bbbk_i} + |S_i|).$$

For $H_{L_i}(F'^{(j)}), g_{L_i/\Bbbk_i}, |S_i|$, we have precisely the same estimates as (9.2.4), (9.2.5), and (9.2.6). Then a similar computation as in the proof of (9.1.14) leads to

$$H_{L_i}(x^{(j)}), mH_{L_i}(y^{(j)}) \leq \Delta_i(nd)^{\exp O(r)}. \tag{9.2.8}$$

Now applying Lemma 7.3.3 and ignoring m for the moment, we get, similar to the proof of (9.1.14)

$$\overline{\deg}\, x, \overline{\deg}\, y \leq (nd)^{\exp O(r)}$$

which was to be proved. □

We proceed to deduce the upper bound (9.1.18) for m in the Schinzel–Tijdeman equation. We need the following lemma, originally proved by Brindza (1993). We have included another proof.

Lemma 9.2.1 *We have*

$$\bigcap_{i=1}^{q} \Bbbk_i = \overline{\mathbb{Q}}.$$

Proof Clearly, $\overline{\mathbb{Q}} \subseteq \cap_{i=1}^{q} \Bbbk_i$. To prove the reverse inclusion, let $\alpha \in \cap_{i=1}^{q} \Bbbk_i$. Define the fields $\mathbb{F}_i := \mathbb{Q}(X_1, \ldots, X_{i-1}, X_{i+1}, \ldots, X_q)$ $(i = 1, \ldots, q)$ and as before $K_0 := \mathbb{Q}(X_1, \ldots, X_q)$. For $i = 1, \ldots, q$, let $P_i \in \mathbb{F}_i(X)$ be the monic minimal polynomial of α over $\mathbb{F}_i$, and let $P \in K_0(X)$ be the monic minimal polynomial of α over K_0. Then for $i = 1, \ldots, q$, P divides P_i in $K_0(X)$. But this is possible only if the coefficients of P lie in $\mathbb{F}_i$ for $i = 1, \ldots, q$. So the coefficients of P lie in $\cap_{i=1}^{q} \mathbb{F}_i$, implying $\alpha \in \overline{\mathbb{Q}}$. □

Proof of (9.1.18) Assume that in (9.1.16) $y \notin \overline{\mathbb{Q}}$. Then $y \notin \Bbbk_i$ for at least one index i by Lemma 9.2.1. Since $y \in B \subset \Bbbk_i(X_i, w)$ and $[\Bbbk_i(X_i, w) : \Bbbk_i(X_i)] \leq D$, it follows that

$$H_{L_i}(y) = [L_i : \Bbbk_i(X_i, w)]H_{\Bbbk_i(X_i, w)}(y) \geq [L_i : \Bbbk_i(X_i, w)]$$
$$\geq \Delta_i/D.$$

Together with (9.2.8) and $D \leq d^r$ (see Lemma 7.2.3 (i)), this gives

$$m \leq (nd)^{\exp O(r)},$$

which is (9.1.18). □

9.3 Bounding the Heights and Specializations

Combining our degree bounds established in Section 9.2 with the effective specialization method from Chapter 7 and the corresponding effective results from Chapter 4 over number fields, we derive effective bounds for the heights $\overline{h}$ of the solutions of unit equations, Thue equations, and hyper- and superelliptic equations. As was seen in Section 9.1, this will complete our effective proofs for the general version of our equations considered over finitely generated domains.

Before proving the height bounds, we recall again some notation and collect some preparatory results from Chapters 7 and 4. Let $A = \mathbb{Z}[z_1, \ldots, z_r]$ be an integral domain of characteristic 0 finitely generated over $\mathbb{Z}$, K its quotient field, q the transcendence degree of K, $z_1 = X_1, \ldots, z_q = X_q$ algebraically independent of $\mathbb{Q}$, and $A_0 = \mathbb{Z}[X_1, \ldots, X_q]$ if $q > 0$, $A_0 = \mathbb{Z}$ otherwise. Let $w \in A$ with minimal polynomial $\mathcal{F}(X) = X^D + \mathcal{F}_1 X^{D-1} + \cdots + \mathcal{F}_D \in A_0[X]$ be as in Proposition 7.2.5, $\Delta_{\mathcal{F}}$ the discriminant of $\mathcal{F}, g \in A_0 \setminus \{0\}$, $A \subseteq B := A_0[w, g^{-1}]$ as in Proposition 7.2.7, and $\mathcal{T} = \Delta_{\mathcal{F}}\mathcal{F}_D g$ as in (7.4.7). Moreover, in case of the Thue equation (9.1.13) $F(x, y) = \delta$ and the superelliptic equation (9.1.16) $F(x) = \delta y^m$, we apply Proposition 7.2.9 with δ and the discriminant D_F of F belonging to B^*.

For $\mathbf{u} \in \mathbb{Z}^q$ with $\mathcal{T}(\mathbf{u}) \neq 0$, the polynomial $\mathcal{F}_{\mathbf{u}}(X) = X^D + \mathcal{F}_1(\mathbf{u})X^{D-1} + \cdots + \mathcal{F}_D(\mathbf{u})$ has distinct zeros $w_1(\mathbf{u}), \ldots, w_D(\mathbf{u})$ in $\overline{\mathbb{Q}}$ which are all nonzero. Consequently, for $j = 1, \ldots, D$ the substitution $X_1 \mapsto u_1, \ldots, X_q \mapsto u_q, w \mapsto w_j(\mathbf{u})$ defines a ring homomorphism $\varphi_{\mathbf{u}, j}$ from B to $\overline{\mathbb{Q}}$. The image of $\alpha \in B$ under $\varphi_{\mathbf{u}, j}$ is denoted by $\alpha_j(\mathbf{u})$. Then $\varphi_{\mathbf{u}, j}(B)$ is contained in the algebraic number field $K_{\mathbf{u}, j} := \mathbb{Q}(w_j(\mathbf{u}))$.

For a fixed $j \in \{1, \ldots, D\}$ and a suitably chosen finite extension L of $K_{\mathbf{u}, j}$, we let S denote the set of places of L which consists of all infinite places and all finite places lying above the rational prime divisors of $g(\mathbf{u})$. Note that $w_j(\mathbf{u})$ is an algebraic integer and $g(\mathbf{u}) \in \mathcal{O}_S^*$. Thus, $\varphi_{\mathbf{u},j}(B) \subseteq \mathcal{O}_S$ and $\varphi_{\mathbf{u},j}(B^*) \subseteq \mathcal{O}_S^*$. Further, since $\delta, D_F \in B^*$, we have $\delta_j(\mathbf{u}) \neq 0$ and $D_{F,j}(\mathbf{u}) \neq 0$.

As in Chapter 4, $d_L, \mathcal{O}_L, \mathcal{M}_L, D_L, h_L, r_L$, and R_L denote the degree, ring of integers, set of places, discriminant, class number, unit rank, and regulator of L. The absolute norm of an ideal $\mathfrak{a}$ of $\mathcal{O}_L$ is denoted by $N(\mathfrak{a})$.

If S consists only of the infinite places of L, we put $P_S := 2, Q_S := 2$. If S contains also finite places, we let $\mathfrak{p}_1, \ldots, \mathfrak{p}_w$ denote the prime ideals corresponding to the finite places of S and put

$$P_S := \max(N(\mathfrak{p}_1), \ldots N(\mathfrak{p}_w)), Q_S := N(\mathfrak{p}_1, \ldots, \mathfrak{p}_w). \tag{9.3.1}$$

The S-regulator is denoted by R_S. In the case that S consists only of the infinite places of L, it is just R_L, while otherwise

$$R_S = h_S R_L \prod_{i=1}^{w} \log N(\mathfrak{p}_i),$$

where h_S is a positive divisor of h_L, see (4.1.10). Further,

$$R_S \leq |D_L|^{1/2} (\log^* |D_L|)^{d_L - 1} (\log^* Q_S)^w, \tag{9.3.2}$$

see (4.1.13).

Finally, in Subsections 9.3.2 and 9.3.3, we shall need the discriminant estimates from Lemmas 4.1.10 and 4.1.11.

In the proofs in Subsections 9.3.1, 9.3.2, and 9.3.3, most of the preceding notation and results will be used without any further mention.

9.3.1 Unit Equations

We prove the height bound (9.1.8).

Proof of (9.1.8) Let ε, η be a solution of Eq. (9.1.6). We first consider the case $q > 0$. Pick $\mathbf{u} \in \mathbb{Z}^q$ with $\mathcal{T}(\mathbf{u}) \neq 0$, let $j \in \{1, \ldots, D\}$ and $L := K_{\mathbf{u},j}$. Putting S as above, we have $\varphi_{\mathbf{u},j}(B) \subseteq \mathcal{O}_S$ and $\varphi_{\mathbf{u},j}(B^*) \subseteq \mathcal{O}_S^*$. Hence, it follows from (9.1.6) that

$$\varepsilon_j(\mathbf{u}) + \eta_j(\mathbf{u}) = 1, \ \varepsilon_j(\mathbf{u}), \eta_j(\mathbf{u}) \in \mathcal{O}_S^*, \tag{9.3.3}$$

where $\varepsilon_j(\mathbf{u}), \eta(\mathbf{u})$ are the images of ε, η under $\varphi_{\mathbf{u},j}$. Applying Theorem 4.3.1 with $\alpha = \beta = 1, H = 1$ to Eq. (9.3.3), we get

$$\max(h(\varepsilon_j(\mathbf{u})), h(\eta_j(\mathbf{u}))) \leq c_1 P_S R_S (1 + \log^* R_S / \log P_S) \tag{9.3.4}$$

with

$$c_1 = s^{2s+3.5} 2^{7s+27} (\log(2s)) d_L^{2(s+1)} (\log^*(2d_L))^3,$$

where s is the cardinality of S.

We estimate the upper bound (9.3.4). By (9.1.5), $g \in A_0 \backslash \{0\}$ has degree at most d_4 and height at most h_4. Hence,

$$|g(\mathbf{u})| \le d_4^q e^{h_4} \max(1, |\mathbf{u}|)^{d_4} =: R(\mathbf{u}). \tag{9.3.5}$$

Since $d_L := [L : \mathbb{Q}] \le D$, the cardinality s of S is at most $D(1 + w)$ where w denotes the number of prime divisors of $g(\mathbf{u})$. In view of the inequality from prime number theory $w \le O(\log^* |g(\mathbf{u})| / \log^* \log^* |g(\mathbf{u})|)$, we obtain

$$s \le O\left(\frac{D \log^* R(\mathbf{u})}{\log^* \log^* R(\mathbf{u})} \right). \tag{9.3.6}$$

From this, it is easy to deduce that

$$c_1 \le \exp O(D \log^* D \log^* R(\mathbf{u})). \tag{9.3.7}$$

We now estimate P_S and R_S. By (9.3.5), we have

$$P_S \le Q_S \le |g(\mathbf{u})|^D \le \exp O(D \log^* R(\mathbf{u})). \tag{9.3.8}$$

To estimate R_S, we use (9.3.2). Using Lemma 7.4.5 and $d_3 \le d_4$, we infer that

$$|D_L| \le D^{2D-1} (d_4^q e^{h_4} \max(1, |\mathbf{u}|)^{d_4})^{2D-2} \le \exp O(D \log^* D \log^* R(\mathbf{u})),$$

and this gives

$$|D_L|^{1/2} (\log^* |D_L|)^{D-1} \le \exp O(D \log^* D \log^* R(\mathbf{u})).$$

Together with the estimates in (9.3.6), (9.3.8) for s and Q_S, this yields

$$R_S \le \exp O(D \log^* D \log^* R(\mathbf{u}) + s \log^* \log^* Q)$$
$$\le \exp O(D \log^* D \log^* R(\mathbf{u})). \tag{9.3.9}$$

Collecting now (9.3.7)–(9.3.9), we infer that the right-hand side of (9.3.4) is bounded by $\exp O(D \log^* D \log^* R(\mathbf{u}))$. So we obtain from (9.3.4) that

$$h(\varepsilon_j(\mathbf{u})), h(\eta_j(\mathbf{u})) \le \exp O(D \log^* D \log^* R(\mathbf{u})). \tag{9.3.10}$$

We apply Lemma 7.4.7 with $N := 4D^2(q + d_4 + 1)^2$. From the already established (9.1.7), it follows that $\overline{\deg} \, \varepsilon, \overline{\deg} \, \eta \le N$. Further, in view of $d_4 \ge d_3$, we have $N \ge 2Dd_3 + 2(d_4 + 1)(q + 1)$. Hence, indeed, we can apply Lemma 7.4.7 with this value of N. It follows that the set

$$S := \{\mathbf{u} \in \mathbb{Z}^q : |\mathbf{u}| \le N, \, \mathcal{T}(\mathbf{u}) \ne 0\}$$

is not empty. For $\mathbf{u} \in \mathcal{S}$, $j = 1, \ldots, D$, we deduce from (9.3.5) and (9.3.10) that

$$h(\varepsilon_j(\mathbf{u})) \leq \exp O(D \log^* D(q \log d_4 + h_4 + d_4 \log^* N))$$
$$\leq \exp O(N^{1/2}(\log^* N)^2 + (D \log^* D)h_4),$$

and so by Lemma 7.4.7,

$$\overline{h}(\varepsilon) \leq \exp O(N^{1/2}(\log^* N)^2 h_4).$$

For $\overline{h}(\eta)$, we obtain the same upper bound. This gives (9.1.8) for $q > 0$.

Next, assume that $q = 0$. In this case, $A_0 = \mathbb{Z}$, $K = \mathbb{Q}(w)$ is a number field containing $B = \mathbb{Z}[w, g^{-1}]$, where w is an algebraic integer with minimal polynomial $\mathcal{F}(X) = X^D + \mathcal{F}_1 X^{D-1} + \cdots + \mathcal{F}_D \in \mathbb{Z}[X]$ over $\mathbb{Q}$, and g is a nonzero rational integer. By assumption, $\log |g| \leq h_4$, $\log^* |\mathcal{F}_j| \leq h_4$ for $j = 1, \ldots, D$. Denote by $w^{(1)}, \ldots, w^{(D)}$ the conjugates of w over $\mathbb{Q}$, and let $L := \mathbb{Q}(w^{(j)})$ for some j. By a similar argument as in the proof of Lemma 7.4.5, we obtain $|D_L| \leq D^{2D-1} e^{(2D-2)h_4}$. The isomorphism defined by $w \mapsto w^{(j)}$ maps $\mathbb{Q}(w)$ to L and B to $\mathcal{O}_S$, where S consists of the infinite places of L and of the prime ideals of $\mathcal{O}_L$ that divide g. The estimates in (9.3.5)–(9.3.9) remain valid if we replace $R(\mathbf{u})$ by e^{h_4}. Hence, for any solution ε, η of (9.1.6),

$$h(\varepsilon^{(j)}), h(\eta^{(j)}) \leq \exp O((D \log^* D)h_4),$$

where $\varepsilon^{(j)}, \eta^{(j)}$ are the jth conjugates of ε, η, respectively. Now an application of Lemma 7.4.2 with $G = \mathcal{F}, m = D$, and $\beta_i = \varepsilon$ gives

$$\overline{h}(\varepsilon) \leq \exp O((D \log^* D)h_4).$$

We obtain the same upper bound for $\overline{h}(\eta)$, whence (9.1.8) follows. The proof of Proposition 9.1.1 has been completed. $\qquad \square$

9.3.2 Thue Equations

Concluding the proof of Theorem 2.3.1, it remains to prove (9.1.15) from Proposition 9.1.3.

Proof of (9.1.15) Let x, y be a solution of Eq. (9.1.13) in B. Consider first the case $q > 0$. We keep the notation of Chapter 7 and that of the introduction of Section 9.3. Recall that $\mathcal{T} = \Delta_{\mathcal{F}} \mathcal{F}_D g$ and, by (7.4.5) and (7.4.8),

$$\deg \mathcal{T} \leq (nd)^{\exp O(r)}.$$

Choose $\mathbf{u} \in \mathbb{Z}^q$ with $\mathcal{T}(\mathbf{u}) \neq 0$, choose $j \in \{1, \ldots, D\}$, and denote by $F_{\mathbf{u},j}, \delta_j(\mathbf{u}), x_j(\mathbf{u})$, and $y_j(\mathbf{u})$ the images of F, δ, x, y under $\varphi_{\mathbf{u},j}$. The coefficients of $F_{\mathbf{u},j}$ belong to $K_{\mathbf{u},j}$. Let L denote the splitting field of $F_{\mathbf{u},j}$ over $K_{\mathbf{u},j}$, and S the set of places of L, which consists of all infinite places and all finite places lying above the rational prime divisors of $g(\mathbf{u})$. Note that $w_j(\mathbf{u})$ is an algebraic integer and $g(\mathbf{u}) \in \mathcal{O}_S^*$. Hence, $\varphi_{\mathbf{u},j}(B) \subseteq \mathcal{O}_S$, $\varphi_{\mathbf{u},j}(B^*) \subseteq \mathcal{O}_S^*$, and it follows from (9.1.13) that

$$F_{\mathbf{u},j}(x_j(\mathbf{u}), y_j(\mathbf{u})) = \delta_j(\mathbf{u}), \quad x_j(\mathbf{u}), y_j(\mathbf{u}) \in \mathcal{O}_S. \tag{9.3.11}$$

Since by assumption $\delta, D_F \in B^*$, we have $\delta_j(\mathbf{u}) \neq 0$ and $D_{F,j}(\mathbf{u}) \neq 0$. Consequently, $F_{\mathbf{u},j}$ is without multiple zeros. Then we can apply Theorem 4.4.1 to Eq. (9.3.11), and we get

$$\max(h(x_j(\mathbf{u})), h(y_j(\mathbf{u})))$$

$$\leq c_2 P_S R_S (1 + \log^* R_S / \log^* P_S) \cdot (c_3 R_L + \frac{h_L}{d_L} \log Q_S + 2nd_L H_1 + H_2) \tag{9.3.12}$$

where $H_1 = \max(1, h(F_{\mathbf{u},j}))$, $H_2 = \max(1, h(\delta_{\mathbf{u},j}))$,

$$c_2 = 250 n^6 s^{2s+3.5} \cdot 2^{7s+29} (\log 2s) d_L^{2s+4} (\log^*(2d_L))^3$$

and $c_3 = 0$ if $r_L = 0$, $1/d_L$ if $r_L = 1$ and $29 e r_L! r_L \sqrt{r_L - 1} \log d_L$ if $r_L \geq 2$.

We already proved in Subsection 9.2.2 that (9.1.14) in Proposition 9.1.3 holds, i.e.,

$$\overline{\deg} \, x, \overline{\deg} \, y \leq (nd)^{\exp O(r)}. \tag{9.3.13}$$

Thus, we can apply Lemma 7.4.7 with $\alpha = x$ resp. y and

$$N = \max((nd)^{\exp O(r)}, 2Dd_3 + 2(q+1)(d_4+1))$$

to get an upper bound for $\overline{h}(x)$, $\overline{h}(y)$, which still depends on $h(x_j(\mathbf{u})), h(y_j(\mathbf{u}))$. Then, to prove (9.1.15), we have to bound from above the parameters occurring in (9.3.12).

In view of (7.4.5), $D \leq d^r$ (see Lemma (7.2.3) (i)), and $q \leq r$, we obtain

$$N \leq (nd)^{\exp O(r)}. \tag{9.3.14}$$

Applying Lemma 7.4.7, inserting $D \leq d^r$ and the upper bound $h_4 \leq (nd)^{\exp O(r)} h$ from (7.4.5), it follows that there are $\mathbf{u} \in \mathbb{Z}^q$, $j \in \{1, \dots, D\}$ with

$$|\mathbf{u}| \leq (nd)^{\exp O(r)}, \quad \mathcal{T}(\mathbf{u}) \neq 0 \tag{9.3.15}$$

and

$$\max(\overline{h}(x), \overline{h}(y))$$

$$\leq (nd)^{\exp O(r)} \max(h(x_j(\mathbf{u})), h(y_j(\mathbf{u}))). \tag{9.3.16}$$

We proceed further with these $\mathbf{u}, j$ to derive upper bounds for the parameters corresponding to those which occur in (9.3.12).

Write $F(X, Y) = a_0 X^n + a_1 X^{n-1} Y + \cdots + a_n Y^n$ and put

$$\overline{\deg} F := \max_{0 \le k \le n} \overline{\deg} a_k, \ \overline{h}(F) := \max_{0 \le k \le n} \overline{h}(a_k).$$

Notice that by Lemma 7.2.6 applied to δ and the coefficients of F with the choice $d_0 = d$, $h_0 = h$, we have

$$\overline{\deg} F, \overline{\deg} \delta \le (2d)^{\exp O(r)}, \tag{9.3.17}$$

$$\overline{h}(F), \overline{h}(\delta) \le (2d)^{\exp O(r)} h. \tag{9.3.18}$$

It follows from Lemma 7.4.6, $q \le r, D \le d^r$, and (7.4.5) (9.3.17), (9.3.18), and (9.3.15) that

$$h(F_{\mathbf{u}, j}) \le D^2 + q(D \log d_3 + \log \overline{\deg} F) + D h_3 + \overline{h}(F)$$
$$+ (D d_3 + \overline{\deg} F) \log \max(1, |\mathbf{u}|)$$
$$\le (nd)^{\exp O(r)} h. \tag{9.3.19}$$

Similarly, replacing F by δ, we obtain

$$h(\delta_j(\mathbf{u})) \le (nd)^{\exp O(r)} h. \tag{9.3.20}$$

We recall that d_L and D_L denote the degree and discriminant of L over $\mathbb{Q}$. Since $[K_{\mathbf{u}, j} : \mathbb{Q}] \le D$, we have $d_L \le Dn!$. Let $G(X) := F(X, 1)$ and let $\vartheta_1, \ldots, \vartheta_n$ be the zeros of G. We have $n' = n$ if $a_0 \ne 0$ and $n' = n-1$ otherwise. Then $L = K_{\mathbf{u}, j}(\vartheta_1, \ldots, \vartheta_{n'})$. Let $d_{K_{\mathbf{u}, j}}$, d_{L_i} denote the degree and $D_{K_{\mathbf{u}, j}}$, D_{L_i} the discriminant of the number field $K_{\mathbf{u}, j}$ resp. $L_i := K_{\mathbf{u}, j}(\vartheta_i)$, $i = 1, \ldots, n'$. Then by Lemma 4.1.10, we have

$$|D_L| \le \prod_{i=1}^{n'} |D_{L_i}|^{d_L / d_{L_i}}. \tag{9.3.21}$$

We now estimate $|D_L|$. Notice that by Lemma 7.4.5, using the estimates $q \le r, D \le d^r$, and (7.4.4), (7.4.5), and (9.3.15), we obtain

$$|D_{K_{\mathbf{u}, j}}| \le D^{2D-1} (d_3^q e^{h_3} \max(1, |\mathbf{u}|)^{d_3})^{2D-2}$$
$$\le \exp((nd)^{\exp O(r)} h). \tag{9.3.22}$$

Further, by Lemma 4.1.11 and the estimates $D \le d^r$, (9.3.19), (9.3.22), we get

$$|D_{L_i}| \le n^{(2n-1)D} e^{(2n^2-2)h(F_{\mathbf{u}, j})} |D_{K_{\mathbf{u}, j}}|^{[L_i : K_{\mathbf{u}, j}]}$$
$$\le \exp\{[L_i : K_{\mathbf{u}, j}](nd)^{\exp O(r)} h\}.$$

Inserting this into (9.3.21) and using $d_{L_i} = [L_i : K_{\mathbf{u}, j}] \cdot d_{K_{\mathbf{u}, j}}$, we have

$$|D_L| \le \exp\{(nd)^{\exp O(r)} h n d_L / d_{K_{\mathbf{u}, j}}\}$$
$$\le \exp\{n! (nd)^{\exp O(r)} h\}. \tag{9.3.23}$$

We follow now similar arguments as in the preceding proof of (9.1.8) concerning unit equations. The polynomial g has degree at most d_4 and logarithmic height at most h_4, which satisfy (7.4.5). Further, $g(\mathbf{u}) \neq 0$ and by $q \leq r$ and (9.3.15),

$$|g(\mathbf{u})| \leq d_4^q e^{h_4} \max(1, |\mathbf{u}|)^{d_4} \leq \exp\{(nd)^{\exp O(r)} h\}. \tag{9.3.24}$$

The cardinality s of S is at most $d_L(1 + w)$, where w denotes the number of distinct prime divisors of $g(\mathbf{u})$. By prime number theory,

$$s = O(d_L \log^* |g(\mathbf{u})| / \log^* \log^* |g(\mathbf{u})|). \tag{9.3.25}$$

Together with (9.3.24), $D \leq d^r$, and $d_L \leq n! d^r$, this implies that c_2 can be estimated as

$$c_2 \leq \exp\{n!(nd)^{\exp O(r)} h\}. \tag{9.3.26}$$

Next, we estimate P_S, Q_S and R_S. In view of (9.3.24), $d_L \leq n! d^r$, we have

$$P_S \leq Q_S \leq |g(u)|^{d_L} \leq \exp\{n!(nd)^{\exp O(r)} h\}. \tag{9.3.27}$$

To estimate R_S, we use (9.3.2). Then, by (9.3.23) and $d_L \leq n! d^r$, we obtain

$$|D_L|^{1/2} (\log^* |D_L|)^{d_L - 1} \leq \exp\{n!(nd)^{\exp O(r)} h\}. \tag{9.3.28}$$

Further, (9.3.25) and (9.3.27) imply

$$(\log Q_S)^s \leq \exp\left\{ O\left(d_L \frac{\log^* |g(\mathbf{u})|}{\log^* \log^* |g(\mathbf{u})|} (\log d_L + \log^* \log^* |g(\mathbf{u})|) \right) \right\}.$$

Together with (9.3.24), this yields

$$R_S \leq \exp\{n!(nd)^{\exp O(r)} h\}. \tag{9.3.29}$$

Combining (9.3.2) with R_S replaced by R_L (when $\log Q_S < 1$) with (9.3.28) and $R_L > 0.2052$ (see Friedman [1989] or Section 4.1), we obtain

$$\max(h_L, R_L) \leq \exp\{n!(nd)^{\exp O(r)} h\}. \tag{9.3.30}$$

Finally, using $r_L < d_L \leq n! d^r$, we deduce that

$$c_3 \leq \exp O(d_L \log^* d_L) \leq \exp\{n!(nd)^{\exp O(r)}\}. \tag{9.3.31}$$

From the estimates of (9.3.21), (9.3.22), (9.3.26), (9.3.28), (9.3.29), (9.3.30), it follows that the upper bound in (9.3.12) is a sum and product of terms, which are all bounded above by $\exp\{n!(nd)^{\exp O(r)} h\}$. Consequently,

$$h(x_j(\mathbf{u})), \ h(y_j(\mathbf{u})) \leq \exp\{n!(nd)^{\exp O(r)} h\}.$$

Inserting this into (9.3.16), we get the upper bound (9.1.15) for $q > 0$.

Now, assume that $q = 0$. Then $A_0 = \mathbb{Z}, K = \mathbb{Q}(w)$, and $B = \mathbb{Z}[w, g^{-1}]$, where w is an algebraic integer with minimal polynomial $\mathcal{F}(X) = X^D + \mathcal{F}_1 X^{D-1} + \cdots + \mathcal{F}_D \in \mathbb{Z}[X]$ over $\mathbb{Q}$, and g is a nonzero rational integer. By (7.4.4), (7.4.5) we may assume that $\log|g| \leq h_4$, $\log^* \mathcal{F}_j \leq h_4$ for $j = 1, \ldots, D$. Denote by $w^{(1)}, \ldots, w^{(D)}$ the conjugates of w over $\mathbb{Q}$, and let $K_j := \mathbb{Q}(w^{(j)})$ for $j = 1, \ldots, D$. One can prove by a similar argument as in the proof of Lemma 7.4.5 that $|D_{K_j}| \leq D^{2D-1} e^{(2D-2)h_4}$. For $\alpha \in K$, we denote by $\alpha^{(j)}$ the conjugates of α over $\mathbb{Q}$, corresponding to $w^{(j)}$.

Instead of Lemma 7.4.7, we use Lemma 7.4.2, applied with $G = \mathcal{F}$, $m = D$ and $\beta^{(j)} = x^{(j)}$ or $y^{(j)}$. Inserting (7.4.4), (7.4.5), this yields the estimate

$$\max(\overline{h}(x), \overline{h}(y)) \leq (nd)^{\exp O(r)}(h + \max_{1 \leq j \leq D} \max(h(x^{(j)}), h(y^{(j)}))). \quad (9.3.32)$$

We proceed further with the index j for which the maximum is attained.

Now we can follow the preceding argument for the case $q > 0$, except that in all estimates we have to take $q = 0$, and replace $\max(1, |\mathbf{u}|)$ by 1, $K_{\mathbf{u},j}$ by K_j, $g(\mathbf{u})$ by g, and $F_{\mathbf{u},j}$ by F_j, where F_j is the binary form obtained by taking the jth conjugates of the coefficients of F, and $g(\mathbf{u})$ by g. This leads to an estimate

$$h(x^{(j)}), h(y^{(j)}) \leq \exp\{n!(nd)^{\exp O(r)}h\},$$

and combined with (9.3.32), this implies again (9.1.15), which completes the proof of Proposition 9.1.3. □

9.3.3 Hyper- and Superelliptic Equations

It remains to prove (9.1.19) from Proposition 9.1.4. The computations will be similar to those as in the above proof of (9.1.15) but with some simplifications.

Proof of (9.1.19) *of Proposition 9.1.4* Take a solution x, y of Eq. (9.1.16) in B. Consider, again, first the case $q > 0$. We use once more the polynomial $\mathcal{T} := \Delta_{\mathcal{F}} \mathcal{F}_D g$ as in (7.4.7). Take again $\mathbf{u} \in \mathbb{Z}^q$ with $\mathcal{T}(\mathbf{u}) \neq 0$, choose $j \in \{1, \ldots, D\}$, and denote by $F_{\mathbf{u},j}, \delta_j(\mathbf{u}), x_j(\mathbf{u}), y_j(\mathbf{u})$ the images of F, δ, x, y under the specialization $\varphi_{\mathbf{u},j}$. In contrast to our arguments for Thue equations, now we do not have to deal with the splitting field of F. Put $L := K_{\mathbf{u},j}$, and choose for S the set of places of L, which consists of all infinite places and the finite places lying above the rational prime divisors of $g(\mathbf{u})$. Then $\varphi_{\mathbf{u},j}(B) \subseteq \mathcal{O}_S$, and

$$F_{\mathbf{u},j}(x_j(\mathbf{u})) = \delta_j(\mathbf{u}) y_j(\mathbf{u})^m, \text{ where } x_j(\mathbf{u}), y_j(\mathbf{u}) \in \mathcal{O}_S. \quad (9.3.33)$$

We note that by assumption, $\delta, D_F \in B^*$, hence $\delta_j(\mathbf{u}) \neq 0$ and $F_{\mathbf{u},j}$ has nonzero discriminant. Since $F_{\mathbf{u},j}$ has the same number of zeros and the same degree as

F, the degree of $F_{u,j}$ is $n \geq 3$ if $m = 2$ and $n \geq 2$ if $m \geq 3$. Thus, we can apply Theorems 4.5.1 and 4.5.2 to Eq. (9.3.33) according as $m = 2$ or $m \geq 3$. Then we obtain

$$
h(x_j(\mathbf{u})), h(y_j(\mathbf{u})) \leq \begin{cases} c_4 |D_L|^{8n^2} Q_S^{20n^3} e^{50n^4 d_L \widehat{h}} & \text{if } n \geq 3, \\ c_5^{m^3} |D_L|^{2m^2 n^2} Q_S^{3m^2 n^2} e^{8m^2 n^3 d_L \widehat{h}} & \text{if } n \geq 2, m \geq 3, \end{cases}
$$

$$(9.3.34)$$

where $c_4 = (4ns)^{212n^4 s}$, $c_5 = (6ns)^{14m^3 n^3 s}$, and $\widehat{h}$ is defined by (4.5.3).

It follows by precisely the same argument as in the case of Thue equations that there are $\mathbf{u} \in \mathbb{Z}^q$ and $j \in \{1, \ldots, D\}$, which satisfy (9.3.15) and (9.3.16). We proceed further with these $\mathbf{u}, j$.

We estimate the parameters occurring in the bounds in (9.3.34). First, we obtain the same estimates as in (9.3.19) and (9.3.20). These imply

$$
\widehat{h} \leq (n+1)h(F_{\mathbf{u},j}) + h(\delta_j(\mathbf{u})) \leq (nd)^{\exp O(r)} h. \tag{9.3.35}
$$

Second we have, similar to (9.3.23),

$$
|D_L| \leq \exp\{(nd)^{\exp O(r)} h\}, \tag{9.3.36}
$$

and similarly to (9.3.24),

$$
|g(\mathbf{u})| \leq \exp\{(nd)^{\exp O(r)} h\}. \tag{9.3.37}
$$

Now the set S consists of places of L instead of the splitting field of $F_{\mathbf{u},j}$ over K. Because of $[L : \mathbb{Q}] \leq D$, we have $s \leq D(1 + w)$, where w is the number of distinct prime divisors of $g(\mathbf{u})$. This implies, instead of (9.3.25),

$$
s = O(D \log^* |g(\mathbf{u})| / \log^* \log^* |g(\mathbf{u})|). \tag{9.3.38}
$$

Inserting (9.3.37) and $D \leq d^r$, we obtain for the quantities of c_4, c_5 in (9.3.34),

$$
c_4, c_5 \leq \exp\{(nd)^{\exp O(r)} h\}. \tag{9.3.39}
$$

Lastly, by $D \leq d^r$ and (9.3.37), we have

$$
P_S \leq Q_S \leq |g(\mathbf{u})|^D \leq \exp\{(nd)^{\exp O(r)} h\}. \tag{9.3.40}
$$

We now use (9.3.34). By inserting (9.3.35), (9.3.36), and (9.3.39) and $d_L \leq D \leq d^r$ into (9.3.34), we get

$$
h(x_j(\mathbf{u})), h(y_j(\mathbf{u})) \leq \exp\{m^3 (nd)^{\exp O(r)} h\}. \tag{9.3.41}
$$

Finally, inserting this into (9.3.16), we obtain (9.1.19) in the case $q > 0$.

Now let $q = 0$. For $\alpha \in K$, write $\alpha^{(j)}$ for the conjugate of α corresponding to $w^{(j)}$, and let F_j be the polynomial obtained by taking the jth conjugates of the coefficients of F. We simply follow the preceding arguments, replacing everywhere q by 0, $\max(1, \mathbf{u})$ by 1, $K_{\mathbf{u},j}$ by $K_j = \mathbb{Q}(w^{(j)})$, $F_{\mathbf{u},j}$ by F_j, $x_j(\mathbf{u})$ and $y_j(\mathbf{u})$ by $x^{(j)}$ and $y^{(j)}$, and $g(\mathbf{u})$ by $g \in \mathbb{Z}$. Instead of (9.3.16), we have to use (9.3.32). Thus, we obtain the same estimate as (9.3.41), but with $x^{(j)}, y^{(j)}$ instead of $x_j(\mathbf{u}), y_j(\mathbf{u})$. Via (9.3.32), we obtain (9.1.19). This completes our proof for Proposition 9.1.4. □

Proof of Proposition 9.1.5 Assume first that $q > 0$. Let $x \in B$, $y \in B \cap \overline{\mathbb{Q}}$, and $m \in \mathbb{Z}_{\geq 2}$ be a solution of Eq. (9.1.16) such that $y \neq 0$ and y is not a root of unity. Choose again $\mathbf{u}, j$ such that they satisfy (9.3.15), (9.3.16). Note that $y_j(\mathbf{u})$ is a conjugate of y since $y \in \overline{\mathbb{Q}}$. Hence, it is not 0 or a root of unity.

By applying Theorem 4.5.3 to Eq. (9.3.33), we get

$$m \leq c_6 |D_L|^{6n} P_S^{n^2} e^{11 n d_L \widehat{h}}, \tag{9.3.42}$$

where $c_6 = (10n^2 s)^{40ns}$. By (9.3.37) and (9.3.38), the constant c_6 satisfies

$$c_6 \leq \exp\{(nd)^{\exp O(r)} h\}.$$

Further, we have the upper bounds (9.3.35) for $\widehat{h}$, (9.3.36) for $|D_L|$, and (9.3.40) for P_S. Inserting these estimates into the upper bound in (9.3.42), we get $m \leq \exp\{(nd)^{\exp O(r)} h\}$. In the case of $q = 0$, we obtain the same estimates, by making the same modifications as in the proof of Proposition 9.1.4. Our proof is complete. □

9.4 The Catalan Equation

In this section, we complete the proof of Theorem 2.5.1 on the Catalan equation (2.5.1) $x^m - y^n = 1$ in $x, y \in A$ and integers m, n with $m, n > 1$ and $mn > 4$. We follow mostly the proofs of Brindza (1993) and Koymans (2017).

As before, $A = \mathbb{Z}[z_1, \ldots, z_r]$ denotes an integral domain finitely generated over $\mathbb{Z}$, with quotient field K. Then $A \cong \mathbb{Z}[X_1, \ldots, X_r]/\mathcal{I}$, where $\mathcal{I}$ is the ideal of polynomials $f \in \mathbb{Z}[X_1, \ldots, X_r]$ such that $f(z_1, \ldots, z_r) = 0$. Let $d \geq 1, h \geq 1$, and assume again that $\mathcal{I} = (f_1, \ldots, f_M)$ with $\deg f_i \leq d$, $h(f_i) \leq h$ for $i = 1, \ldots, M$.

Proof of Theorem 2.5.1 Let x, y, m, n be an arbitrary solution of Eq. (2.5.1) with nonzero $x, y \in A$, not roots of unity, and with integers m, n such that $m, n > 1$ and $mn > 4$. We keep the notation and assumptions from the beginning of Sections 9.3 and 9.4. Further, by Proposition 7.2.7, there exists a nonzero $g \in A_0$ such that

$$A \subseteq B = A_0[w, g^{-1}]$$

and

$$\deg g \le (2d)^{\exp O(r)}, \ h(g) \le (2d)^{\exp O(r)} h. \tag{9.4.1}$$

We shall work in this larger ring B to bound m and n.

We start with the case $q = 0$. Then $A_0 = \mathbb{Z}$, $K_0 = \mathbb{Q}$, K is a number field of degree $D \le d^r$, and D_K is the discriminant of K. Since $|D_K| \le |D(\mathcal{F})|$ and, as was seen in the proof of Lemma 7.4.5, $|D(\mathcal{F})| \le D^{2D-1} H(\mathcal{F})^{2D-2}$, where $H(\mathcal{F})$ denotes the maximum of the absolute values of the coefficients of $\mathcal{F}$, we infer that

$$|D_K| \le \exp((2d)^{\exp O(r)} h).$$

Let S denote the set of infinite places and of the finite places of K corresponding to the prime ideal divisors, say, $\mathfrak{p}_1, \ldots, \mathfrak{p}_w$, of g. Write $s = |S|$, and let P_S, Q_S be as in (9.3.1). In view of (9.4.1), it follows $s \le (2d)^{\exp O(r)} h$, $P_S \le \exp((2d)^{\exp O(r)} h)$, and $Q_S \le |g|^d \le \exp((2d)^{\exp O(r)} h)$. Using the estimates for s, P_S, Q_S and combining them with Theorem 4.6.1, we obtain (2.5.2).

Consider now the case $q > 0$. Fix an algebraic closure $\overline{K}_0$ of K_0, and let $\Bbbk_i$ be the algebraic closure of $\mathbb{Q}(X_1, \ldots, X_{i-1}, X_{i+1}, \ldots, X_q)$ in $\overline{K}_0$. Thus, A_0 is contained in $\Bbbk_i[X_i]$. Define

$$L_i := \Bbbk_i(X_i, w^{(1)}, \ldots, w^{(D)}),$$

where $w^{(1)}, \ldots, w^{(D)}$ are the conjugates of w over K_0.

We start with the case $x \in \Bbbk_i$ for $i = 1, \ldots, q$. Then by Lemma 9.2.1, x and y belong to the algebraic number field $\overline{\mathbb{Q}} \cap K$. We are now going to apply Theorem 4.6.1.

Let $\mathbf{u} = (u_1, \ldots, u_q) \in \mathbb{Z}^q$, and put $|\mathbf{u}| = \max(|u_1|, \ldots, |u_q|)$. As in Section 7.4, we extend the ring homomorphism from a subring of K_0 to $\mathbb{Q}$,

$$\varphi_{\mathbf{u}} : \alpha \mapsto \alpha(\mathbf{u}) : \{g_1/g_2 : g_1, g_2 \in A_0, g_2(\mathbf{u}) \ne 0\} \to \mathbb{Q},$$

defined by the substitution $X_1 \mapsto u_1, \ldots, X_q \mapsto u_q$, to a ring homomorphism from B to $\overline{\mathbb{Q}}$ for which we need to impose some restriction on $\mathbf{u}$. Denote by $\Delta_{\mathcal{F}}$ the discriminant of $\mathcal{F}$, and let $\mathcal{T} = \Delta_{\mathcal{F}} \mathcal{F}_D g$. Obviously, $\mathcal{T} \in A_0$. Since $\Delta_{\mathcal{F}}$ is a polynomial of degree $2D - 2$ with integer coefficients in $\mathcal{F}_1, \ldots, \mathcal{F}_D$, it follows that $\deg \mathcal{T} \le (2d)^{\exp O(r)}$.

Lemma 7.4.4 implies that

$$\mathcal{S} := \{\mathbf{u} \in \mathbb{Z}^q : |\mathbf{u}| \le N, \ \mathcal{T}(\mathbf{u}) \ne 0\}$$

is nonempty, provided that $N = (2d)^{\exp O(r)}$ and the constant implied by the O-symbol is sufficiently large. Take $\mathbf{u} \in \mathcal{S}$ and consider the polynomial $\mathcal{F}_{\mathbf{u}}(X) :=$

$X^D + \mathcal{F}_1(\mathbf{u})X^{D-1} + \cdots + \mathcal{F}_D(\mathbf{u})$. It has distinct zeros, say, $w_1(\mathbf{u}), \ldots, w_D(\mathbf{u})$, which are all different from 0. Then, for $j = 1, \ldots, D$,

$$X_1 \mapsto u_1, \ldots, X_q \mapsto u_q, \mathbf{u} \mapsto w_j(\mathbf{u})$$

defines, as was claimed in Section 7.4, a ring homomorphism $\varphi_{\mathbf{u},j}$ from B to $\overline{\mathbb{Q}}$. The image of $\alpha \in B$ under $\varphi_{\mathbf{u},j}$ is again denoted by $\alpha_j(\mathbf{u})$. It is clear that $\varphi_{\mathbf{u},j}$ is the identity on $B \cap \mathbb{Q}$. Consequently, if $\alpha \in B \cap \overline{\mathbb{Q}}$, then $\varphi_{\mathbf{u},j}$ has the same minimal polynomial as α, and hence it is conjugate to α.

Consider the algebraic number field $K_{\mathbf{u},j} := \mathbb{Q}(w_j(\mathbf{u}))$ and denote by $D_{K_{\mathbf{u},j}}$ its discriminant for $j = 1, \ldots, D$. By Lemma 7.4.5 we have

$$[K_{\mathbf{u},j} : \mathbb{Q}] \le D, \ |D_{K_{\mathbf{u},j}}| \le D^{2D-1}(d_3^q e^{h_3} \max(1, |\mathbf{u}|)^{d_3})^{2D-2},$$

where

$$d_3 = \max(d, \deg \mathcal{F}_1, \ldots, \deg \mathcal{F}_D), \ h_3 = \max(d, h(\mathcal{F}_1), \ldots, h(\mathcal{F}_D)).$$

We have $d_3 \le (2d)^{\exp O(r)}$, $h_3 \le (2d)^{\exp O(r)} h$, which, together with $D \le d^r$, gives

$$|D_{K_{\mathbf{u},j}}| \le \exp((2d)^{\exp O(r)} h).$$

Fix now any of $j = 1, \ldots, D$. In $K_{\mathbf{u},j}$, denote by S the set consisting of the infinite places and of the finite places corresponding to the prime ideal divisors of $g(\mathbf{u})$. Then $\varphi_{\mathbf{u},j}$ maps B to the ring of S-integers of $K_{\mathbf{u},j}$. To apply Theorem 4.6.1, we still need to bound P_S, Q_S, and s.

It is easy to see that for any $\mathbf{u} \in \mathbb{Z}^q$,

$$\log |g(\mathbf{u})| \le q \log \deg g + h(g) + \deg g \log \max(1, |\mathbf{u}|)$$

which together with (9.4.1) and with the choice of N gives

$$|g(\mathbf{u})| \le (2d)^{q \exp O(r)} \cdot \exp((2d)^{\exp O(r)} \cdot h) \cdot (2d)^{(2d)^{\exp O(r)}}$$
$$\le \exp((2d)^{\exp O(r)} \cdot h).$$

Thus, we get

$$Q_S \le |g(\mathbf{u})|^D \le \exp((2d)^{\exp O(r)} h), \tag{9.4.2}$$
$$P_S \le \exp((2d)^{\exp O(r)} h) \tag{9.4.3}$$

and

$$s \le (2d)^{\exp O(r)} h. \tag{9.4.4}$$

Combining Theorem 4.6.1 with the bounds in (9.4.2) to (9.4.4), we obtain again (2.5.3).

Finally, we deal with the case $x \notin \Bbbk_i$ for some i. Fix such an i. Let S denote the set of valuations of L_i over $\Bbbk_i$ such that $v(X_i) < 0$, $v(g) > 0$. Let v be any valuation of L_i over $\Bbbk_i$ with $v \notin S$. Then $v(X_i) \geq 0$. We recall that w is integral over $\Bbbk_i[X_i]$. This implies that $v(w) \geq 0$. We also have that $v(g) \leq 0$, whence $v(g^{-1}) \geq 0$. Consequently, the ring $B = A_0[w, g^{-1}]$ is a subring of the ring of S-integers $\mathcal{O}_S$ of L_i.

Since $x, y \in A$, and $A \subseteq B$, it follows that $x, y \in \mathcal{O}_S \backslash \Bbbk_i$. Now in view of Theorem 5.3.1 (i), we get

$$mH_{L_i}(x), \, nH_{L_i}(y) \leq 6(|S| + 2g_{L_i/\Bbbk_i} - 2). \tag{9.4.5}$$

Putting $K_i = \Bbbk_i(X_i, w)$, $\Delta_i = [L_i : \Bbbk(X_i)]$ and using the fact that $x, y \in K_i$ and $[K_i : \Bbbk_i(X_i)] \leq D$, we infer that

$$H_{L_i}(x) = [L_i : K_i]H_{K_i}(x) \geq [L_i : K_i] \geq \Delta_i/[K_i : \Bbbk_i(X_i)] \geq \Delta_i/D$$

and similarly for $H_{L_i}(y)$. Together with (9.4.5), this gives

$$\max(m, n) \leq \frac{6D}{\Delta_i}(|S| + 2g_{L_i/\Bbbk_i} - 2). \tag{9.4.6}$$

We are now going to estimate $|S|$ and $g_{L_i/\Bbbk_i}$. Every valuation of $\Bbbk_i(X_i)$ can be extended to at most Δ_i valuations of L_i. Thus, L_i has at most Δ_i valuations with $v(X_i) < 0$ and at most $\Delta_i \deg_{X_i} g$ valuations with $v(g) > 0$. Using also $\deg g \leq (2d)^{\exp O(r)}$ from Proposition 7.2.7, we deduce that

$$|S| \leq \Delta_i + \Delta_i \deg_{X_i} g \leq \Delta_i(1 + \deg g) \leq \Delta_i(2d)^{\exp O(r)}. \tag{9.4.7}$$

Further, L_i being the splitting field of $\mathcal{F}$ over $\Bbbk_i(X_i)$, by Lemma 5.1.1 and Proposition 7.2.5, we obtain

$$g_{L_i/\Bbbk_i} \leq \Delta_i D \max_j \deg_{X_i} \mathcal{F}_j \leq \Delta_i D \max_j \deg \mathcal{F}_j \leq \Delta_i D(2d)^{\exp O(r)}. \tag{9.4.8}$$

Thus, it follows from $D \leq d^r$, (9.4.6), (9.4.7), and (9.4.8) that

$$\max(m, n) \leq (2d)^{\exp O(r)},$$

which completes the proof of (2.5.3) and hence that of Theorem 2.5.1. $\qquad\square$

10

Proofs of the Results from Sections 2.6 to 2.8
Reduction to Unit Equations

In Section 10.1, we prove our central results of Theorem 2.6.1 and Corollary 2.6.2 on *decomposable form equations*, stated in Section 2.6, together with the corollaries stated in that section. Their proofs are based on Theorem 2.2.1 on unit equations. In fact, we apply Győry's method to reduce the decomposable form equation under consideration to unit equations in two unknowns, which was originally developed in an effective form over number fields, e.g. in Győry (1976, 1981a), Győry and Papp (1978), and, in an ineffective form, over arbitrary finitely generated domains in Győry (1982). We make this fully effective and quantitative by employing some of the degree-height estimates from Chapter 8.

In Section 10.2, we prove the results for *norm form equations* stated in Section 2.7 and in Section 10.3 the results for *discriminant form equations* and *discriminant equations* stated in Section 2.8. These are all consequences of Theorem 2.6.1.

10.1 Proofs of the Central Results on Decomposable Form Equations

Keeping the notation of Sections 2.6 and 8.1, we assume that A, δ, $\mathcal{L} = (\ell_1, \ldots, \ell_n)$ satisfy the conditions of Theorem 2.6.1. Denote as before by $\mathcal{G}(\mathcal{L})$ the triangular graph of $\mathcal{L}$, defined by (2.6.4), and let $\mathcal{L}_1, \ldots, \mathcal{L}_k$ be the vertex systems of the connected components of $\mathcal{G}(\mathcal{L})$, and $[\mathcal{L}_j]$ the $\overline{K}$-vector space generated by $\mathcal{L}_j$, for $j = 1, \ldots, k$.

Proof of Theorem 2.6.1 Take $\mathbf{x} \in A^m$ such that

$$F(\mathbf{x}) = \ell_1(\mathbf{x}) \cdots \ell_n(\mathbf{x}) = \delta, \quad \text{there is } \ell \in [\mathcal{L}_1] \cap \cdots \cap [\mathcal{L}_k] \text{ with } \ell(\mathbf{x}) \neq 0.$$
$$(2.6.7)$$

We have to show that the coset $\mathbf{x} + \mathcal{Z}_{A,F}$ is represented by $\widetilde{\mathbf{x}} \in \mathbb{Z}[X_1, \ldots, X_r]^m$ with (2.6.8). The proof is divided into a couple of steps.

Step 1. *Construction of certain scalar multiples ℓ_i' of ℓ_i, for $i = 1, \ldots, n$, and a finitely generated domain $A' \supset A$, such that $\ell_1'(\mathbf{x}), \ldots, \ell_n'(\mathbf{x})$ are units of A'.*

Write

$$\ell_i = \alpha_{i,1} X_1 + \cdots + \alpha_{i,m} X_m \text{ for } i = 1, \ldots, n.$$

Put

$$\mathcal{R} := 2mn \cdot v^{mn} d, \quad \mathcal{R}' := 2mn \cdot v^{vmn} d.$$

Let G be the extension of K generated by the $\alpha_{i,j}$ $(i = 1, \ldots, n, j = 1, \ldots, m)$. We may assume that also $\delta \in G$, since otherwise (2.6.7) cannot hold. By Corollary 8.3.4, there is $\theta \in G$, such that $G = K(\theta)$, θ has monic minimal polynomial $F_\theta \in A[X]$ over K, and

$$\theta \overset{\text{int}}{\prec} \left(\mathcal{R}^{\exp O(r)}, \mathcal{R}^{\exp O(r)} h \right). \tag{10.1.1}$$

Further, letting $\mathcal{E} := [G : K]$, we have

$$\alpha_{i,j} = \sum_{t=0}^{\mathcal{E}-1} a_{i,j,t} \theta^t, \quad \delta = \sum_{t=0}^{\mathcal{E}-1} b_t \theta^t,$$

with $b_{i,j,t}, b_t \in K$, $b_t, b_{i,j,t} \prec \left(\mathcal{R}'^{\exp O(r)}, \mathcal{R}'^{\exp O(r)} h \right)$ \quad (10.1.2)

for $i = 1, \ldots, n$, $j = 1, \ldots, m$, and $t = 0, \ldots, \mathcal{E} - 1$. We clear the denominators of the $a_{i,j,t}, b_t$. According to the definitions, we have for all i, j, t that

$$a_{i,j,t} = \frac{g_{i,j,t}'(z_1, \ldots, z_r)}{g_{i,j,t}''(z_1, \ldots, z_r)}, \quad b_t = \frac{g_t'(z_1, \ldots, z_r)}{g_t''(z_1, \ldots, z_r)},$$

where $g_t', g_{i,j,t}', g_t'', g_{i,j,t}''$ are polynomials in $\mathbb{Z}[X_1, \ldots, X_r]$ of total degree at most $\mathcal{R}'^{\exp O(r)}$ and logarithmic height at most $\mathcal{R}'^{\exp O(r)} h$. Now define

$$\gamma_0 := \prod_{t=0}^{\mathcal{E}-1} g_t''(z_1, \ldots, z_r),$$

$$\gamma_i := \prod_{t=0}^{\mathcal{E}-1} \prod_{j=1}^{m} g_{i,j,t}''(z_1, \ldots, z_r) \text{ for } i = 1, \ldots, n.$$

Then by Lemma 4.1.7 and $\mathcal{E} \leq v^{mn}$,

$$\gamma_0, \gamma_1, \ldots, \gamma_n \overset{\text{int}}{\prec} \left(\mathcal{R}'^{\exp O(r)}, \mathcal{R}'^{\exp O(r)} h \right). \tag{10.1.3}$$

Define the quantities

$$
\begin{cases}
a'_{i,j,t} := \gamma_0 \gamma_i a_{i,j,t}, \\
\alpha'_{i,j} := \sum_{t=0}^{\mathcal{E}-1} a'_{i,j,t} \theta^t = \gamma_0 \gamma_i \alpha_{i,j}, \quad \ell'_i := \sum_{j=1}^{m} \alpha'_{i,j} X_j = \gamma_0 \gamma_i \ell_i
\end{cases}
\tag{10.1.4}
$$

for $i = 1, \ldots, n$, $j = 1, \ldots, m$, $t = 0, \ldots, \mathcal{E} - 1$, and

$$
\begin{cases}
b'_t := \gamma_0^n \gamma_1 \cdots \gamma_n b_t \quad \text{for } t = 0, \ldots, \mathcal{E} - 1, \\
\delta' := \displaystyle\sum_{t=0}^{\mathcal{E}-1} b'_t \theta^t = \gamma_0^n \gamma_1 \cdots \gamma_n \delta.
\end{cases}
\tag{10.1.5}
$$

With this construction, we have

$$
a'_{i,j,t}, \; b'_t \in A \quad \text{for all } i, j, t
\tag{10.1.6}
$$

and

$$
\ell'_1(\mathbf{x}) \cdots \ell'_n(\mathbf{x}) = \delta'.
\tag{10.1.7}
$$

Further, by Lemma 4.1.7,

$$
a'_{i,j,t}, \; b'_t \overset{\text{int}}{\prec} \left(\mathcal{R}'^{\exp O(r)}, \mathcal{R}'^{\exp O(r)} h \right).
\tag{10.1.8}
$$

Now, let $A' := A[\theta, \delta'^{-1}] = \mathbb{Z}[z_1, \ldots, z_r, \theta, \delta'^{-1}]$. Then from (10.1.6), it is clear that $\ell'_1(\mathbf{x}), \ldots, \ell'_n(\mathbf{x}) \in A'$ and subsequently by (10.1.7),

$$
\ell'_i(\mathbf{x}) \in A'^* \quad \text{for } i = 1, \ldots, n.
\tag{10.1.9}
$$

To apply Theorem 2.2.1, we need an ideal representation for A'. By (10.1.1), the generator θ of G has monic minimal polynomial over A,

$$
F_\theta = X^{\mathcal{E}} + p_1(z_1, \ldots, z_r) X^{\mathcal{E}-1} + \cdots + p_{\mathcal{E}}(z_1, \ldots, z_r),
$$

where $p_1, \ldots, p_{\mathcal{E}}$ are polynomials in $\mathbb{Z}[X_1, \ldots, X_r]$ of total degree at most $\mathcal{R}^{\exp O(r)}$ and logarithmic height at most $\mathcal{R}^{\exp O(r)} h$. Let

$$
f_{M+1} := X_{r+1}^{\mathcal{E}} + p_1 X_{r+1}^{\mathcal{E}-1} + \cdots + p_{\mathcal{E}}.
$$

Then using $\mathcal{E} \leq v^{mn}$, we get

$$
\deg f_{M+1} \leq \mathcal{R}^{\exp O(r)}, \quad h(f_{M+1}) \leq \mathcal{R}^{\exp O(r)} h.
\tag{10.1.10}
$$

Further,

$$
A[\theta] \cong \mathbb{Z}[X_1, \ldots, X_r, X_{r+1}]/\mathcal{I}', \quad \text{with } \mathcal{I}' := (f_1, \ldots, f_M, f_{M+1}).
$$

The quantity δ' corresponds to the residue class modulo $\mathcal{I}'$ of $\sum_{t=0}^{\mathcal{E}-1} \widetilde{b_t} X_{r+1}^t$, where $\widetilde{b_t} \in \mathbb{Z}[X_1, \ldots, X_r]$ is a representative for b_t, of total degree at most $\mathcal{R}'^{\exp O(r)}$, and logarithmic height at most $\mathcal{R}'^{\exp O(r)} h$. By Lemma 9.1.2, we have

$$A' \cong \mathbb{Z}[X_1, \ldots, X_r, X_{r+1}, X_{r+2}]/(f_1, \ldots, f_M, f_{M+1}, f_{M+2}), \qquad (10.1.11)$$

where

$$f_{M+2} := X_{r+2}\left(\sum_{t=0}^{\mathcal{E}-1} \widetilde{b_t} X_{r+1}^t\right) - 1.$$

Using again $\mathcal{E} \le v^{mn}$, we infer that f_{M+2} has total degree at most $\mathcal{R}'^{\exp O(r)}$ and logarithmic height at most $\mathcal{R}'^{\exp O(r)} h$. Combined with (10.1.10) and our assumptions $\deg f_i \le d$, $h(f_i) \le h$, this gives

$$\deg f_i \le \mathcal{R}'^{\exp O(r)}, \quad h(f_i) \le \mathcal{R}'^{\exp O(r)} h \text{ for } i = 1, \ldots, M+2. \qquad (10.1.12)$$

Step 2. *Let ℓ_i, ℓ_j with $i \ne j$ be connected by an edge in $\mathcal{G}(\mathcal{L})$. Then*

$$\frac{\ell_i(\mathbf{x})}{\ell_j(\mathbf{x})} \prec \left(\exp\left(\mathcal{R}'^{\exp O(r)} h\right), \exp\left(\mathcal{R}'^{\exp O(r)} h\right)\right). \qquad (10.1.13)$$

To prove this, we first assume that ℓ_i, ℓ_j are linearly dependent on $\overline{K}$. Then $\ell_i(\mathbf{x})/\ell_j(\mathbf{x})$ is equal to the quotient of a coefficient of ℓ_i and a coefficient of ℓ_j. So by Corollary 8.3.3 and (8.3.2), we have

$$\frac{\ell_i(\mathbf{x})}{\ell_j(\mathbf{x})} \prec \left((2v^2 d)^{\exp O(r)}, (2v^2 d)^{\exp O(r)} h\right),$$

which is much stronger than what we want to prove.

Next, assume that ℓ_i, ℓ_j are linearly independent of $\overline{K}$. Here, we have to apply our Theorem 2.2.1 on unit equations. There is $q \ne i, j$ such that ℓ_i, ℓ_j, ℓ_q are linearly dependent on $\overline{K}$. Then clearly, $\ell_i', \ell_j', \ell_q'$ are also linearly dependent on $\overline{K}$. In fact, we have

$$\lambda_i \ell_i' + \lambda_j \ell_j' + \lambda_q \ell_q' = 0,$$

where $\lambda_i, \lambda_j, \lambda_q$ are certain 2×2-determinants of the coefficients of $\ell_i', \ell_j', \ell_k'$. By (10.1.4), we have

$$\lambda_i = \widetilde{\lambda_i}\left(z_1, \ldots, z_r, \theta, \delta'^{-1}\right),$$

where $\widetilde{\lambda_i} \in \mathbb{Z}[X_1, \ldots, X_r, X_{r+1}, X_{r+2}]$ is a polynomial of total degree at most $\mathcal{R}'^{\exp O(r)}$ and logarithmic height at most $\mathcal{R}'^{\exp O(r)} h$. We have something similar for λ_j and λ_q. Clearly,

$$\lambda_i \cdot \frac{\ell_i'(\mathbf{x})}{\ell_j'(\mathbf{x})} + \lambda_q \cdot \frac{\ell_q'(\mathbf{x})}{\ell_j'(\mathbf{x})} = -\lambda_j,$$

while $\ell'_i(\mathbf{x})/\ell'_j(\mathbf{x})$, $\ell'_q(\mathbf{x})/\ell'_j(\mathbf{x}) \in A'^*$ by (10.1.9). Now invoking (10.1.11), (10.1.12) and applying Theorem 2.2.1, we obtain

$$\frac{\ell'_i(\mathbf{x})}{\ell'_j(\mathbf{x})} = g(z_1, \ldots, z_r, \beta, \delta'^{-1}),$$

where g is a polynomial in $\mathbb{Z}[X_1, \ldots, X_r, X_{r+1}, X_{r+2}]$ of total degree and logarithmic height at most $\exp(\mathcal{R}'^{\exp O(r)} h)$. An application of (8.3.2) and Corollary 8.3.3 yields

$$\frac{\ell'_i(\mathbf{x})}{\ell'_j(\mathbf{x})} \prec \left(\exp(\mathcal{R}'^{\exp O(r)} h), \exp(\mathcal{R}'^{\exp O(r)} h) \right).$$

Finally, using $\ell_i(\mathbf{x})/\ell_j(\mathbf{x}) = (\gamma_j/\gamma_i)(\ell'_i(\mathbf{x})/\ell'_j(\mathbf{x}))$, estimate (10.1.3), and again (8.3.2) and Corollary 8.3.3, we arrive at (10.1.13).

Step 3. *Let ℓ_i, ℓ_j belong to the same connected component of $\mathcal{G}(\mathcal{L})$. Then we have again* (10.1.13).

There is a sequence $\ell_{i_0}, \ell_{i_1}, \ldots, \ell_{i_s}$ with $i_0 = i$, $i_s = j$ of length $s \leq n$ of which any two consecutive linear forms are connected by an edge in $\mathcal{G}(\mathcal{L})$. Now apply Step 2 and Corollary 8.3.3 to

$$\frac{\ell_j(\mathbf{x})}{\ell_i(\mathbf{x})} = \prod_{t=0}^{s-1} \frac{\ell_{i_{t+1}}(\mathbf{x})}{\ell_{i_t}(\mathbf{x})}$$

to finish Step 3.

Step 4. *Let ℓ_i, ℓ_j be any two distinct linear forms from $\mathcal{L}$ with $i \neq j$. Then we have again* (10.1.13).

This is clear if $\mathcal{G}(\mathcal{L})$ is connected, so assume that its number k of connected components is >1. By assumption, there is $\ell \in [\mathcal{L}_1] \cap \cdots \cap [\mathcal{L}_k]$ such that $\ell(\mathbf{x}) \neq 0$. We can partition $\{1, \ldots, n\}$ into $I_1 \cup \cdots \cup I_k$ such that

$$\mathcal{L}_t = (\ell_s : s \in I_t) \text{ for } t = 1, \ldots, k.$$

Notice that $[\mathcal{L}_1] \cap \cdots \cap [\mathcal{L}_k]$ consists of the linear forms ℓ of the shape

$$\sum_{s \in I_1} c_s \ell_s = \cdots = \sum_{s \in I_k} c_s \ell_s \qquad (10.1.14)$$

for certain $c_1, \ldots, c_n \in \overline{K}$. We recall that in general, if $\mathcal{A}$ is an $a \times b$-matrix of rank t, say, with elements from a field L, then the solution space of vectors $\mathbf{x} \in L^b$ with $\mathcal{A}\mathbf{x} = \mathbf{0}$ has a basis consisting of vectors whose coordinates are $t \times t$-subdeterminants of $\mathcal{A}$. In particular, the linear subspace of $\overline{K}^n$, consisting of the vectors $\mathbf{c} = (c_1, \ldots, c_n)$ satisfying (10.1.14), has a basis, consisting of vectors whose coordinates are determinants of order at most m, with elements from the coefficients of $\pm\ell_i$ $(i = 1, \ldots, n)$. So by Corollary 8.3.3, these coordinates have degree-height estimates

$$\prec ((2mv^{m^2}d)^{\exp O(r)}, (2mv^{m^2}d)^{\exp O(r)}h).$$

This basis contains a vector $\mathbf{c} = (c_1, \ldots, c_n)$ such that the linear form ℓ given by (10.1.14) does not vanish at $\mathbf{x}$. Now suppose for instance that $\ell_i \in \mathcal{L}_1, \ell_j \in \mathcal{L}_2$. Then

$$\frac{\ell(\mathbf{x})}{\ell_i(\mathbf{x})} = \sum_{s \in I_1} c_s \frac{\ell_s(\mathbf{x})}{\ell_i(\mathbf{x})}.$$

By Corollary 8.3.3 and what we established in Step 3, we get

$$\frac{\ell(\mathbf{x})}{\ell_i(\mathbf{x})} \prec \left(\exp(\mathcal{R}'^{\exp O(r)}h), \exp(\mathcal{R}'^{\exp O(r)}h) \right).$$

We get a similar estimate for $\ell(\mathbf{x})/\ell_j(\mathbf{x})$. Another application of Corollary 8.3.3, in combination with (8.3.2), completes Step 4.

Step 5. *For $i = 1, \ldots, n$, we have*

$$\ell_i(\mathbf{x}) \prec \left(\exp(\mathcal{R}'^{\exp O(r)}h), \exp(\mathcal{R}'^{\exp O(r)}h) \right). \tag{10.1.15}$$

To prove this, observe that

$$\ell_i(\mathbf{x})^n = \delta \prod_{j=1}^{n} \frac{\ell_i(\mathbf{x})}{\ell_j(\mathbf{x})}.$$

Now apply Step 4 and Proposition 8.3.2.

Step 6. *Completion of the proof.*

Write $\beta_i := \ell_i(\mathbf{x})$ for $i = 1, \ldots, n$. Note that $\beta_i \in G$, so $\deg_K \beta_i \leq v^{mn}$ for $i = 1, \ldots, n$. By Corollary 8.3.5, there is $\mathbf{x}' = (x'_1, \ldots, x'_m) \in A^m$ such that

$$\ell_i(\mathbf{x}') = \beta_i \text{ for } i = 1, \ldots, n$$

and

$$x'_i \overset{\text{int}}{\prec} \left(\exp(\mathcal{R}'^{\exp O(r)}h), \exp(\mathcal{R}'^{\exp O(r)}h) \right) \text{ for } i = 1, \ldots, m,$$

which means precisely that $\mathbf{x}'$ has a representative $\widetilde{\mathbf{x}}' \in \mathbb{Z}[X_1, \ldots, X_r]^m$ with $s(\widetilde{\mathbf{x}}') \leq \exp(\mathcal{R}'^{\exp O(r)}h)$, which is (2.6.8). Further, $\mathbf{x}' - \mathbf{x} \in \mathcal{Z}_{A,F}$, so in fact, $\widetilde{\mathbf{x}}'$ represents the coset $\mathbf{x} + \mathcal{Z}_{A,F}$. This completes the proof of Theorem 2.6.1. $\quad\square$

Proof of Corollary 2.6.2 We prove only the effective part of the statement of Corollary 2.6.2.

Let α be either δ or one of the coefficients of $\ell_1, \ldots, \ell_n$. Then an effective representation for α is given, and from this, we can compute effective representations for $\alpha^2, \ldots, \alpha^{[G:K]}$. There is a divisor v of $[G : K]$ such that $1, \alpha, \ldots, \alpha^v$ are linearly dependent over K. Using the effective representations for the α^i, we can determine the smallest such v, a K-linear relation between $1, \alpha, \ldots, \alpha^v$, the monic minimal polynomial of α over K, and finally, a degree–height estimate for α. Now Theorem 2.6.1 gives an effectively computable number C such that every $\mathcal{Z}_{A,F}$-coset of solutions of (2.6.7) is represented by some $\widetilde{\mathbf{x}} \in \mathbb{Z}[X_1, \ldots, X_r]^m$ with $s(\widetilde{\mathbf{x}}) \leq C$.

There are only a finite number of such tuples $\widetilde{\mathbf{x}}$, which can be determined effectively, say, $\widetilde{\mathbf{x}}_1, \ldots, \widetilde{\mathbf{x}}_s$. These represent tuples $\mathbf{x}_1, \ldots, \mathbf{x}_s \in A^m$. Now, using the effective representations of δ and the coefficients of $\ell_1, \ldots, \ell_n$, one can check for each $\mathbf{x}_i$ whether $F(\mathbf{x}_i) = \delta$.

Using the effective representations for G and the coefficients of $\ell_1, \ldots, \ell_n$, one can compute the triangular graph $\mathcal{G}(\mathcal{L})$ and, thus, the vertex systems $\mathcal{L}_1, \ldots, \mathcal{L}_k$ of its connected components. The intersection $\mathcal{V} := [\mathcal{L}_1] \cap \cdots \cap [\mathcal{L}_k]$ was defined as a $\overline{K}$-vector space, but since the $\mathcal{L}_i$ consist of linear forms with coefficients from G, the space $\mathcal{V}$ is defined by a system of linear equations with coefficients from G. By solving this system using standard linear algebra, one can decide whether $\mathcal{V}$ is nonzero and, if so, compute a basis $\mathcal{B}$ of $\mathcal{V}$, consisting of linear forms from G. If for one of the $\mathbf{x}_i$ previously mentioned there is $\ell \in \mathcal{V}$ with $\ell(\mathbf{x}_i) \neq 0$, then there is such $\ell \in \mathcal{B}$, and one can find it by computing $\ell(\mathbf{x}_i)$ for all $\ell \in \mathcal{B}$. Thus, for each $\mathbf{x}_i$, it can be checked whether it satisfies (2.6.7).

Finally, using the effective representations, one can decide for each pair $\mathbf{x}_i$, $\mathbf{x}_j$ with $i, j = 1, \ldots, s$ whether $\mathbf{x}_i$, $\mathbf{x}_j$ lie in the same $\mathcal{Z}_{A,F}$-coset, i.e., $\ell_t(\mathbf{x}_i) = \ell_t(\mathbf{x}_j)$ for $t = 1, \ldots, n$. This shows that one can compute a finite set, consisting of one representative for each $\mathcal{Z}_{A,F}$-coset of solutions of (2.6.7). □

Proof of Corollary 2.6.3 By dividing F and δ by one of the coefficients of F, we can transform (2.3.1) into an equation

$$F'(x, y) = \ell_1(x, y) \cdots \ell_n(x, y) = \delta',$$

where δ' and the coefficients of F' lie in K, and are all represented by pairs in $\mathbb{Z}[X_1, \ldots, X_r]^2$ of total degree at most d and logarithmic height at most h, and where each ℓ_i is either of the form Y, or of the form $X - \alpha Y$, with $\alpha \in \overline{K}$. Since F is divisible by at least three pairwise nonproportional linear forms, the system $\mathcal{L} = (\ell_1, \ldots, \ell_n)$ is triangularly connected. Further, the module of $(x, y) \in A^2$ with $\ell_i(x, y) = 0$ for $i = 1, \ldots, n$ is equal to $\{\mathbf{0}\}$. Lastly, each of the preceding α has a degree at most n over K, and by Proposition 8.2.3, it is represented by a tuple of degree at most $(nd)^{\exp O(r)}$ and logarithmic height at most $(nd)^{\exp O(r)}$. Now an application of Theorem 2.6.1 immediately gives the bound (2.6.9), and then using Proposition 2.1.1, one can determine the solutions. □

Proof of Corollary 2.6.4 Recall that every solution $\mathbf{x} = (x_1, x_2, x_3) \in A^3$ of the system of double Pell equations (2.6.10) is a solution of the decomposable form equation (2.6.12), with the decomposable form F given by (2.6.14) and δ given by (2.6.13). It is easily shown that the linear factors of the decomposable form F in (2.6.12) form a triangularly connected system, and moreover, $\mathcal{Z}_{A,F} = \{\mathbf{0}\}$. Further, by Proposition 8.3.2 and Corollary 8.3.3, both δ and the coefficients of the linear factors of F are represented by tuples of degree at most $(2d)^{\exp O(r)}$ and logarithmic height at most $(2d)^{\exp O(r)} h$. An application of Theorem 2.6.1 directly gives the bound (2.6.15), and the solutions can be found by means of Proposition 2.1.1. □

10.2 Proofs of the Results for Norm Form Equations

Proof of Theorem 2.7.1 Denote by $\mathcal{L}$ the set of conjugates of the linear form $\ell := \alpha_1 X_1 + \cdots + \alpha_m X_m$ with respect to K'/K. The coefficients of the linear forms in $\mathcal{L}$ all have degree at most $n = [K' : K]$ over K. Further, $\alpha_1, \ldots, \alpha_m$ and their conjugates over K are all represented by tuples of degree at most d and height at most h. The rank of $\mathcal{L}$ is equal to m, since $\alpha_1, \ldots, \alpha_m$ are assumed to be linearly independent of K. So with $F := N_{K'/K}(\alpha_1 X_1 + \cdots + \alpha_m X_m)$, the module $\mathcal{Z}_{A,F}$ is $\{\mathbf{0}\}$. Letting $\mathcal{L}_1, \ldots, \mathcal{L}_k$ denote the vertex systems of the connected components of $\mathcal{G}(\mathcal{L})$, we verify below that $X_m \in [\mathcal{L}_1] \cap \cdots \cap [\mathcal{L}_k]$. Once this has been done, Theorem 2.7.1 follows directly from Theorem 2.6.1, taking $v = n$.

Observe that any two distinct conjugates of ℓ are linearly independent, since $K' = K(\alpha_1, \ldots, \alpha_m)$. Partition the linear forms in $\mathcal{L}$ into subsets such that ℓ', ℓ'' belong to the same subset if the coefficients of $X_1, \ldots, X_{m-1}$ in ℓ', ℓ'' coincide. This leads to a partition $\mathcal{L}'_1, \ldots, \mathcal{L}'_{k'}$ of $\mathcal{L}$ with k' denoting the degree of $K'' := K(\alpha_1, \ldots, \alpha_{m-1})$ over K. Further, in view of the condition that α_m be of degree ≥ 3 over K'', each of the sets $\mathcal{L}'_1, \ldots, \mathcal{L}'_{k'}$ has cardinality of at least 3. As is easily seen, any three linear forms from the same set $\mathcal{L}'_i$ are linearly dependent; hence any two linear forms from the same set $\mathcal{L}'_i$ are connected by an edge in $\mathcal{G}(\mathcal{L})$. That is, each set $\mathcal{L}'_i$ is contained in the vertex system of one of the connected components of $\mathcal{G}(\mathcal{L})$. Next, the difference of any two linear forms from the same set $\mathcal{L}'_i$ is proportional to X_m. This shows

$$X_m \in [\mathcal{L}'_1] \cap \cdots \cap [\mathcal{L}'_{k'}] \subseteq [\mathcal{L}_1] \cap \cdots \cap [\mathcal{L}_k].$$

As previously mentioned, this completes the proof of Theorem 2.7.1. □

Proof of Corollary 2.7.2 By assumption, an irreducible monic polynomial $P \in K[X]$ is given such that $K' \cong K[X]/(P)$. Denote by G the splitting field of P

and put $\mathcal{E} := [G : K]$. Notice that G is the normal closure of K'/K. By Corollaries 6.2.5 and 6.2.6, we can compute θ such that $G = K(\theta)$, together with the monic minimal polynomial of θ over K, and express $\alpha_1, \ldots, \alpha_m$ and their powers as K-linear combinations of $1, \theta, \ldots, \theta^{\mathcal{E}-1}$. With these expressions, we can compute which K-linear combinations of powers of α_i are 0 and, thus, compute the monic minimal polynomial of α_i over K for $i = 1, \ldots, m$. These monic minimal polynomials have all their roots in G, and we can compute these using Theorem 6.2.3. In this way, we can compute the conjugates of $\alpha_1, \ldots, \alpha_m$ over K. Consequently, the conjugates of the linear form ℓ are effectively given. Now Corollary 2.6.2 applies to Eq. (2.7.1), and the assertion follows. □

Proof of Corollary 2.7.3 Let $\mathbf{x} = (x_1, \ldots, x_m) \in A^m$ be a solution of (2.7.1), and denote by m' the greatest integer with $x_{m'} \neq 0$. If $m' \geq 2$, Corollary 2.7.2 applies with m' instead of m, while for $m' = 1$ we get $N(\alpha_1)x_1^n = \delta$, whence, by Theorems 6.2.3 and 6.3.2, we can check that $x_1 \in K$ and $x_1 \in A$. □

10.3 Proofs of the Results for Discriminant Form Equations and Discriminant Equations

Let Ω be a finite étale K-algebra with $[\Omega : K] = n$. Recall that Ω is given in the form $K[X]/(P)$, where $P \in K[X]$ is monic and separable of degree n. Thus, $\Omega = K[\theta]$, with $P(\theta) = 0$. Let G be the splitting field of P and $\theta^{(i)}$ $(i = 1, \ldots, n)$ the zeros of P in G. We can express $\alpha \in \Omega$ as

$$\alpha = \sum_{j=0}^{n-1} a_j \theta^j \quad \text{with } a_j \in K \text{ for } j = 0, \ldots, n-1.$$

Thus, the images of α under the n K-homomorphisms of Ω are given by

$$\alpha^{(i)} = \sum_{j=0}^{n-1} a_j (\theta^{(i)})^j \quad (i = 1, \ldots, n).$$

By Vandermonde's identity, we have $\det(\theta^{(i)})^{j-1})_{i,j=1,\ldots,n} \neq 0$. This implies that

$$\alpha^{(1)} = \cdots = \alpha^{(n)} \iff a_1 = \cdots = a_{n-1} = 0$$

$$\iff \alpha \in K. \tag{10.3.1}$$

In what follows, let $\omega_1, \ldots, \omega_m \in \Omega$.

Proof of Theorem 2.8.1 Let $\ell^{(i)} := \omega_1^{(i)} X_1 + \cdots + \omega_m^{(i)} X_m$ for $i = 1, \ldots, n$, and $\ell_{i,j} := \ell^{(i)} - \ell^{(j)}$ for $i, j = 1, \ldots, n$, $i \neq j$. We want to apply Theorem 2.6.1 to Eq. (2.8.3) in the form

$$F(\mathbf{x}) := \prod_{\substack{i,j=1 \\ i \neq j}}^{n} \ell_{i,j}(\mathbf{x}) = (-1)^{n(n-1)/2}\delta \quad \text{in} \quad \mathbf{x} \in A^m. \tag{10.3.2}$$

Let $\mathcal{L}$ denote the system of the linear forms $\ell_{i,j}$. We first show that $\mathcal{L}$ is triangularly connected. Indeed, using $\ell_{i,i'} + \ell_{i',i''} + \ell_{i'',i} = 0$ for any three distinct $i, i', i'' \in \{1, \ldots, n\}$, one infers that if $\{i,j\}$, $\{i',j'\}$ are any two distinct subsets of $\{1, \ldots, n\}$, then $\ell_{i,j}, \ell_{i',j'}$ are connected by an edge in $\mathcal{G}(\mathcal{L})$ if $\{i,j\} \cap \{i',j'\} \neq \emptyset$. If $\{i,j\} \cap \{i',j'\} = \emptyset$, then there are edges from $\ell_{i,j}$ to $\ell_{i',j}$ and from $\ell_{i',j}$ to $\ell_{i',j'}$, implying that $\ell_{i,j}$ and $\ell_{i',j'}$ belong to the same connected component of $\mathcal{G}(\mathcal{L})$.

Next, using (10.3.1), it follows that

$$\mathbf{x} \in \mathcal{Z}_{A,F} \iff \ell_{i,j}(\mathbf{x}) = 0 \text{ for all } i, j$$
$$\iff \ell^{(1)}(\mathbf{x}) = \cdots = \ell^{(n)}(\mathbf{x}) \iff \ell(\mathbf{x}) \in K \iff \mathbf{x} \in \mathcal{Z}_{A,D}.$$

Notice that there are $n(n-1)$ linear forms $\ell_{i,j}$ and that their coefficients $\omega_t^{(i)} - \omega_t^{(j)}$ have degree at most n^2 over K. Now Theorem 2.8.1 is proved by applying Theorem 2.6.1 with $k = 1$, with $n(n-1)$ instead of n and with n^2 instead of ν to Eq. (10.3.2). $\qquad \square$

Proof of Corollary 2.8.2 Recall that Ω is given in the form $K[X]/(P)$, with P an effectively given, monic separable polynomial in $K[X]$. Further, a set of A-module generators $\omega_1, \ldots, \omega_m \in \Omega$ of $\mathcal{M}$ is effectively given. Similar to the proof of Corollary 2.7.2, we can compute an effective representation for the splitting field G of P, as well as effective representations for $\omega_j^{(i)}$, for $i = 1, \ldots, n$, $j = 1, \ldots, m$. So, the coefficients of the linear forms $\ell_{i,j}$ defined in the proof of Theorem 2.8.1 are effectively given. So, by Corollary 2.6.2, we can compute a finite set of solutions $\mathbf{x} \in A^n$ of (2.8.2), consisting of one representative from each $\mathcal{Z}_{A,D}$-coset. But this means precisely that we can compute a finite set of solutions $\xi \in \mathcal{M}$ of Eq. (2.8.2), consisting of one element from each $\mathcal{M} \cap K$-coset. $\qquad \square$

Proof of Corollary 2.8.3 By Corollary 2.8.2, we can effectively compute a full system of representatives, say $\{\xi_1, \ldots, \xi_s\}$, for the $\mathcal{O} \cap K$-cosets of solutions of

$$D_{\Omega/K}(\xi) = \delta \quad \text{in} \quad \xi \in \mathcal{O}. \tag{10.3.3}$$

Further, by Corollary 6.3.9, we can compute a full system of representatives, say, $\{\eta_1, \ldots, \eta_t\}$, for the cosets of $\mathcal{O} \cap K$ modulo A. Then the finite set $\{\xi_i + \eta_j : i = 1, \ldots, s, j = 1, \ldots, t\}$ is an effectively computable full system of representatives for the A-cosets of solutions of (10.3.3). $\qquad \square$

In the proof of Theorem 2.8.4, we shall need the following.

Lemma 10.3.1 *For every integral domain A of characteristic 0 that is finitely generated over $\mathbb{Z}$, and every two monic polynomials $f, f' \in A[X]$, all effectively given, we can determine effectively whether f, f' are strongly A-equivalent.*

Proof It suffices to consider the case when f, f' have equal degrees. Write $f(X) = X^n + a_1 X^{n-1} + \cdots$, $f'(X) = X^n + b_1 X^{n-1} + \cdots$. We have to check whether there is an $a \in A$ such that $f'(X) = f(X + a)$. Comparing the coefficients of X^{n-1}, we see that for such a we must have $na = b_1 - a_1$. Using Theorem 6.3.2, we can check whether $a \in A$ and then whether indeed $f'(X) = f(X + a)$. $\qquad\square$

Proof of Theorem 2.8.4 Let A, G, δ, n be effectively given and satisfy the conditions of Theorem 2.8.4. Denote by A_K the integral closure of A in K and by A_G the integral closure of A in G. Let f be a polynomial with

$$D(f) = \delta, \quad f \text{ is monic}, f \in A[X],$$
$$\deg f = n, \ f \text{ has all its zeros in } G. \qquad (2.8.6)$$

Write

$$f = (X - \xi_1) \cdots (X - \xi_n).$$

Then

$$\xi_1, \ldots, \xi_n \in A_G, \quad \xi_1 + \cdots + \xi_n \in A, \qquad (10.3.4)$$

$$\prod_{1 \leq i < j \leq n} (\xi_i - \xi_j)^2 = \delta. \qquad (10.3.5)$$

For arbitrary $\xi_1, \ldots, \xi_n$ with (10.3.4) and (10.3.5), we write $\xi := (\xi_1, \ldots, \xi_n)$, $f_\xi := (X - \xi_1) \cdots (X - \xi_n)$. It is important to notice that conversely, if ξ satisfies (10.3.4) and (10.3.5), then f_ξ satisfies all conditions in (2.8.6), except that it need not belong to $A[X]$.

Two tuples $\xi' = (\xi'_1, \ldots, \xi'_n)$, $\xi'' = (\xi''_1, \ldots, \xi''_n)$ with (10.3.4) and (10.3.5) are said to lie in the same A-coset if

$$\xi'_1 - \xi''_1 = \cdots = \xi'_n - \xi''_n \in A.$$

Notice that in this case, the corresponding polynomials $f_{\xi'}$, $f_{\xi''}$ are strongly A-equivalent. We show that there are only finitely many A-cosets of tuples ξ with (10.3.4) and (10.3.5), and determine a full system of representatives, and subsequently select those tuples ξ from this system for which $f_\xi \in A[X]$. Then every polynomial f with (2.8.6) is strongly A-equivalent to one of these f_ξ.

By Theorem 6.3.6, A_G is finitely generated as an A-module, and we can effectively determine a system of A-module generators for A_G, say, $\{\omega_1, \ldots, \omega_m\}$.

Thus, we can express $\xi_1, \ldots, \xi_n$ with (10.3.4) as

$$\begin{aligned}
\xi_i &= \ell_i(\mathbf{x}) := x_{i,1}\omega_1 + \cdots + x_{i,m}\omega_m \quad (i = 1, \ldots, n-1), \\
\xi_n &= \ell_n(\mathbf{x}) := x_0 - \ell_1(\mathbf{x}) - \cdots - \ell_{n-1}(\mathbf{x}),
\end{aligned} \tag{10.3.6}$$

where

$$\mathbf{x} = (x_{1,1}, \ldots, x_{1,m}, \ldots, x_{n-1,1}, \ldots, x_{n-1,m}, x_0) \in A^{m(n-1)+1},$$

and then (10.3.5) translates into the decomposable form equation

$$F(\mathbf{x}) := \prod_{\substack{1 \le i,j \le n \\ i \ne j}} (\ell_i(\mathbf{x}) - \ell_j(\mathbf{x})) = (-1)^{n(n-1)/2}\delta \text{ in } \mathbf{x} \in A^{m(n-1)+1}. \tag{10.3.7}$$

Completely similar to the proof of Theorem 2.8.1, one shows that the system of linear forms

$$\mathcal{L} := (\ell_i - \ell_j : 1 \le i, j \le n, i \ne j)$$

is triangularly connected. Further, we have

$$\mathbf{x} \in \mathcal{Z}_{A,F} \iff \ell_1(\mathbf{x}) = \cdots = \ell_n(\mathbf{x}).$$

By Corollary 2.6.2, Eq. (10.3.7) has only a finite number of $\mathcal{Z}_{A,F}$-cosets of solutions, and a full system of representatives of these can be determined effectively. Notice that if $\mathbf{x} = (\ldots, x_0) \in \mathcal{Z}_{A,F}$, then

$$\ell_1(\mathbf{x}) = \cdots = \ell_n(\mathbf{x}) = \tfrac{1}{n}x_0 \in \tfrac{1}{n}A \cap A_G = \tfrac{1}{n}A \cap A_K.$$

Translating this back to (10.3.4) and (10.3.5), we see that the tuples ξ with (10.3.4), (10.3.5) lie in only a finite number of $(\tfrac{1}{n}A \cap A_K)$-cosets. Moreover, a full system of representatives for these cosets can be determined effectively. Let $\mathcal{C}$ be such a full system of representatives. By Corollary 6.3.7, there are only a finite number of cosets of $\tfrac{1}{n}A \cap A_K$ modulo A, and a full system of representatives of these can be determined effectively. Let $\mathcal{C}'$ be such a system. From $\mathcal{C}$ and $\mathcal{C}'$, we compute

$$\mathcal{C}'' := \{\xi + a^* : \xi \in \mathcal{C}, a \in \mathcal{C}'\}, \text{ where } a^* := \underbrace{(a, \ldots, a)}_{n \text{ times}},$$

which is a full system of representatives for the A-cosets of tuples ξ with (10.3.4) and (10.3.5). From the tuples $\xi \in \mathcal{C}''$, we effectively select those for which $f_\xi \in A[X]$; namely, using the effective representations of the coefficients of f_ξ, we can first decide whether $f_\xi \in K[X]$, and then subsequently whether $f_\xi \in A[X]$ by means of Theorem 6.3.2. Further, using Lemma 10.3.1, from the f_ξ, with the remaining ξ, we select a maximal set of pairwise not strongly A-equivalent polynomials. What is left is a finite, full system of representatives for the strong A-equivalence classes of polynomials f with (2.8.6). □

References

Aschenbrenner M. (2004), *Ideal membership in polynomial rings over the integers*, J. Amer. Math. Soc. **17**, 407–442.

Baker A. (1966), *Linear forms in the logarithms of algebraic numbers*, Mathematika **13**, 204–216.

Baker A. (1967a), *Linear forms in the logarithms of algebraic numbers, II*, Mathematika **14**, 102–107.

Baker A. (1967b), *Linear forms in the logarithms of algebraic numbers, III*, Mathematika **14**, 220–228.

Baker A. (1968a), *Linear forms in the logarithms of algebraic numbers, IV*, Mathematika **15**, 204–216.

Baker A. (1968b), *Contributions to the theory of Diophantine equations*, Philos. Trans. Roy. Soc. London. Ser. A **263**, 173–208.

Baker A. (1968c), *The Diophantine equation $y^2 = ax^3 + bx^2 + cx + d$*, J. London Math. Soc. **43**, 1–9.

Baker A. (1969), *Bounds for the solutions of the hyperelliptic equation*, Proc. Camb. Philos. Soc. **65**, 439–444.

Baker A. (1975), Transcendental Number Theory, Cambridge University Press.

Baker A., ed. (1988), New Advances in Transcendence Theory, Cambridge University Press.

Baker A. and Coates J. (1970), *Integer points on curves of genus 1*, Proc. Camb. Philos. Soc. **67**, 595–602.

Baker A. and Masser D. W., eds. (1977), Transcendence Theory: Advances and Applications, Academic Press.

Baker A. and Wüstholz G. (2007), Logarithmic Forms and Diophantine Geometry, Cambridge University Press.

Bennett M. A. and Skinner C. S. (2004), *Ternary Diophantine equations via Galois representations and modular forms*, Canad. J. Math. **56**, 23–54.

Bérczes A. (2015a), *Effective results for unit points on curves over finitely generated domains*, Math. Proc. Camb. Philos. Soc. **158**, 331–353.

Bérczes A. (2015b), *Effective results for division points on curves in $\mathbb{G}_m^2$*, J. Théorie Nombres Bordeaux **27**, 405–437.

Bérczes A., Evertse J.-H., and Győry K. (2009), *Effective results for linear equations in two unknowns from a multiplicative division group*, Acta Arith. **136**, 331–349.

206

Bérczes A., Evertse J.-H., and Győry K. (2013), *Effective results for hyper- and superelliptic equations over number fields*, Publ. Math. Debrecen **82**, 727–756.

Bérczes A., Evertse J.-H., and Győry K. (2014), *Effective results for Diophantine equations over finitely generated domains*, Acta Arith. **163**, 71–100.

Bérczes A., Evertse J.-H., Győry K., and Pontreau C. (2009), *Effective results for points on certain subvarieties of a tori*, Math. Proc. Camb. Phil. Soc. **147**, 69–94.

Bilu Yu. F. (1995), *Effective analysis of integral points on algebraic curves*, Israel J. Math. **90**, 235–252.

Birch B. J. and Merriman J. R. (1972), *Finiteness theorems for binary forms with given discriminant*, Proc. London Math. Soc. (3) **24**, 385–394.

Bombieri E. (1993), *Effective Diophantine approximation on* $\mathbb{G}_M$, Ann. Scuola Norm. Sup. Pisa (IV) **20**, 61–89.

Bombieri E. and Cohen P. B. (1997), *Effective Diophantine approximation on* $\mathbb{G}_m$, *II*, Ann. Scuola Norm. Sup. Pisa (IV) **24**, 205–225.

Bombieri E. and Cohen P. B. (2003), *An elementary approach to effective Diophantine approximation on* $\mathbb{G}_m$, in: *Number Theory and Algebraic Geometry*, Cambridge University Press, pp. 41–62.

Bombieri E. and Gubler W. (2006), Heights in Diophantine Geometry, Cambridge University Press.

Borevich Z. I. and Shafarevich I. R. (1967), Number Theory, 2nd ed., Academic Press.

Borosh I., Flahive M., Rubin D., and Treybig B. (1989), *A sharp bound for solutions of linear Diophantine equations*, Proc. Amer. Math. Soc. **105**, 844–846.

Brindza B. (1984), *On S-integral solutions of the equation* $y^m = f(x)$, Acta Math. Hungar. **44**, 133–139.

Brindza B. (1987), *On S-integral solutions of the Catalan equation*, Acta Arith. **48**, 397–412.

Brindza B. (1989), *On the equation* $f(x) = y^n$ *over finitely generated domains*, Acta Math. Hungar. **53**, 377–383.

Brindza B. (1993), *The Catalan equation over finitely generated integral domains*, Publ. Math. Debrecen **42**, 193–198.

Brindza B., Győry K., and Tijdeman R. (1986), *On the Catalan equation over algebraic number fields*, J. Reine Angew. Math. **367**, 90–102.

Brownawell W. D. and Masser D. W. (1986), *Vanishing sums in function fields*, Math. Proc. Camb. Phil. Soc. **100**, 427–434.

Bugeaud Y. (1998), *Bornes effectives pour les solutions des équations en S-unités et des équations de Thue–Mahler*, J. Number Theory **71**, 227–244.

Bugeaud Y. (2018), *Linear Forms in Logarithms and Applications*, European Mathematical Society.

Bugeaud Y. and Győry K. (1996a), *Bounds for the solutions of unit equations*, Acta Arith. **74**, 67–80.

Bugeaud Y. and Győry K. (1996b), *Bounds for the solutions of Thue–Mahler equations and norm form equations*, Acta Arith. **74**, 273–292.

Cassels J.W.S. (1959), An Introduction to the Geometry of Numbers, Springer.

Catalan E. (1844), *Note extraite d'une lettre adressée à l'éditeur*, J. Reine Angew. Math. **27**, 192.

Coates J. (1969), *An effective p-adic analogue of a theorem of Thue*, Acta Arith. **15**, 279–305.

Dobrowolski E. (1979), *On a question of Lehmer and the number of irreducible factors of a polynomial*, Acta Arith. **34**, 391–401.

Dvornicich R. and Zannier U. (1994), *A note on Thue's equation over function fields*, Monatsh. Math. **118**, 219–230.

Evertse J.-H. (1983), *Upper bounds for the numbers of solutions of Diophantine equations*, PhD thesis, Leiden University. Also published as Math. Centre Tracts No. **168**, CWI, Amsterdam.

Evertse J.-H. (1984), *On sums of S-units and linear recurrences*, Compos. Math. **53**, 225–244.

Evertse J.-H. and Győry K. (1988a), *On the number of solutions of weighted unit equations*, Compos. Math. **66**, 329–354.

Evertse J.-H. and Győry, K. (1988b), *Finiteness criteria for decomposable form equations*, Acta Arith. **50**, 357–379.

Evertse J.-H. and Győry K. (2013), *Effective results for unit equations over finitely generated integral domains*, Math. Proc. Camb. Phil. Soc. **154**, 351–380.

Evertse J.-H. and Győry K. (2014), *Effective results for Diophantine equations over finitely generated domains: A survey*, in: Turán Memorial, Number Theory, Analysis and Combinatorics, (J. Pintz, A. Biró, K. Győry, G. Harcos, M. Simonovits, and J. Szabados, eds.). De Gruyter, 63–74.

Evertse J.-H. and Győry K. (2015), Unit Equations in Diophantine Number Theory, Cambridge University Press.

Evertse J.-H. and Győry K. (2017a), Discriminant Equations in Diophantine Number Theory, Cambridge University Press.

Evertse J.-H. and Győry K. (2017b), *Effective results for discriminant equations over finitely generated integral domains*, in: Number Theory, Diophantine Problems, Uniform Distribution and Applications (C. Elsholtz and P. Grabner, eds.). Springer, 237–256.

Evertse J.-H., Győry K., Stewart C. L., and Tijdeman R. (1988a), *On S-unit equations in two unknowns*, Invent. Math. **92**, 461–477.

Evertse J.-H., Győry K., Stewart C. L., and Tijdeman R. (1988b), *S-unit equations and their applications*, in: New Advances in Transcendence Theory, Proc. Conf. Durham 1986 (A. Baker, ed.). Cambridge University Press, 110–174.

Faltings G. (1983), *Endlichkeitssätze für abelsche Varietäten über Zahlkörpern*, Invent. Math. **73**, 349–366, Erratum: ibid. **75** (1984), 381.

Faltings G. and Wüstholz G. (1984), *Rational points,* Seminar Bonn/Wuppertal 1983/84, Aspects. Math. E6, Vieweg.

Feldman N. I. and Nesterenko Y. V. (1998), *Transcendental Numbers*, Springer Verlag. Vol. 44 of Encyclopaedia of Mathematical Sciences.

Le Fourn S. (2020), *Tubular approaches to Baker's method for curves and varieties*, Algebra Number Theory **14**, 763–785.

Freitas N., Kraus A., and Siksek S. (2020a), *On the unit equation over cyclic number fields of prime degree*, arXiv:2012.06445v1 [math.NT] 11 December 2020.

Freitas N., Kraus A., and Siksek S. (2020b), *On local criteria for the unit equation and the asymptotic Fermat's last theorem*, arXiv:2012.12666v1 [math.NT] 23 December 2020.

Friedman E. (1989), *Analytic formulas for regulators of number fields,* Invent. Math. **98**, 599–622.

Gel'fond A. O. (1934), *Sur le septième problème de Hilbert*, Izv. Akad. Nauk SSSR **7**, 623–630.

Gel'fond A. O. (1935), *On approximating transcendental numbers by algebraic numbers*, Dokl. Akad. Nauk SSSR **2**, 177–182.

Győry K. (1972), *Sur l'irreducibilité d'une classe des polynômes II.*, Publ. Math. Debrecen **19**, 293–326.

Győry K. (1973), *Sur les polynômes à coefficients entiers et de discriminant donné*, Acta Arith. **23**, 419–426.

Győry K. (1974), *Sur les polynômes à coefficients entiers et de discriminant donné II*, Publ. Math. Debrecen **21**, 125–144.

Győry K. (1976), *Sur les polynômes à coefficients entiers et de discriminant donné III*, Publ. Math. Debrecen **23**, 141–165.

Győry K. (1978a), *On polynomials with integer coefficients and given discriminant IV*, Publ. Math. Debrecen **25**, 155–167.

Győry K. (1978b), *On polynomials with integer coefficients and given discriminant V, P-adic generalizations*, Acta Math. Acad. Sci. Hungar. **32**, 175–190.

Győry K. (1979), *On the number of solutions of linear equations in units of an algebraic number field*, Comment. Math. Helv. **54**, 583–600.

Győry K. (1980a), *Explicit upper bounds for the solutions of some Diophantine equations*, Ann. Acad. Sci. Fenn. Ser. A I. Math. **5**, 3–12.

Győry K. (1980b), *Résultats effectifs sur la représentation des entiers par des formes désomposables*, Queen's Papers in Pure and Applied Math., No. **56**, Kingston, Canada.

Győry K. (1981a), *On S-integral solutions of norm form, discriminant form and index form equations*, Studia Sci. Math. Hungar. **16**, 149–161.

Győry K. (1981b), *On discriminants and indices of integers of an algebraic number field*, J. Reine Angew. Math. **324**, 114–126.

Győry K. (1982), *On certain graphs associated with an integral domain and their applications to Diophantine problems*, Publ. Math. Debrecen **29**, 79–94.

Győry K. (1983), *Bounds for the solutions of norm form, discriminant form and index form equations in finitely generated integral domains*, Acta Math. Hungar. **42**, 45–80.

Győry K. (1984a), *On norm form, discriminant form and index form equations*, in: Topics in Classical Number Theory (A. Baker, ed.). North-Holland Publ. Comp., 617–676.

Győry K. (1984b), *Effective finiteness theorems for polynomials with given discriminant and integral elements with given discriminant over finitely generated domains*, J. Reine Angew. Math. **346**, 54–100.

Győry K. (1990), *On arithmetic graphs associated with integral domains*, in: A Tribute to Paul Erdős (A. Baker, B. Bollobás and A. Hajnal, eds.). Cambridge University Press, 207–222.

Győry K. (1992), *Some recent applications of S-unit equations*, Astérisque **209**, 17–38.

Győry K. (1993), *On the numbers of families of solutions of systems of decomposable form equations*, Publ. Math. Debrecen **42**, 65–101.

Győry K. (1998), *Bounds for the solutions of decomposable form equations*, Publ. Math. Debrecen **52**, 1–31.

Győry K. (2002), *Solving Diophantine equations by Baker's theory*, in: A Panorama of Number Theory (G. Wüstholz, ed.). Cambridge University Press, 38–72.

Győry K. (2019), *Bounds for the solutions of S-unit equations and decomposable form equations II*, Publ. Math. Debrecen **94**, 507–526.

Győry K. and Papp Z. Z. (1977), *On discriminant form and index form equations*, Studia Sci. Math. Hungar. **12**, 47–60.

Győry K. and Papp Z. Z. (1978), *Effective estimates for the integer solutions of norm form and discriminant form equations*, Publ. Math. Debrecen **25**, 311–325.

Győry K. and Yu K. (2006), *Bounds for the solutions of S-unit equations and decomposable form equations*, Acta Arith. **123**, 9–41.

Hartshorne R. (1977), Algebraic Geometry, Springer Verlag.

Hermann G. (1926), *Die Frage der endlich vielen Schritte in der Theorie der Polynomideale*, Math. Ann. **95**, 736–788.

de Jong T. (1998), *An algorithm for computing the integral closure*, J. Symb. Comput. **26**, 273–277.

von Känel R. (2014), *Modularity and integral points on moduli schemes*, arXiv: 1310.7263v2 [math.NT].

von Känel R. and Matschke B. (2016), *Solving S-unit, Mordell, Thue, Thue–Mahler and generalized Ramanujan–Nagell equations via Shimura–Taniyama conjecture*, arXiv:1605.06079.

Kim D. (2017), *A modular approach to cubic Thue–Mahler equations*, Math. Comp. **86**, 1435–1471.

Kotov S. V. and Sprindžuk V. G. (1973), *An effective analysis of the Thue–Mahler equation in relative fields*, Dokl. Akad. Nauk BSSR **17**, 393–395 (Russian).

Kotov S. V. and Trelina L. A. (1979), *S-ganze Punkte auf elliptischen Kurven*, J. Reine Angew. Math. **306**, 28–41.

Koymans P. (2016), *The Catalan equation*, Master thesis, Leiden University.

Koymans P. (2017), *The Catalan equation*, Indag. Math. (N.S.) **28**, 321–352.

Lang S. (1960), *Integral points on curves*, Inst. Hautes Études Sci. Publ. Math. **6**, 27–43.

Lang S. (1962), Diophantine Geometry, Wiley.

Lang S. (1965a), *Division points on curves*, Ann. Mat. Pura Appl. (4) **70**, 229–234.

Lang S. (1965b), *Report on Diophantine approximations*, Bull. Soc. Math. France **93**, 177–192.

Lang S. (1978), Elliptic Curves: Diophantine Analysis, Springer.

Lang S. (1983), Fundamentals of Diophantine Geometry, Springer.

Lang S. (1994), Algebraic Number Theory, 2nd ed., Springer.

Laurent M. (1984), *Équations diophantiennes exponentielles*, Invent. Math. **78**, 299–327.

LeVeque W. J. (1964), *On the equation $y^m = f(x)$*, Acta Arith. **9**, 209–219.

Liardet P. (1974), *Sur une conjecture de Serge Lang*, C.R. Acad. Sci. Paris **279**, 435–437.

Liardet P. (1975), *Sur une conjecture de Serge Lang*, Astérisque **24–25**, Soc. Math. France, 187–210.

Loher T. and Masser D. (2004), *Uniformly counting points of bounded height*, Acta Arith. **111**, 277–297.

Louboutin S. (2000), *Explicit bounds for residues of Dedekind zeta functions, values of L-functions at s = 1, and relative class numbers*, J. Number Theory **85**, 263–282.

Loxton J. H. and van der Poorten A. J. (1983), *Multiplicative dependence in number fields*, Acta Arith. **42**, 291–302.

Mahler K. (1933), *Zur Approximation algebraischer Zahlen I: Über den grössten Primteiler binärer Formen*, Math. Ann. **107**, 691–730.

Mahler K. (1934), *Über die rationalen Punkte auf Kurven vom Geschlecht Eins*, J. Reine Angew. Math. **170**, 168–178.

Mason R. C. (1981), *On Thue's equation over function fields*, London Math. Soc. **24**, 414–426.

Mason R. C. (1983), *The hyperelliptic equation over function fields*, Math. Proc. Camb. Phil. Soc. **93**, 219–230.

Mason R. C. (1984), Diophantine Equations over Function Fields, Cambridge University Press.

Mason R. C. (1986), *Norm form equations I.*, J. Number Theory **22**, 190–207.

Mason R. C. (1988), *The study of Diophantine equations over function fields*, in: New Advances in Transcendence Theory (A. Baker, ed.). Cambridge University Press, 229–244.

Matsumura H. (1986), Commutative Ring Theory, Cambridge University Press.

Matsumoto R. (2000), *On computing the integral closure*, Commun. Algebra **28**, 401–405.

Matveev E. M. (2000), *An explicit lower bound for a homogeneous rational linear form in logarithms of algebraic numbers, II*. Izvestiya: Mathematics **64**, 1217–1269.

Mihâilescu P. (2004), *Primary cyclotomic units and a proof of Catalan's conjecture*, J. Reine Angew. Math. **572**, 167–195.

Mordell L. J. (1922a), *On the rational solutions of the indeterminate equations of the third and fourth degrees*, Proc. Camb. Phil. Soc. **21**, 179–192.

Mordell L. J. (1922b), *Note on the integer solutions of the equation $Ey^2 = Ax^3 + Bx^2 + Cx + D$*, Messenger Math. **51**, 169–171.

Mordell L. J. (1923), *On the integer solutions of the equation $ey^2 = ax^3 + bx^2 + cx + d$*, Proc. London Math. Soc. (2) **21**, 415–419.

Mordell L. J. (1969), Diophantine Equations, Academic Press.

Moriwaki A. (2000), *Arithmetic height functions over finitely generated fields*, Invent. Math. **140**, 101–142.

Murty M. R. and Pasten H. (2013), *Modular forms and effective Diophantine approximation*, J. Number Theory **133**, 3739–3754.

Nagata M. (1956), *A general theory of algebraic geometry over Dedekind domains I.*, Amer. J. Math. **78**, 78–116.

Osgood C. F. (1973), *An effective lower bound on the 'Diophantine approximation' of algebraic functions by rational functions*, Mathematika **20**, 4–15.

Osgood C. F. (1975), *Effective bounds on the 'Diophantine approximation' of algebraic functions over fields of arbitrary characteristic and applications to differential equations*, Indag. Math. **37**, 105–119.

Parry C. J. (1950), *The p-adic generalization of the Thue-Siegel theorem*, Acta Math. **83**, 1–100.

Pasten H. (2017), *Shimura curves and the abc conjecture*, arXiv:1705.09251v1 [math.NT], 25 May 2017.

Poonen B. (2019), *The S-integral points on the projective line minus three points via étale covers and Skolem method.* Available at https://math.mit.edu/~poonen/papers/siegel_for_Q.pdf.

Roquette P. (1957), *Einheiten und Divisorenklassen in endlich erzeugbaren Körpern*, Jahresber. Deutsch. Math. Verein **60**, 1–21.

Schlickewei H. P. (1977), *On norm form equations*, J. Number Theory **9**, 370–380.

Schinzel A. and Tijdeman R. (1976), *On the equation $y^m = P(x)$*, Acta Arith. **31**, 199–204.

Schmidt W. M. (1971), *Linearformen mit algebraischen Koeffizienten II*, Math. Ann. **191**, 1–20.

Schmidt W. M. (1972), *Norm form equations*, Ann. Math. **96**, 526–551.

Schmidt W. M. (1976), *On Osgood's effective Thue theorem for algebraic functions*, Commun. Pure Applied Math. **29**, 759–773.

Schmidt W. M. (1978), *Thue's equation over function fields*, J. Austral. Math. Soc. Ser A **25**, 385–422.

Schmidt W. M. (1991), Diophantine Approximations and Diophantine Equations, Lecture Notes Math. 1467, Springer.

Schneider T. (1934), *Transzendenzuntersuchungen periodischer Funktionen: I Transzendenz von Potenzen; II Transzendenzeigenschaften elliptischer Funktionen*, J. Reine Angew. Math. **172**, 65–74.

Seidenberg A. (1974), *Constructions in algebra*, Trans. Amer. Math. Soc. **197**, 273–313.

Shorey T. N. and Tijdeman R. (1986), Exponential Diophantine Equations, Cambridge University Press.

Siegel C. L. (1921), *Approximation algebraischer Zahlen*, Math. Z. **10**, 173–213.

Siegel C. L. (1926), *The integer solutions of the equation $y^2 = ax^n + bx^{n-1} + \cdots + k$*, J. London Math. Soc. **1**, 66–68.

Siegel C. L. (1929), *Über einige Anwendungen diophantischer Approximationen*, Abh. Preuss. Akad. Wiss., Phys. Math. Kl., No. **1**.

Siksek S. (2013), *Explicit Chabauty over number fields*, Algebra Number Theory **7**, 765–793.

Simmons H. (1970), *The solution of a decision problem for several classes of rings*, Pacific J. Math. **34**, 547–557.

Smart N. P. (1998), The Algorithmic Resolution of Diophantine Equations, Cambridge University Press.

Sprindžuk V. G. (1982), Classical Diophantine Equations in Two Unknowns (Russian), Nauka, Moskva.

Sprindžuk V. G. (1993), Classical Diophantine Equations, Lecture Notes Math. 1559, Springer.

Stark H. M. (1974), *Some effective cases of the Brauer–Siegel theorem*, Invent. Math. **23**, 135–152.

Stothers W. W. (1981), *Polynomial identities and Hauptmoduln*, Quart. J. Math., Oxford Ser. (2) **32**, 349–370.

Thue A. (1909), *Über Annäherungswerte algebraischer Zahlen*, J. Reine Angew. Math. **135**, 284–305.

Tijdeman R. (1976), *On the equation of Catalan*, Acta Arith. **29**, 197–209.

Triantafillou N. (2020), *The unit equation has no solutions in number fields of degree prime to 3 where 3 splits completely*, arXiv:2003.02414.

van der Poorten A. J. and Schlickewei H. P. (1982), *The growth condition for recurrence sequences*, Macquarie Univ. Math. Rep. 82–0041.

van der Poorten A. J. and Schlickewei H. P. (1991), *Additive relations in fields*, J. Austral. Math. Soc. (Ser. A) **51**, 154–170.

Végső J. (1994), *On superelliptic equations*, Publ. Math. Debrecen **44**, 183–187.

Voutier P. (1996), *An effective lower bound for the height of algebraic numbers*, Acta Arith. **74**, 81–95.

Waldschmidt M. (2000), Diophantine Approximation on Linear Algebraic Groups, Springer.

Wüstholz G., ed. (2002), A Panorama of Number Theory or the View from Baker's Garden, Cambridge University Press.

Yu K. (2007), *P-adic logarithmic forms and group varieties III*, Forum Mathematicum **19**, 187–280.

Zannier U. (2009), Lecture Notes on Diophantine Analysis, Lecture Notes, Scuola Normale Superiore di Pisa (New Series), **8** Edizioni della Normale.

Index

214

Printed in the United States
by Baker & Taylor Publisher Services